CENTRIFUGAL PUMPS
Design & Application

Second Edition

Gulf Professional Publishing
An Imprint of Elsevier

CENTRIFUGAL PUMPS
Design & Application
Second Edition

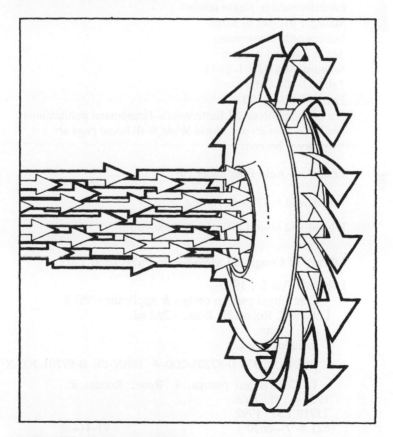

Val S. Lobanoff
Robert R. Ross

CENTRIFUGAL PUMPS
Design & Application

Second Edition

Originally published by Gulf Professional Publishing, Houston, TX.

For information, please contact:
Manager of Special Sales
Butterworth–Heinemann
225 Wildwood Avenue
Woburn, MA 01801–2041
Tel: 781-904-2500
Fax: 781-904-2620
For information on all Butterworth–Heinemann publications available, contact our World Wide Web home page at: http://www.bh.com

Printed on Acid-Free Paper (∞)

Transferred to Digital Printing, 2010

Printed and bound in the United Kingdom

Library of Congress Cataloging-in-Publication Data

Lobanoff, Val S., 1910–
    Centrifugal pumps: design & application/Val S. Lobanoff, Robert R. Ross.—2nd ed.
        p.   cm.
    Includes index.
        ISBN-13: 978-0-87201-200-4   ISBN-10: 0-87201-200-X
        1. Centrifugal pumps. I. Ross, Robert R., 1934– . II. Title.
    TJ919.L52 1992
    621.6'7—dc20   91-41458
                                                CIP

        ISBN-13: 978-0-87201-200-4
        ISBN-10: 0-87201-200-X

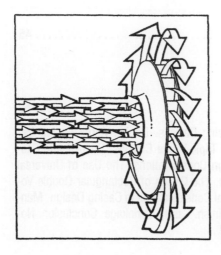

Contents

ern Pumps. Mine Dewatering Pumps. Wet Pit Pumps. Water Supply Pumps. Condenser Cooling Water Pumps. Cooling Tower Pumps. Flood Control Pumps. Transfer Pumps. Barrel-Mounted or Can-Mounted Pumps. Condensate and Heater Drain Pumps. Process Pumps. Small Boiler Feed Pumps. Cryogenic Pumps. Loading Pumps. Pipeline Booster Pumps. Design Features. The Bowl Assembly. The Column Assembly. Outer Column. Column Shaft. Shaft Enclosing Tube. The Head Assembly. Pump Vibration. References.

10 Pipeline, Waterflood, and CO$_2$ Pumps 139

Pipeline Pumps. Condition Changes. Destaging. Bi-rotors. Slurry Pipelines. Example of Pipeline Pump Selection. Series vs. Parallel. Waterflood Pumps. CO$_2$ Pumps. Mechanical Seals. Horsepower Considerations. Notation. References.

11 High Speed Pumps 173

by Edward Gravelle
History and Description of an Unconventional Pump Type. Terminology. Partial Emission Formulae. Specific Speed. Suction Specific Speed. Inducers. Partial Emission Design Evolution. Design Configuration Options. Other High-Speed Considerations. References.

12 Double-Case Pumps 206

by Erik B. Fiske
Configurations. Pump Casing. Volute Casing with Opposed Impellers. Diffuser Casings with Balance Drum. Diffuser Casings with Balance Disk. Applications. Boiler Feed Pumps. Charge Pumps. Waterflood Pumps. Pipeline Pumps. Design Features. Removable Inner Case Subassembly. Auxiliary Take-off Nozzles. Double-Suction First-Stage Impellers. Mounting of the Impellers. Impeller Wear Rings. Shaft Seals. Radial Bearings. Thrust Bearings. Baseplates and Foundations. Mounting of the Barrel. Design Features for Pumping Hot Oil with Abrasives. Double-Case Pump Rotordynamic Analysis. The Effect of Stage Arrangement on Rotordynamics. The Effect of Impeller Growth from Centrifugal Forces. Comparison of Diffuser Casings with Volute Casings. Diffuser Casings. Volute Casings. References.

13 Slurry Pumps 226

by George Wilson
Slurry Abrasivity. Pump Materials to Resist Abrasive Wear. Slurry Pump Types. Specific Speed and Wear. Areas of Wear. Casing. Impeller. Wear Plates. Bearing Frames. Sealing. Sump Design. Pump Drive. The Effect of Slurries on Pump Performance.

Stage Pumps. Single-Suction Single-Stage Overhung Pumps. Multi-Stage Pumps. Notation. References.

17 Mechanical Seals 354

18 Vibration and Noise in Pumps 422

Part 4: Extending Pump Life   495

19 Alignment 497

20 Rolling Element Bearings and Lubrication 524

Oxidation Resistance. Emulsification. Rust Prevention. Additives. General Lubricant Considerations. Application Limits for Greases. Life-Time Lubricated, "Sealed" Bearings. Oil Viscosity Selection. Applications of Liquid Lubricants in Pumps. Oil Bath Lubrication. Drip Feed Lubrication. Forced Feed Circulation. Oil Mist Lubrication. Selecting Rolling Element Bearings for Reduced Failure Risk. Magnetic Shaft Seals in the Lubrication Environment. References.

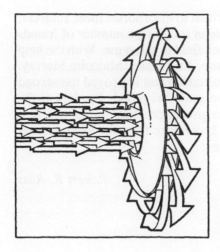

Preface

When Val and I decided to collaborate and write the first edition, our goal was to produce an easy-to-read, easy-to-understand, practical text-book stressing hydraulic design, that could be of hands-on use to the pump designer, student, and rotating equipment engineer. Although feedback from readers indicates that we achieved our desired goal, we did recognize that we had omitted several important topics. We had said little about the design of chemical pumps and touched only lightly on the extensive range of composite materials and the manufacturing techniques used in nonmetallic pump applications. We had totally ignored the subject of mechanical seals, yet we fully recognized that a knowledge of seal fundamentals and theory of operation is essential to the pump designer and rotating equipment engineer.

Another major omission was the subject of vibration and noise in centrifugal pumps. With today's high energy pumps operating at ever increasing speeds, it is essential that we understand the sources of pump noise and causes of vibration that result from installation, application, cavitation, pulsation, or acoustic resonance.

Although we had touched lightly on rotor dynamics, we felt this subject deserved to be expanded, particularly in the areas of bearing stiffness and damping, seal effects, and the evaluation of critical speed calculations. Finally, we had said nothing about the knowledge necessary to extend pump life during installation and operation, which requires a deep understanding of bearings, lubrication, mechanical seal reliability, and the external alignment of pump and driver.

This second edition was therefore written to incorporate these subjects, and in this regard, I have been fortunate in soliciting a number of friends and colleagues, each expert in his chosen field to assist me. With the help of Heinz Bloch, Gordon Buck, Fred Buse, Erik Fiske, Malcolm Murray, Jim Netzel, and Fred Szenasi, I have expanded and improved the second edition in a manner I know would have made Val proud. Readers of the first edition will find this book of even greater practical value, and new readers will gain an in-depth practical knowledge from the extensive experience of the authors and contributors.

Robert R. Ross

CENTRIFUGAL PUMPS
Design & Application

Second Edition

Part 1

Elements of Pump Design

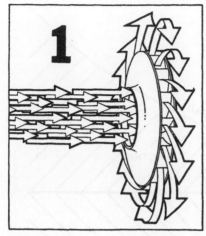

Introduction

System Analysis for Pump Selection

Before a pump can be selected or a prototype designed, the application must be clearly defined. Whether a simple recirculation line or a complex pipeline is needed, the common requirement of all applications is to move liquid from one point to another. As pump requirements must match system characteristics, analysis of the overall system is necessary to establish pump conditions. This is the responsibility of the user and includes review of system configuration, changes in elevation, pressure supply to the pump, and pressure required at the terminal. Relevant information from this analysis is passed on to the pump manufacturer in the form of a pump data sheet and specification. From the information given, the following will ultimately determine pump selection.

- Capacity range of liquid to be moved
- Differential head required
- NPSHA
- Shape of head capacity curve
- Pump speed
- Liquid characteristics
- Construction

Differential Head Required

The head to be generated by the pump is determined from the system head curve. This is a graphical plot of the total static head and friction

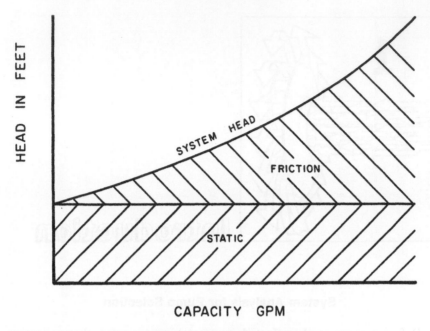

Figure 1-1. System head curve.

losses for various flow rates. For any desired flow rate, the head to be generated by the pump or pumps, can be read directly (Figures 1-1 and 1-2).

NPSHA

Net positive suction head available (NPSHA) is of extreme importance for reliable pump operation. It is the total suction head in feet of liquid absolute, determined at the suction nozzle and referred to datum, less the vapor pressure of the liquid in feet absolute. This subject is discussed in detail in Chapter 9.

Shape of Head Capacity Curve

The desired shape of the head capacity (H-Q) curve is determined during analysis of the system. Most specifications call for a continuously rising curve (Figure 1-3) with the percentage rise from the best efficiency point (BEP) determined by system limits and mode of operation. Unsta-

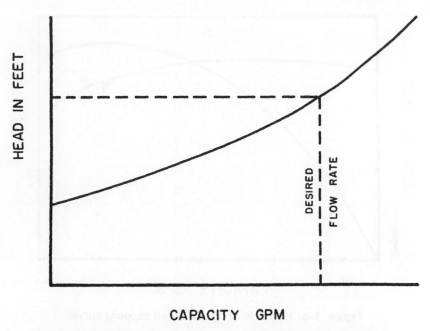

HEAD IN FEET

DESIRED

FLOW RATE

CAPACITY GPM

Figure 1-2. The system head curve establishes pump conditions.

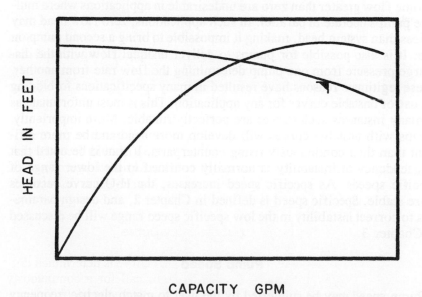

HEAD IN FEET

CAPACITY GPM

Figure 1-3. Continuously rising head capacity curve.

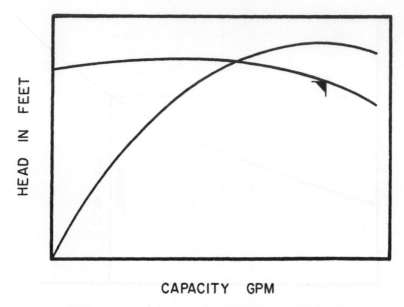

Figure 1-4. Unstable or hooked head capacity curve.

ble or hooked curves (Figure 1-4) where the maximum developed head is at some flow greater than zero are undesirable in applications where multiple pumps operate in parallel. In such applications, zero flow head may be less than system head, making it impossible to bring a second pump on line. It is also possible for pumps to deliver unequal flow with the discharge pressure from one pump determining the flow rate from another. These legitimate reasons have resulted in many specifications forbidding the use of unstable curves for any application. This is most unfortunate as in many instances such curves are perfectly suitable. More importantly, pumps with unstable curves will develop more head and be more efficient than their continuously rising counterparts. It should be noted that this tendency of instability is normally confined to the lower range of specific speeds. As specific speed increases, the H-Q curve becomes more stable. Specific speed is defined in Chapter 2, and design parameters to correct instability in the low specific speed range will be discussed in Chapter 3.

Pump Speed

Pump speed may be suggested by the user to match electric frequency or available driver speed. The pump manufacturer, however, has the ultimate responsibility and must confirm that the desired speed is compatible with NPSHA and satisfies optimum efficiency selection.

Liquid Characteristics

To have reasonable life expectancy, pump materials must be compatible with the liquid. Having intimate knowledge of the liquid to be pumped, the user will often specify materials to the pump manufacturer. When the pump manufacturer is required to specify materials, it is essential that the user supply all relevant information. Since liquids range from clear to those that contain gases, vapors, and solid material, essential information includes temperature, specific gravity, pH level, solid content, amount of entrained air and/or dissolved gas, and whether the liquid is corrosive. In determining final material selection the pump manufacturer must also consider operating stresses and effects of corrosion, erosion, and abrasion.

Viscosity

As liquid flows through a pump, the hydrodynamic losses are influenced by viscosity and any increase results in a reduction in head generated and efficiency, with an increase in power absorbed (Figure 1-5). Centrifugal pumps are routinely applied in services having viscosities below 3,000 SSU and have been used in applications with product viscosities up to 15,000 SSU. It is important to realize that the size of the internal flow passages has a significant effect on the losses, thus the smaller the pump is, the greater are the effects of viscosity. As the physical size of a pump increases, the maximum viscosity it can handle increases. A pump with a 3-in. discharge nozzle can handle 500 SSU, while a pump with a 6-in. discharge nozzle can handle 1,700 SSU. Centrifugal pumps can handle much higher viscosities, but beyond these limits, there is an increasing penalty loss. When viscosity is too high for a particular size pump, it will be necessary to go to a larger pump. A reasonable operating range of viscosity versus pump size is shown in Figure 1-6. Methods to predict pump performance with viscous liquids are clearly defined in the *Hydraulic Institute Standards*.

Specific Gravity

When pumping a nonviscous liquid, pumps will generate the same head uninfluenced by the specific gravity of the liquid. Pressure will change with specific gravity and can be calculated from:

$$\text{Differential pressure (psi)} = \frac{\text{Differential head (ft)} \times \text{sp gr}}{2.31}$$

VISCOUS PERFORMANCE CHANGE

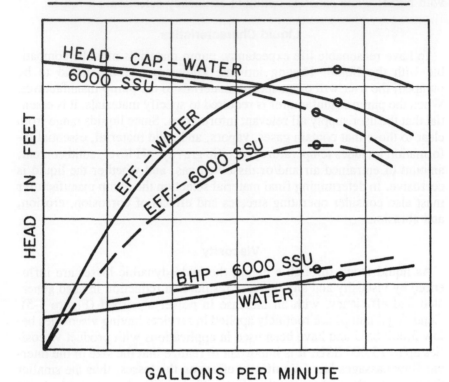

Figure 1-5. Viscous performance change.

Thus, pumps with a change in product density generating the same head will show a change in pressure, and horsepower absorbed by the pump will vary directly with the change in specific gravity (Figures 1-7 and 1-8). A pump being purchased to handle a hydrocarbon of 0.5 specific gravity will normally have a motor rating with some margin over end of curve horsepower. During factory testing on water with 1.0 specific gravity, the absorbed horsepower will be two times that of field operation, thus preventing use of the contract motor during the test. In such instances the pump manufacturers standard test motor is used.

Construction

Pump construction is quite often spelled out on the pump data sheet. General terms like horizontal, vertical, radial split, and axial split are

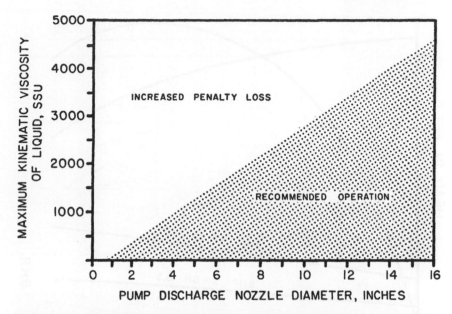

Figure 1-6. Maximum liquid viscosity for centrifugal pumps (from C.E. Petersen, Marmac, "Engineering and System Design Considerations for Pump Systems and Viscous Service," presented at Pacific Energy Association, October 15, 1982).

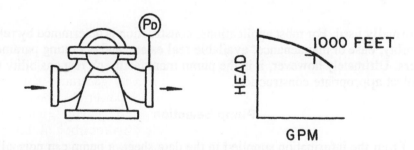

LIQUID	SPECIFIC GRAVITY	P_D
ETHANE	.5	216 PSI
CRUDE OIL	.85	368 PSI
WATER	1.0	433 PSI

Figure 1-7. Pressure vs. specific gravity.

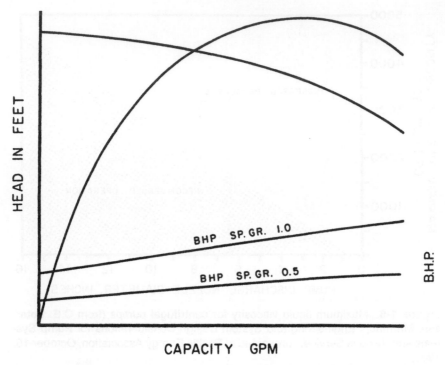

Figure 1-8. Horsepower change with specific gravity.

normally used. For most applications, construction is determined by reliability, ease of maintenance, available real estate, and operating parameters. Ultimately however, it is the pump manufacturer's responsibility to select appropriate construction.

Pump Selection

From the information supplied in the data sheet, a pump can normally be selected from the pump manufacturer's sales book. These are normally divided into sections, each representing a particular construction. Performance maps show the range of capacity and head available, while individual performance curves show efficiency and NPSHR. If the pump requirements fall within the performances shown in the sales book, the process of selection is relatively simple. When the required pumping conditions, however, are outside the existing range of performance, selection is no longer simple and becomes the responsibility of the pump designer.

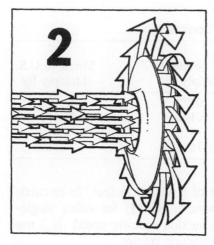

Specific Speed and Modeling Laws

Specific speed and suction specific speed are very useful parameters for engineers involved in centrifugal pump design and/or application. For the pump designer an intimate knowledge of the function of specific speed is the only road to successful pump design. For the application or product engineer specific speed provides a useful means of evaluating various pump lines. For the user specific speed is a tool for use in comparing various pumps and selecting the most efficient and economical pumping equipment for his plant applications.

A theoretical knowledge of pump design and extensive experience in the application of pumps both indicate that the numerical values of specific speed are very critical. In fact, a detailed study of specific speed will lead to the necessary design parameters for all types of pumps.

Definition of Pump Specific Speed and Suction Specific Speed

Pump specific speed (N_s) as it is applied to centrifugal pumps is defined in U.S. units as:

$$N_s = \frac{RPM \times GPM^{.5}}{H^{.75}}$$

Specific speed is always calculated at the best efficiency point (BEP) with maximum impeller diameter and single stage only. As specific speed can be calculated in any consistent units, it is useful to convert the calculated number to some other system of units. See Table 2-1. The suction specific speed (N_{ss}) is calculated by the same formula as pump specific speed (N_s)

11

Table 2-1
Specific Speed Conversion

Capacity	UNITS Head/ Stage	U.S. to Metric Multiply By	Metric to U.S. Multiply By
Ft³/Sec	Feet	.0472	21.19
M³/Sec	Meters	.0194	51.65
M³/Min	Meters	.15	6.67
M³/Hr	Meters	1.1615	.8609

but uses NPSHR values in feet in place of head (H) in feet. To calculate pump specific speed (N_s) use full capacity (GPM) for either single- or double-suction pumps. To calculate suction specific speed (N_{ss}) use one half of capacity (GPM) for double-suction pumps.

$$N_{ss} = \frac{RPM \times GPM^{.5}}{NPSHR^{.75}}$$

It is well known that specific speed is a reference number that describes the hydraulic features of a pump, whether radial, semi-axial (Francis type), or propeller type. The term, although widely used, is usually considered (except by designers) only as a characteristic number without any associated concrete reference or picture. This is partly due to its definition as the speed (RPM) of a geometrically similar pump which will deliver one gallon per minute against one foot of head.

To connect the term specific speed with a definite picture, and give it more concrete meaning such as GPM for rate of flow or RPM for rate of speed, two well known and important laws of centrifugal pump design must be borne in mind—the affinity law and the model law (the model law will be discussed later).

The Affinity Law

This is used to refigure the performance of a pump from one speed to another. This law states that for similar conditions of flow (i.e. substantially same efficiency) the capacity will vary directly with the ratio of speed and/or impeller diameter and the head with the square of this ratio at the point of best efficiency. Other points to the left or right of the best efficiency point will correspond similarly. The flow cut-off point is usually determined by the pump suction conditions. From this definition, the rules in Table 2-2 can be used to refigure pump performance with impeller diameter or speed change.

Table 2-2
Formulas for Refiguring Pump Performance with
Impeller Diameter or Speed Change

Diameter Change Only	Speed Change Only	Diameter and Speed Change
$Q_2 = Q_1 \left(\dfrac{D_2}{D_1}\right)$	$Q_2 = Q_1 \left(\dfrac{N_2}{N_1}\right)$	$Q_2 = Q_1 \left(\dfrac{D_2}{D_1} \times \dfrac{N_2}{N_1}\right)$
$H_2 = H_1 \left(\dfrac{D_2}{D_1}\right)^2$	$H_2 = H_1 \left(\dfrac{N_2}{N_1}\right)^2$	$H_2 = H_1 \left(\dfrac{D_2}{D_1} \times \dfrac{N_2}{N_1}\right)^2$
$bhp_2 = bhp_1 \left(\dfrac{D_2}{D_1}\right)^3$	$bhp_2 = bhp_1 \left(\dfrac{N_2}{N_1}\right)^3$	$bhp_2 = bhp_1 \left(\dfrac{D_2}{D_1} \times \dfrac{N_2}{N_1}\right)^3$

Q_1, H_1, bhp_1, D_1, and N_1 = Initial capacity, head, brake horsepower, diameter, and speed.
Q_2, H_2, bhp_2, D_2, and N_2 = New capacity, head, brake horsepower, diameter, and speed.

Example

A pump operating at 3,550 RPM has a performance as shown in solid lines in Figure 2-1. Calculate the new performance of the pump if the operating speed is increased to 4,000 RPM.

Step 1

From the performance curve, tabulate performance at 3,550 RPM (Table 2-3).

Step 2

Establish the correction factors for operation at 4,000 RPM.

$4,000/3,550 = $ $f = 1.13.$
$f^2 = 1.27.$
$f^3 = 1.43.$

Step 3

Calculate new conditions at 4,000 RPM from:

$Q_2 = Q_1 \times 1.13.$
$H_2 = H_1 \times 1.27.$
$bhp_2 = bhp_1 \times 1.43.$

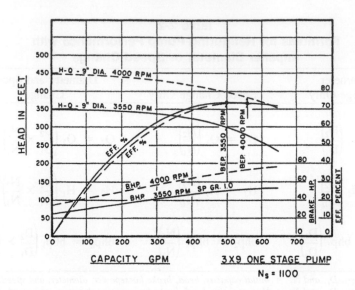

Figure 2-1. New pump factored from model pump—different speed.

Table 2-3
Tabulated Performance at 3,550 RPM

GPM	H(ft)	Eff. %	bhp
0	350	0	25
100	349	28	31
200	345	48	36
300	337	52	42
400	325	70	46
500	300	74	51
600	260	73	54
650	235	72	53

Results are tabulated in Table 2-4 and shown as a dotted line, in Figure 2-1. Note that the pump efficiency remains the same with the increase in speed.

Specific Speed Charts

We have prepared a nomograph (Figure 2-2) relating pump specific speed and suction specific speed to capacity, speed, and head. The nomograph is very simple to use: Locate capacity at the bottom of the graph, go vertically to the rotating speed, horizontally to TDH, and vertically to

Table 2-4
Tabulated Performance at 4,000 RPM

GPM	H(ft)	Eff. %	bhp
0	445	0	37
113	443	28	45
226	438	48	52
337	427	52	60
452	412	70	66
565	381	74	73
678	330	73	77
732	298	72	76

the pump specific speed. To obtain suction specific speed continue from the rotating speed to NPSHR and vertically to the suction specific speed. Pump specific speed is the same for either single-suction or double-suction designs.

For estimating the expected pump efficiencies at the best efficiency points, many textbooks have plotted charts showing efficiency as a function of specific speed (N_s) and capacity (GPM). We have prepared similar charts, but ours are based on test results of many different types of pumps and many years of experience.

Figure 2-3 shows efficiencies vs. specific speed as applied to end-suction process pumps (API-types). Figure 2-4 shows them as applied to single-stage double-suction pumps, and Figure 2-5 shows them as applied to double-volute-type horizontally split multi-stage pumps.

Figure 2-5 is based on competitive data. It is interesting to note that although the specific speed of multi-stage pumps stays within a rather narrow range, the pump efficiencies are very high, equal almost to those of the double-suction pumps. The data shown are based on pumps having six stages or less and operating at 3,560 RPM. For additional stages or higher speed, horsepower requirements may dictate an increase in shaft size. This has a negative effect on pump performance and the efficiency shown will be reduced.

As can be seen, efficiency increases very rapidly up to N_s 2,000, stays reasonably constant up to N_s 3,500, and after that begins to fall off slowly. The drop at high specific speeds is explained by the fact that hydraulic friction and shock losses for high specific speed (low head) pumps contribute a greater percentage of total head than for low specific speed (high head) pumps. The drop at low specific speeds is explained by the fact that pump mechanical losses do not vary much over the range of specific speeds and are therefore a greater percentage of total power consumption at the lower specific speeds.

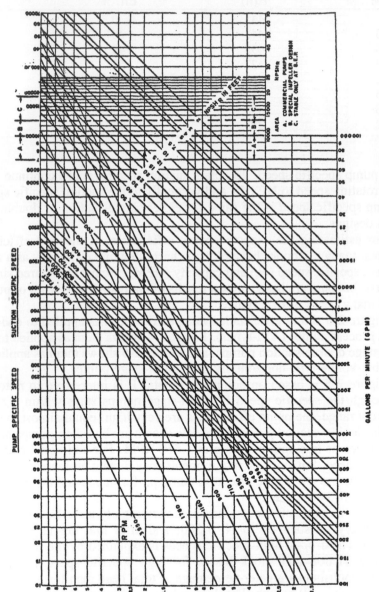

Figure 2-2. Specific speed and suction specific speed nomograph.

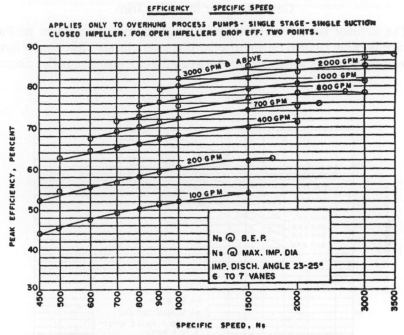

Figure 2-3. Efficiency for overhung process pumps.

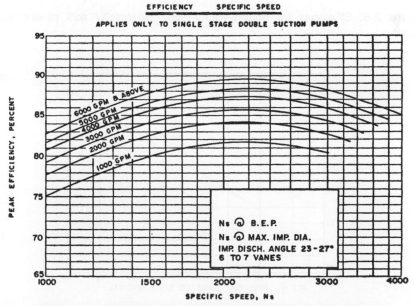

Figure 2-4. Efficiency for single-stage double-suction pumps.

Correction for Impeller Trim

The affinity laws described earlier require correction when performance is being figured on an impeller diameter change. Test results have shown that there is a discrepancy between the calculated impeller diame-

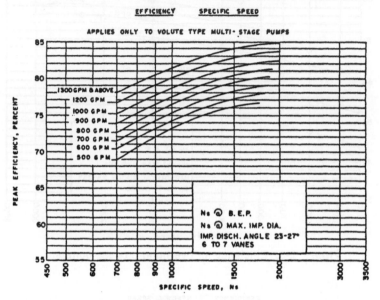

Figure 2-5. Efficiency for double-volute-type horizontally split multi-stage pumps.

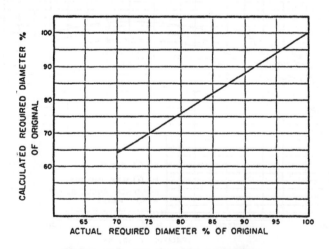

Figure 2-6. Impeller trim correction.

ter and the achieved performance. The larger the impeller cut, the larger the discrepancy as shown in Figure 2-6.

Example

What impeller trim is required on a 7-in. impeller to reduce head from 135 ft to 90 ft?

Step 1.

From affinity laws:

$$H_2 = H_1 \left(\frac{D_2}{D_1}\right)^2$$

$$90 = 135 \left(\frac{D_2}{7}\right)^2$$

$$D_2 = 5.72 \text{ in.}$$

Calculated percent of original diameter $= 5.72/7 = .82$

Step 2.

Establish correction factor:

From Figure 2-6 calculated diameter $.82 =$ Actual required diameter $.84$.

Trim diameter $= 7 \times .84 = 5.88$ in.

Impeller trims less than 80% of original diameter should be avoided since they result in a considerable drop in efficiency and might create unstable pump performance. Also, for pumps of high specific speed (2,500–4,000), impeller trim should be limited to 90% of original diameter. This is due to possible hydraulic problems associated with inadequate vane overlap.

Model Law

Another index related to specific speed is the pump modeling law. The "model law" is not very well known and usually applies to very large pumps used in hydroelectric applications. It states that two geometrically similar pumps working against the same head will have similar flow conditions (same velocities in every pump section) if they run at speeds inversely proportional to their size, and in that case their capacity will vary with the square of their size. This is easily understood if we realize that the peripheral velocities, which are the product of impeller diameter and RPM, will be the same for the two pumps if the diameter increase is inversely proportional to the RPM increase. The head, being proportional to the square of the peripheral velocity, will also be the same. If the velocities are the same, the capacities will be proportional to the areas, i.e. to the square of the linear dimensions.

As a corollary, the linear dimensions of similar pumps working against the same head will change in direct proportion to the ratio of the square root of their capacities, and the RPM in inverse proportion to the same ratio. This permits selection of a model pump for testing as an alternative to building a full-size prototype. The selected model must agree with the following relationship:

$$\frac{N_1}{N} = \left(\frac{D}{D_1}\right)\left(\frac{H_1}{H}\right)^{.5}$$

$$\frac{Q_1}{Q} = \left(\frac{D_1}{D}\right)^2 \left(\frac{H_1}{H}\right)^{.5}$$

$$\frac{1 - \eta_1}{1 - \eta} = \left(\frac{D}{D_1}\right)^n$$

where D_1, N_1, H_1, and η_1, are model diameter, speed, head, and efficiency, and D, N, H and η, are prototype diameter, speed, head, and efficiency. n will vary between zero and 0.26, depending on relative surface roughness.

Other considerations in the selection of a model are:

1. Head of the model pump is normally the same as the prototype head. However, successful model testing has been conducted with model head as low as 80% prototype head.
2. Minimum diameter of the model impeller should be 12 in.
3. Model speed should be such that the specific speed remains the same as that of the prototype.
4. For meaningful evaluation prototype pump and model pump must be geometrically similar and flow through both kinematically similar.
5. Suction requirements of model and prototype should give the same value of sigma (see Chapter 9).

An example of model selection is described in detail in the *Hydraulic Institutes Standards, 14th Edition.*

Factoring Law

In addition to the affinity law and model law, there are some other principles of similarity that are very useful to the pump designer. The "factoring law" can be used to design a new pump that has the same specific speed and running speed as the existing model but which is larger or smaller in size. Factoring the new pump can be determined by using capacity or impeller diameter ratio. From BEP condition, a new pump head and flow are calculated from:

$$Q_1 = Q_m \times f^3$$
$$H_1 = H_m \times f^2$$

$$\frac{D_1}{D_m} = f$$

where Q_1 = Capacity of new pump
H_1 = Head of new pump
D_1 = Impeller diameter of new pump
Q_m = Capacity of model pump
H_m = Head of model pump
D_m = Impeller diameter of model pump

Thus, the capacity will change with factor cubed (f^3). The head and all areas will change with factor squared (f^2), and all linear dimensions will change directly with factor (f).

The specific speed of the model pump should be the same as the new pump or within ± 10% of the new pump's specific speed. The model pump should ideally be of the same type as the new pump.

Example

If we take the 3-in. × 9-in. pump shown in solid lines on Figure 2-1, with a performance at the BEP of 500 GPM; 300 ft; 74% efficiency; 55 maximum HP; 3,550 RPM; and 9-in. impeller diameter, and increase pump size to 700 GPM at BEP and running at 3,550 RPM, the new performance would be:

$$\frac{Q_1}{Q_m} = \frac{700}{500} = f^3 = 1.40$$

$$f^2 = 1.25$$

$$f = 1.12$$

Table 2-5
Tabulated Performance of Model Pump
(Model 9-In. Diameter Impeller)

Q_m	H_m	$Eff._m$ %	bhp_m
0	350	0	25
100	349	28	31
200	345	48	36
300	337	52	42
400	325	70	46
500	300	74	57
600	260	73	54
650	235	72	54

Table 2-6
Tabulated Performance of New Pump
(New Pump 10¹/₈-In. Diameter Impeller)

Q_1	H_1	$Eff._1$ %	bhp_1
0	437	0	50
140	436	28	55
280	431	49	62
420	421	53	84
560	406	71	81
700	375	75	88
840	325	74	93
910	293	72	94

Applying these factors to the model, the new pump performance at BEP will be: 700 GPM; 375 ft; 75% efficiency; 94 maximum HP; 3,550 RPM; and 10¹/₈-in. impeller diameter.

To obtain complete H-Q performance refer to Tables 2-5 and 2-6.

The new pump size will be a 4-in. × 6-in. × 10-in. with performance as shown in Figure 2-7.

All major pump manufacturers have in their files records of pump performance tests covering a wide range of specific speeds. Each test can be used as a model to predict new pump performance and to design same. In the majority of cases, the required model specific speed can be found having the same running speed as required by the new pump. There are cases, however, where the model pump running speed is different than required by pump in question.

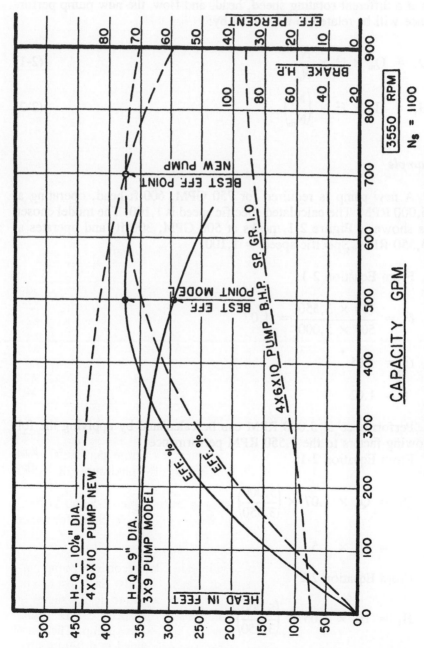

Figure 2-7. New pump factored from model pump—same speed.

If a new pump has the same specific speed (N_s) as the model but is to run at a different rotating speed, head, and flow, the new pump performance will be related to the model by:

$$Q_1 = Q_m \times f^3 \times \frac{N_1}{N_m} \tag{2-1}$$

$$H_1 = H_m \times f^2 \times \left(\frac{N_1}{N_m}\right)^2 \tag{2-2}$$

Example

A new pump is required for 750 GPM, 600-ft head, operating at 5,000 RPM. The calculated specific speed is 1,100. The model chosen is shown in Figure 2-1, peaks at 500 GPM, 300 ft, and operates at 3,550 RPM. Specific speed = 1,100.

From Equation 2-1:

$$f^3 = \frac{750 \times 3,550}{500 \times 5,000} = 1.07$$

$$f^2 = 1.05$$

$$f = 1.02$$

Performance at 5,000 RPM can be calculated by applying the following factors to the 3,550 RPM performance,
From Equation 2-1:

$$Q_1 = Q_m \times 1.07 \times \left(\frac{5,000}{3,550}\right)$$

$$= Q_m \times 1.5$$

From Equation 2-2:

$$H_1 = H_m \times 1.05 \times \left(\frac{5,000}{3,550}\right)^2$$

$$= H_m \times 2.07$$

Table 2-7
Tabulated Performance at 3,550 RPM

GPM	H(ft)	Eff. %
0	350	0
100	349	28
200	345	48
300	337	52
400	325	70
500	300	74
600	260	73
650	235	72

Table 2-8
Tabulated Performance at 5,000 RPM

GPM	H(ft)	Eff. %	bhp
0	700	0	91
150	698	28	94.4
300	690	48	109
450	674	52	142
600	650	70	140
750	600	74	153
900	520	73	161
975	470	72	160

For complete performance conversion, see Tables 2-7 and 2-8. The hydraulic performance of the new 4-in. × 6-in. × 9-in. pump operating at 5,000 RPM is shown in Figure 2-8.

Conclusion

There is no question that specific speed is the prime parameter for evaluating design of centrifugal pumps, evaluating pump selections, and predicting possible field problems. It is obvious, however, that no single parameter can relate to all aspects of final pump design. Different pump specifications covering a wide variety of applications force the designer to consider additional factors that may have an unfavorable effect on pump hydraulic performance. Predicted performance can be affected by any of the following:

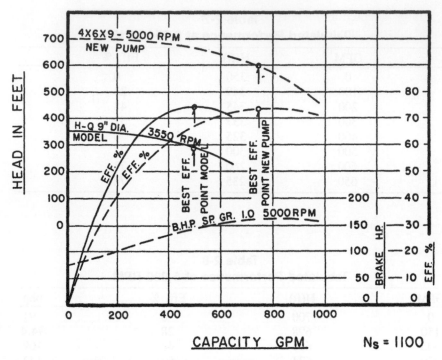

Figure 2-8. Performance change with speed change.

- *Mechanical Considerations*
  High horsepower/large shaft diameter
  High suction pressure
  Operating speed
  Operating temperature
  Running clearances
- *Pump Liquid*
  Slurries
  Abrasives
  High viscosity
  Dissolved gases
- *System Considerations*
  NPSHA
  Suction piping arrangement
  Discharge piping arrangement
  Shape of H-Q curve
  Runout conditions
  Vibration limits
  Noise limits

Pumps selected as models must have reliable hydraulic performance based on accurate instrumentation. Unreliable model tests used as a basis for new development will lead to total disaster. It is strongly suggested that the selected model pump be retested to verify its hydraulic performance.

In the following chapters of this book "specific speed" will be referred to with many additional charts, curves, and technical plots.

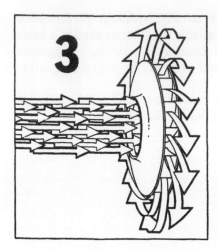

Impeller Design

In this chapter detailed information for designing a centrifugal pump impeller will be presented. This information will apply to a new design where a model in the same specific speed is not available. The design factors shown are based on a theoretical approach and many years of collecting thousands of performance tests in various specific speeds.

The following example, will demonstrate the procedure necessary to design and layout a new impeller.

Example

The requirements for a new pump are shown on curve Figure 3-1, which at Best Efficiency Point (BEP) is:

2,100 GPM—450 ft—3,600 RPM—Liquid—Water

Step 1: Calculate pump specific speed.

$$N_s = \frac{RPM \times (GPM)^{.5}}{(H)^{.75}}$$

$$= \frac{3,600 \times (2,100)^{.5}}{(450)^{.75}} = 1,688$$

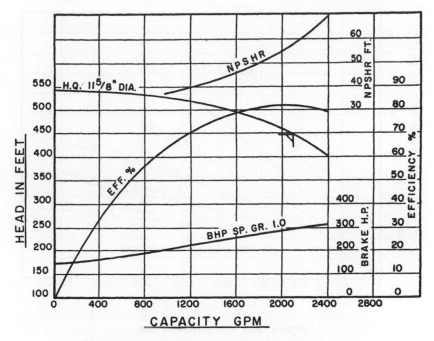

Figure 3-1. Required performance of new pump.

Step 2: Select vane number and discharge angle.

The desired head rise from BEP to zero GPM is 20% continuously rising (Figure 3-1). To produce this head rise, the impeller should be designed with six equally spaced vanes having a 25° discharge angle (Figure 3-2).

Step 3: Calculate impeller diameter.

From Figure 3-3:

Head constant K_u = 1.075

$$D_2 = \frac{1,840 \times K_u \times H^{.5}}{RPM}$$

$$= \frac{1,840 \times 1.075 \times (450)^{.5}}{3,600}$$

$$= 11.66 \text{ in. (say } 11^5/8 \text{ in.)}$$

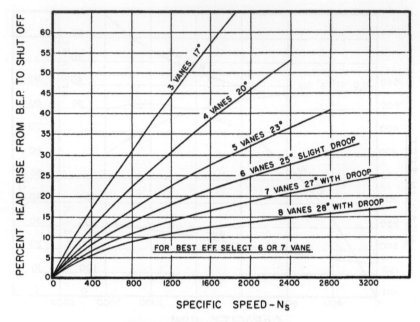

Figure 3-2. Percent head rise.

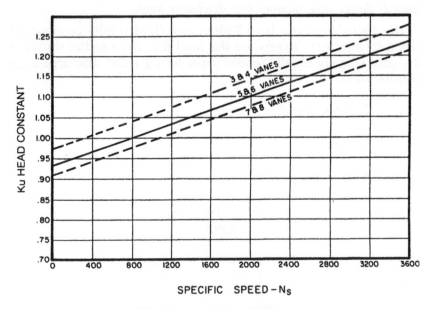

Figure 3-3. Head constant.

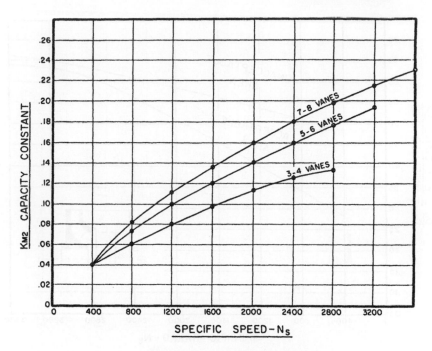

Figure 3-4. Capacity constant.

Step 4: Calculate impeller width b_2.

From Figure 3-4:

$$K_{m2} = .125$$

$$C_{m2} = K_{m2} \times (2gH)^{.5}$$

$$= .125 \times 170 = 21.3 \text{ ft/sec}$$

$$b_2 = \frac{\text{GPM} \times .321}{C_{m2} \times (D_2\pi - ZS_u)}$$

Estimated S_u = ½ in. (This will be confirmed during vane development and the calculation repeated if necessary.)

$$b_2 = \frac{2,100 \times .321}{21.3(11.66\pi - 6 \times .5)} = 1.09 \text{ in.}$$

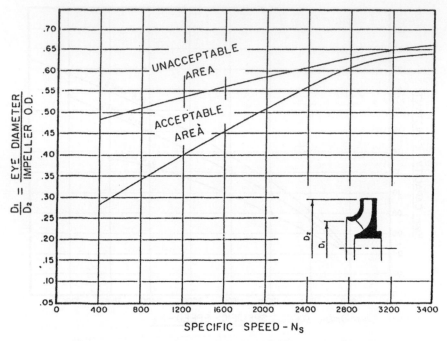

Figure 3-5. Impeller eye diameter/outside diameter ratio.

Step 5: Determine eye diameter.

From Figure 3-5:

D_1/D_2 = .47

D_1 = 11.66 × .47 = 5.5 in.

Step 6: Determine shaft diameter under impeller eye.

Methods for calculating shaft diameter are discussed in detail under shaft design. For this exercise, the shaft diameter under the impeller eye is assumed to be 2 in.

Step 7: Estimate impeller eye area.

Eye area = Area at impeller eye − shaft area

= 23.76 − 3.1 = 20.66 sq in.

Step 8: Estimate NPSHR.

From Figure 3-6:

$$C_{ml} = \frac{2,100 \times .321}{20.66} = 32.63 \text{ ft/sec}$$

$$U_t = \frac{5.5 \times 3,600}{229} = 86.5 \text{ ft/sec}$$

From Figure 3-6:

$$\text{NPSHR} = 59 \text{ ft}$$

$$N_{ss} = \frac{3,600 \times (2,100)^{.5}}{(59)^{.75}} = 7,749$$

Step 9: Determine volute parameters.

To finalize the impeller design, we must consider any mechanical limitations of the pump casing. Designing the impeller alone is not sufficient as we must see how it physically relates to the volute area, cutwater diameter, volute width, etc.

- *Volute area.* Figure 3-7 shows a number of curves for volute velocity constant K_3. These represent the statistical gathering efforts of a number of major pump companies. Figure 3-8 shows the average of these curves and is recommended for estimating volute area.

$$\text{Volute area } A_8 = \frac{0.04 \times \text{GPM}}{K_3 \times (H)^{.5}}$$

$$= \frac{.04 \times 2,100}{.365 \times (450)^{.5}}$$

$$= 10.85 \text{ sq in.}$$

The calculated volute area is the final area for a single volute pump. For double-volute pumps, this area should be divided by 2, and for diffuser type pumps it should be divided by the number of vanes in the diffuser casing.

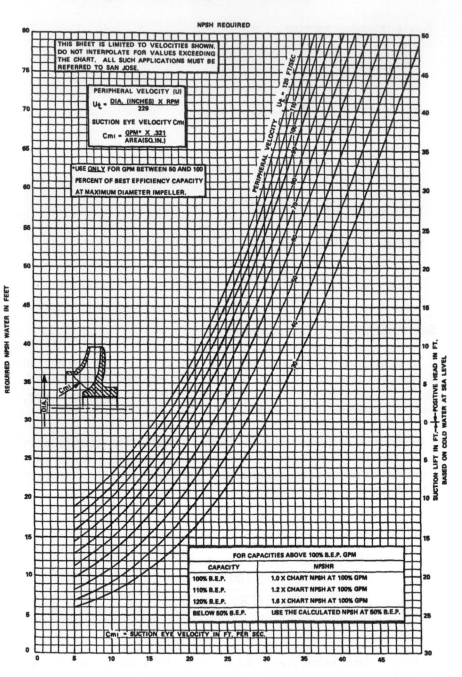

Figure 3-6. NPSHR prediction chart.

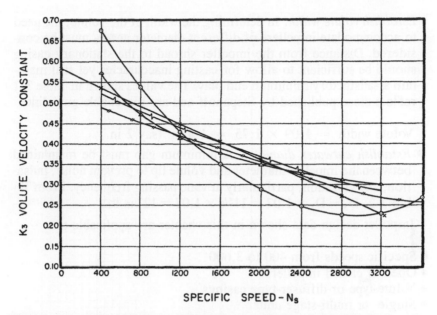

Figure 3-7. Volute velocity constant (data acquisition of different pump companies).

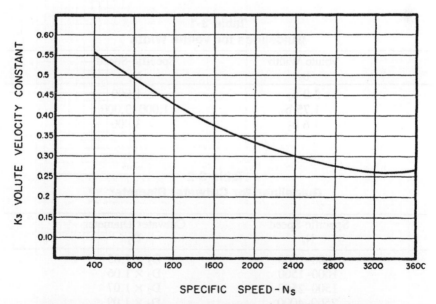

Figure 3-8. Volute velocity constant (author's recommendation).

- *Establish volute width.* In determing the width of the volute, the need to accommodate impellers of different diameter and b_2 must be considered. Distance from the impeller shroud to the stationary casing should be sufficient to allow for casting inaccuracies yet still maintain a satisfactory minimum end play. The values shown in Table 3-1 reflect those published by Stepanoff and are reasonable guidelines.

Volute width $= 1.09 \times 1.75 = 1.9$ in. (say 2 in.)

- *Establish cutwater diameter.* A minimum gap must be maintained between the impeller diameter and volute lip to prevent noise, pulsation, and vibration, particularly at vane passing frequency. From Table 3-2, $D_3 = D_2 \times 1.07 = 11^5/8 \times 1.07 = 12^7/16$ in.

Impeller design data shown in this chapter are applicable to:

- Specific speeds from 400 to 3,600.
- Open or closed impellers.
- Volute-type or diffuser-type casings.
- Single- or multi-stage units.
- Vertical or horizontal pumps.
- Single suction or double suction impellers.

In establishing discharge geometry for double suction impellers, use factors as given in this chapter. Do not divide the specific speed by

Table 3-1
Guidelines for Volute Width

Volute Width b_3	Specific Speed N_s
2.0 b_2	< 1,000
1.75 b_2	1,000–3,000
1.6 b_2	> 3,000

Table 3-2
Guidelines for Cutwater Diameter

Specific Speed N_s	Cutwater Diameter D_3
600–1000	$D_2 \times 1.05$
1000–1500	$D_2 \times 1.06$
1500–2500	$D_2 \times 1.07$
2500–4000	$D_2 \times 1.09$

$(2)^{.5}$, as final results will not be accurate. On the impeller suction side however, divide the inlet flow and eye area by 2 for each side of the impeller centerline.

Impeller Layout

Development of Impeller Profile (Plan View)

Lay out the impeller width b_2 at full diameter (Figure 3-9). Begin to develop the hub and shroud profiles by expanding this width by approximately 5° on each side of the vertical center line toward the suction eye. Complete these profiles by ending in such a way as to produce the required eye area. The area change from eye to discharge should be gradual. To minimize axial and radial thrust, make the impeller as symmetrical as possible about its vertical centerline.

Development of Impeller End View

Draw a circle to the full impeller diameter. Divide the circle into several even-degree segments. In this case 15° segments have been chosen (Points 1–10). The smaller the segments, the more accurate the vane layout will be.

Impeller Inlet Angles

Inlet angles are established from a layout of the velocity triangle, which shows the various component velocities of flow, entering the impeller (Figure 3-10). The vector connecting U_t and C_{M1} represents the angle of flow θ. Vane angle B_1 is drawn to intersect P_{S1} and should be greater than θ to allow for recirculated flow and nonuniform velocity. For reasonable design the prerotation angle should not exceed 30° and for optimum NPSH it is recommended that $P_{S1} = 1.05$ to 1.2 times C_{M1}.

Development of Impeller Vane

Begin the vane development by drawing a line equal to the discharge angle. On the end view estimate and locate the distance a_r. Transmit a_r to discharge angle line on the vane development establishing the distance a. Transmit a to the front shroud on the impeller profile (plan view). Measure R_2, then scribe R_2 on the end view, locating Point 2. Check to see if the estimated a_r was approximately correct, making a new estimate and repeating the process if necessary. Continue this same procedure to the minimum impeller cut diameter, in this case Point 4. Complete the vane development for the shroud contour by drawing several lines equal to the

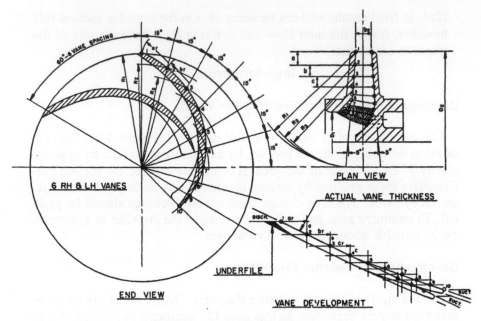

Figure 3-9. Impeller layout.

suction entrance angles to establish match between discharge angle and suction angles. Once a match is established with a smooth curve, lay out the remaining vane points until the eye diameter is reached, in this case at Point 10.

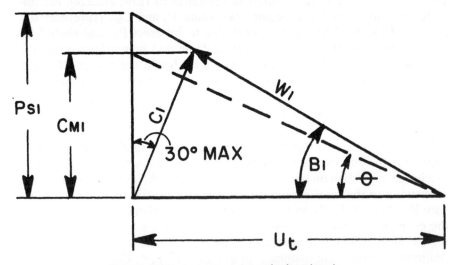

Figure 3-10. Impeller inlet velocity triangle.

The vane development and layout for the back shroud is done in the same manner, taking into account the required hub vane angle. This procedure should result in Point 10 reaching the estimated hub vane diameter. If this diameter is missed by more than 1/4-in., the hub vane angle should be changed and the layout repeated.

Complete the vane layout by adding vane thickness (Figure 3-11) and indicating a slight underfile at the vane OD and a thinning at the suction. This underfile will produce higher head and improved efficiency. The vane spacing location of the second vane on the end view will be determined by the number of vanes.

At this point it is necessary to check the ratio of the vane area A_v to the suction eye area A_e. Referring to Figure 3-12, this ratio should be 0.4 to 0.6. The area A_v is the area which is shaded on the plan view. If the area ratio falls outside the limits shown, it will be necessary to change the plan view profile and/or vane thickness.

Pumps in the higher specific speed range (N_s 1,000–5,000) have suction velocity triangles that dictate different suction vane angularity at the impeller hub and impeller eye. This difference in angularity restricts vane removal from the core box, requiring a segmented core and more costly patterns and casting. Hydraulically, it is very necessary, and we must learn to live with it. On low specific speeds however, (N_s 500–1000) a straight vane is acceptable and will not jeopardize hydraulic performance.

Design Suggestions

The following points should be kept in mind when designing an impeller.

- Standardize the relationship between the number of vanes and discharge vane angle. The relationship shown in Figure 3-13 is suggested. Number and angularity of vanes greatly affect H-Q pump performance. Standardization as shown will lead to more accurate performance prediction.
- Figure 3-14 shows the effect of the number of impeller vanes in the same casing, on H-Q performance. Please note that with less vanes, the lower is the BEP head and efficiency and steeper is the shape of H-Q curve.
- It is possible to maintain same BEP, Q, and efficiency by increasing b_2 where number of vanes and angularity are reduced (Figure 3-15). For best efficiency, the velocity ratio between volute and impeller peripheral must be maintained. Similarly for the same impeller diameter, vane number and discharge angle BEP will change with change in b_2 (Figure 3-16).
- Avoid using even number of vanes in double volute pumps.

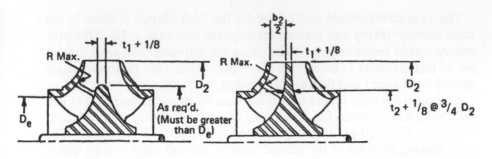

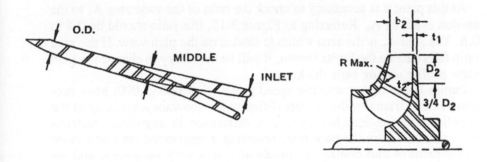

IMP. DIA	MIN. VANE THICKNESS			MIN. SHROUD THICKNESS	
	O.D.	MIDDLE	INLET	t_1	t_2
> 6-11	7/32	5/16	1/8	1/8	3/16
> 11-15	1/4	3/8	3/16	5/32	5/16
> 15-19	9/32	7/16	3/16	3/16	3/8
> 19-23	5/16	1/2	1/4	7/32	7/16
> 23-31	11/32	9/16	1/4	1/4	9/16
> 31-35	3/8	5/8	5/16	5/16	5/8
> 35-39	7/16	11/16	5/16	3/8	11/16
> 39-50	1/2	3/4	3/8	7/16	3/4

Figure 3-11. Recommended minimum impeller vane and shroud thickness for castability (standard cast materials).

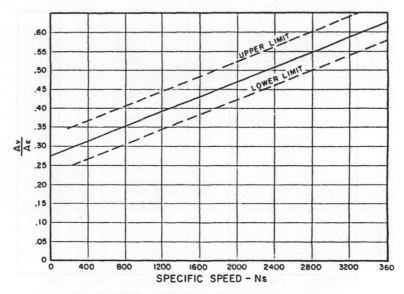

Figure 3-12. Area between vanes/eye area ratio.

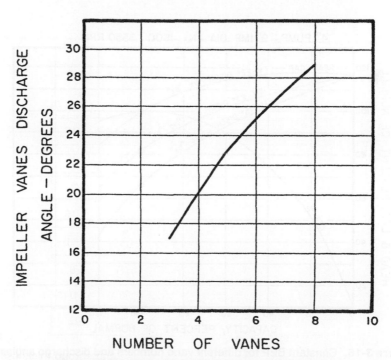

Figure 3-13. Suggested standard for vane number.

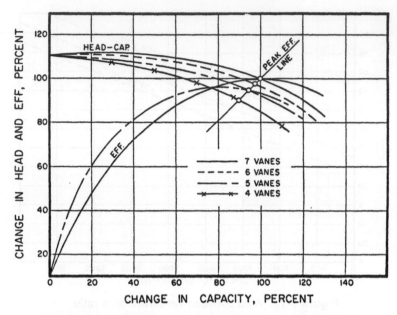

Figure 3-14. Influence of vane number on pump performance.

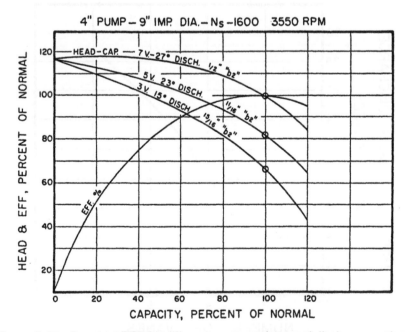

Figure 3-15. Constant BEP for different vane numbers and discharge angles by change in b_2.

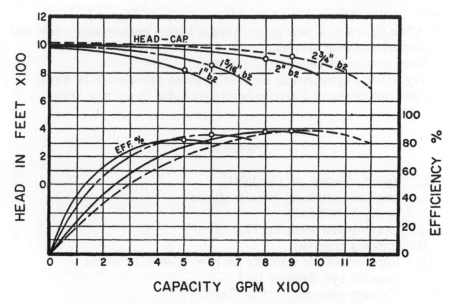

Figure 3-16. Influence of b_2 on pump performance.

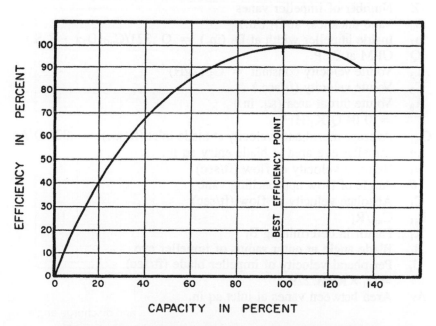

Figure 3-17. Predicting efficiency on both sides of BEP.

- To predict efficiency on both sides of BEP, use Figure 3-17.
- When designing a new pump, always have a performance curve and general sectional drawing.
- Keep neat and accurate design records.
- Try to use a symmetrical impeller to avoid excessive axial thrust.
- Consider mechanical requirements applicable to the new design.

Notation

K_u Speed constant = $U_2/(2gH)^{.5}$

U_2 Impeller peripheral velocity (ft/sec)
= $D_2 \times RPM/229$

g Gravitational constant (32.2 ft/sec²)

H Impeller head (ft)

D_2 Impeller outside diameter (in.) = $1840\ K_u H^{.5}/RPM$

D_1 Impeller eye diameter (in.)

K_{m2} Capacity constant = $C_{m2}/(2gH)^{.5}$

C_{m2} Radial velocity at impeller discharge (ft/sec) = $Q\ .321/A_2$

D_3 Volute cutwater diameter (in.)

A_2 Impeller discharge area (sq in.)
= $(D_2\pi - ZS_u)b_2$

Z Number of impeller vanes

S_u Vane thickness at D_2 (in.)

b_2 Inside impeller width at D_2 (in.) = $Q\ .321/C_{m2}(D_2\pi - Z\ S_u)$

Q GPM at BEP

K_3 Volute velocity constant = $C_3/(2gH)^{.5}$

C_3 Volute velocity (ft/sec.)

A_8 Volute throat area (sq. in.)
= $0.04\ Q/K_3(H)^{.5}$

C_{m1} Average meridianal velocity at blade inlet (ft/sec) = $.321\ Q/Ae$

Ae Impeller eye area at blade entry sq in.

w_1 Relative velocity of flow (ft/sec)

θ Angle of flow approaching vane

C_1 Absolute velocity of flow (ft/sec)

P_{s1} C_{m1}/R_1

R_1 Factor in determining B_1

B_1 Blade angle at outer radius of impeller eye

U_t Peripheral velocity of impeller blade (ft/sec)
= $D_1 \times RPM/229$

Av Area between vanes at inlet sq in.

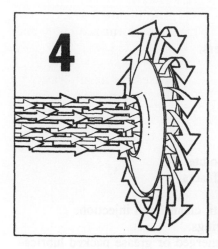

General Pump Design

It is not a difficult task to design a centrifugal pump; however, designing the right pump for a specific application related to a specific industry and service requires an extensive knowledge of hydraulics. Also required is experience with industrial specifications, end users, and contractors' special requirements and many years of practical experience in engineering and marketing.

The variables that exist for pump requirements are so numerous that the design of the right pump in the right service is a complex project. There is no such product as the "universal pump."

As an example, let us take a pump that is required to produce 500 GPM and 200-ft head, rotating at 2 or 4 pole speed. In any or all industries this hydraulic requirement exists; however, the mechanical specifications are entirely different for each and every industry. For instance, the type of pumps used in the pulp and paper industry are entirely different from pumps used in the petroleum industry, petrochemical industry, or chemical industry. Thus, the pump required to deliver 500 GPM and 200-ft head, will be different for each of the following applications:

- Slurry
- Boilerfeed
- Pipeline
- Nuclear
- Municipal
- Agricultural
- Marine
- Cargo

45

Mechanical variables include:

- Open or closed or semi-open impellers.
- Single-stage or multi-stage.
- Vertical or horizontal.
- With water jacket or without.
- Overhung or two-bearings design.
- Close-coupled or coupled units.
- Stiff shaft or flexible shaft design.
- Single volute, double volute, quad volute.
- Short elbow-type or turbine-type diffusers.
- Mechanical seals or packing.
- Stuffing boxes with bleed-offs or with clean flush injection.
- Ball, sleeve, or Kingsbury-type bearings.
- Oil rings, forced feed, oil mist, submerged or grease packed lubrication.

It can be seen from the variables listed that it is a complex job for a pump designer to design the right pump for the right environment.

After complete hydraulic and mechanical specifications are established, the designer should be ready for pump layout documents.

General pump design can be classified in the following categories:

1. Design a new pump to satisfy basic engineering requirements such as shape of H-Q curve, NPSHA, efficiency, etc.
2. Design a new pump to satisfy special applications such as boiler-feed, nuclear coolant, pipeline, large circulator, condensate, secondary recovery, etc.
3. Design a new line of pumps, such as API pumps, ANSI pumps, double-suction pumps, pulp and paper pumps, building trade pumps, boilerfeed pumps, etc.

Performance Chart

For pumps in any category, an overall performance chart should be prepared (if not available) as a first step in the design study. This chart will establish the flow and head for each pump, establish the number and size of pumps required to satisfy the range chart, and avoid overlap or gaps between pump sizes. Even if only one pump is required, the range chart should be confronted to be sure that the new pump fits into the overall planning.

Many old pump lines have poorly planned range charts, resulting in similar pump overlaps and uncovered gaps between pump sizes. Such a

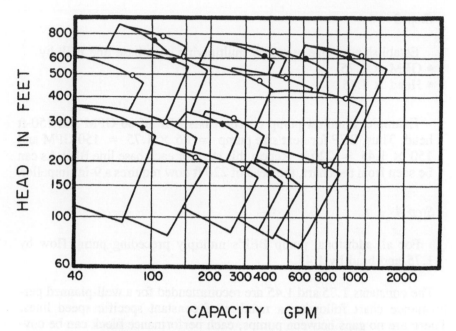

Figure 4-1. Poorly planned performance chart.

chart is shown in Figure 4-1. Black dots show pumps that could be eliminated, permitting a substantial reduction in inventory without hurting overall hydraulic performance. This type of chart is not recommended. A suggested properly layed out performance range chart is shown in Figure 4-2. Steps to develop this chart are now described.

Step 1

Establish BEP of lowest capacity and lowest head pump required. In this example, this is 86 GPM, 150-ft head, which is achieved by a 1-in. pump, and a 7-in. diameter impeller.

Step 2

Extend BEP capacity coverage at a constant 150-ft head by multiples of 1.75. Thus,

Second pump BEP = 86 GPM × 1.75 = 150 GPM.
Third pump BEP = 150 GPM × 1.75 = 260 GPM, etc.

In this manner, the base line BEP is established.

Step 3

Establish next size pump by multiplying each base line BEP by
- GPM × 1.75
- Head × 1.45

Example: Smallest pump on chart has BEP of 86 GPM and 150-ft head. Thus, BEP for next size pump = 86 × 1.75 = 150 GPM and 150 × 1.45 = 220 ft. Repeat this step for each base line BEP. As can be seen from the chart, the head of 220 ft now requires a 9-in. impeller.

Step 4

For all additional pump BEP's multiply preceding pump flow by 1.75 and head by 1.45.

The constants 1.75 and 1.45 are recommended for a well-planned performance chart following a number of constant specific speed lines. There are no gaps between pumps, each performance block can be covered by normal impeller trim, and there are no overlaps. The chart is also helpful to the designer. In this example, only six small pumps have to be

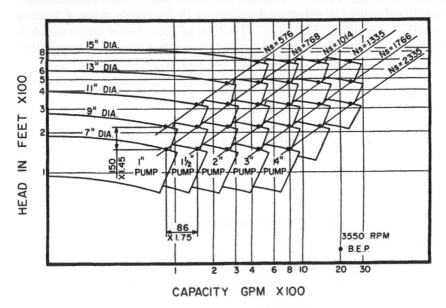

Figure 4-2. Recommended performance chart.

designed and test checked. The other sizes that follow specific speed lines can be factored up and their performance accurately predicted.

designed and test checked. The other sizes that follow specific speed lines can be factored up as described in Chapter 2 and their performance accurately predicted.

After the hydraulic performance range chart is complete, the designer should check the mechanical requirements as outlined by the applicable industrial specification.

If a complete line is being designed, the following mechanical features should be checked.

- Shaft sizes
- Bearing arrangements
- Stuffing box design
- Bolting for maximum pressures
- Suction pressures
- Pump axial and radial balance
- Bed plates, motor supports, etc.
- Gasketing
- Lubrication

The standardization and the use of existing parts should be considered at this time; however hydraulic performance should never be sacrificed for mechanical or cost reasons. If sacrifice becomes necessary, adjust pump hydraulics accordingly. Mechanical design features will be covered in other chapters.

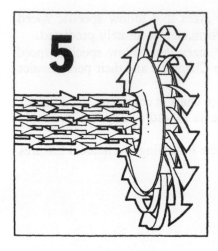

Volute Design

The object of the volute is to convert the kinetic energy imparted to the liquid by the impeller into pressure. The pump casing has no part in the (dynamic) generation of the total head and therefore deals only with minimizing losses.

The absolute velocity of the liquid at the impeller discharge is an important parameter in pump casing design. This velocity is, of course, different from the average liquid velocity in the volute sections, which is the primary casing design parameter. The relationship between these two velocities is given indirectly in Figure 5-1. This relationship is given in a slightly different form in Figure 3-7.

Volutes, like all pump elements, are designed based on average velocities. The average velocity is, of course, that velocity obtained by dividing the flow by the total area normal to that flow. Designs are usually based only on the desired BEP flow, and the performance over the rest of the head-capacity is merely estimated. The results of many tests in which the pressure distribution within the volute casing was measured indicate that:

1. The best volutes are the constant-velocity design.
2. Kinetic energy is converted into pressure only in the diffusion chamber immediately after the volute throat.
3. The most efficient pumps use diffusion chambers with a total divergence angle between 7 and 13 degrees.
4. Even the best discharge nozzle design does not complete the conversion of kinetic energy. This was indicated on the Grand Coulee model pump where the highest pressure was read seven pipe diameters from the discharge flange.

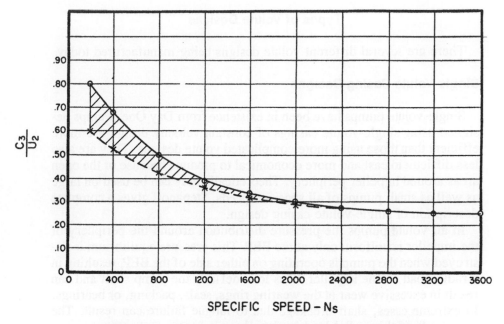

$\dfrac{C_3}{U_2}$

SPECIFIC SPEED – Ns

Figure 5-1. Volute velocity/impeller peripheral velocity ratio.

The hydraulic characteristics of a volute casing are a function of the following design elements:

- *Impeller diameter*
- *Cutwater diameter*—This is directly related to specific speed as shown in Table 3-2.
- *Volute lip angle*—This is selected to suit the absolute flow angle at the impeller discharge. However, considerable deviation is acceptable in low to medium specific speed pumps.
- *Volute areas*—These are so arranged that areas increase gradually from the volute tongue or cutwater, toward the volute nozzle, thus accommodating the discharge along the impeller periphery.
- *Volute width* (b_3)—This is made 1.6 to 2.0 times the impeller width (b_2) (Table 3-1).
- *Discharge nozzle diameter*
- *Throat area*—This area is the most important factor in determining the pump capacity at BEP and is determined using Figure 3-8.

Types of Volute Designs

There are several different volute designs being manufactured today.

Single-Volute Casing Designs

Single-volute pumps have been in existence from Day One. Pumps designed using single-volute casings of constant velocity design are more efficient than those using more complicated volute designs. They are also less difficult to cast and more economical to produce because of the open areas around impeller periphery. Theoretically they can be used on large as well as small pumps of all specific speeds. Stepanoff gives a complete description of single-volute casing design.

In all volute pumps the pressure distribution around the periphery of the impeller is uniform only at the BEP. This pressure equilibrium is destroyed when the pump is operating on either side of the BEP, resulting in a radial load on the impeller. This load deflects the pump shaft and can result in excessive wear at the wearing rings, seals, packing, or bearings. In extreme cases, shaft breakage due to fatigue failure can result. The magnitude of this radial load is given by:

$$P = KHD_2B_2 \text{ sp gr}/2.31$$

Values of the experimental constant K are given in Figure 5-2. For a specific single-volute pump it reaches its maximum at shutoff and will vary between 0.09 and 0.38 depending upon specific speed. The effect of the force will be most pronounced on a single-stage pump with a wide b_2 or a large-sized pump.

It is safe to say that with existing design techniques, single-volute designs are used mainly on low capacity, low specific speed pumps or pumps for special applications such as variable slurries or solids handling.

Double-Volute Casing Designs

A double-volute casing design is actually two single-volute designs combined in an opposed arrangement. The total throat area of the two volutes is identical to that which would be used on a comparable single-volute design.

Double-volute casings were introduced to eliminate the radial thrust problems that are inherent in single-volute designs. Test measurements, however, indicate that while the radial forces in a double volute are greatly reduced, they are not completely eliminated. This is because al-

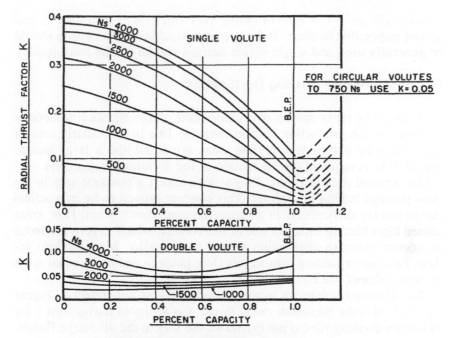

Figure 5-2. Radial thrust factor.

though the volute proper is symmetrical about its centerline, the two passages carrying the liquid to the discharge flange often are not. For this reason, the pressure forces around the impeller periphery do not precisely cancel, and a radial force does exist even in double-volute pumps.

Values of the constant K have been established experimentally by actually measuring the pressure distributions in a variety of double-volute pumps. The data presented in Figure 5-2 apply to conventional single-stage double-volute pumps and indicate substantial reductions in the magnitude of K. Tests on multistage pumps with completely symmetrical double-volute casings indicate that the radial thrust is nearly zero over the full operating range of the pump.

The hydraulic performance of double-volute pumps is nearly as good as that of single-volute pumps. Tests indicate that a double-volute pump will be approximately one to one and one-half points less efficient at BEP, but will be approximately two points more efficient on either side of BEP than a comparable single-volute pump. Thus the double-volute casing produces a higher efficiency over the full range of the head-capacity curve than a single volute.

Double-volute pump casings should not be used in low-flow (below 400 GPM) single-stage pumps. The small liquid passages behind the long dividing rib make this type of casing very difficult to manufacture and almost impossible to clean. In large pumps double-volute casings should be generally used and single-volute designs should not be considered.

The Double-Volute Dividing Rib (Splitter)

The dividing rib or splitter in double-volute pumps causes considerable problems in the production of case castings. This is particularly true on small-capacity pumps where flow areas are small and a large unsupported core is required on the outside of the dividing rib (splitter).

The original double-volute designs maintained a constant area in the flow passage behind the splitter. This concept proved to be impractical due to casting difficulties. In addition the consistently small flow areas caused high friction losses in one of the volutes, which in turn produced an uneven pressure distribution around the impeller. Most modern designs have an expanding area in this flow passage ending in equal areas on both sides of the volute rib.

The effects of volute rib length on radial thrust are shown in Figure 5-3. Note that the minimum radial thrust was achieved during Test 2 for which the dividing rib did not extend all the way to the discharge flange. Also note that even a short dividing rib (Test 4) produced substantially less radial thrust than would have been obtained with a single volute.

Triple-Volute Casings

Some pumps use three volutes symmetrically spaced around the impeller periphery. Total area of the three volutes is equal to that of a comparable single volute. The triple volute casing is difficult to cast, almost impossible to clean, and expensive to produce. We do not see any justification for using this design in commercial pumps.

Quad-Volute Casings

Approximately 15 years ago a 4-vane (quad) volute was introduced. Later this design was applied to large primary nuclear coolant pumps (100,000 GPM, 10–15,000 HP). The discharge liquid passage of these pumps is similar to that of a multi-stage crossover leading to a side discharge. There is no hydraulic advantage to this design.

The only advantage of this design is its reduced material cost. The overall dimensions of quad-volute casing are considerably smaller than those of a comparable double-volute pump.

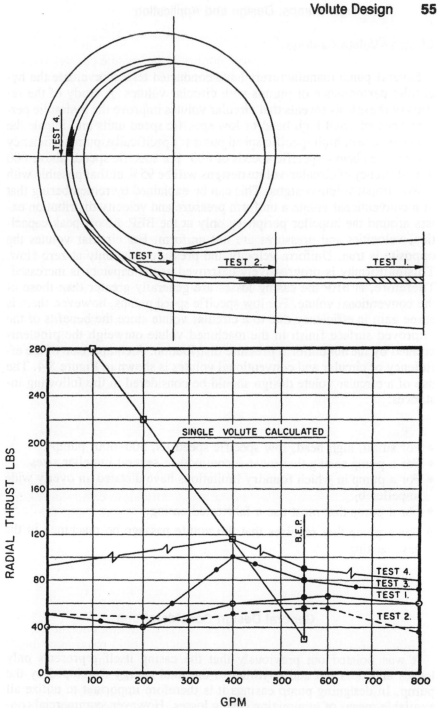

Figure 5-3. Influence of volute rib length on radial thrust.

Circular-Volute Casings

Several pump manufacturers have conducted tests to evaluate the hydraulic performance of pumps with circular volutes. A study of the results of these tests reveals that circular volutes improve the hydraulic performance of small high head or low specific speed units and impair the performance of high specific speed pumps. Specifically, pump efficiency is improved below specific speeds of 600. For specific speeds above 600 the efficiency of circular-volute designs will be 95% of that possible with conventional volute designs. This can be explained by remembering that in a conventional volute a uniform pressure and velocity distribution exists around the impeller periphery only at the BEP. At off-peak capacities, velocities and pressures are not uniform. For circular volutes the opposite is true. Uniform velocity and pressure exist only at zero flow. This uniformity is progressively destroyed as the capacity is increased. Therefore, at BEP the casing losses are generally greater than those of the conventional volute. For low specific speed pumps, however, there is some gain in efficiency due to a circular volute since the benefits of the improved surface finish in the machined volute outweigh the problems created by the nonuniform pressure distribution. A comparison of the efficiency of circular and conventional volutes is shown in Figure 5-4. The use of a circular volute design should be considered in the following instances:

- For small, high head, low specific speed (N_s 500–600) pumps.
- For a pump casing that must accommodate several impeller sizes.
- For a pump in which foundry limitations have dictated an overly wide impeller b_2.
- For a pump that must use a fabricated casing.
- For a pump that requires that the volute passage be machined in the case casting.

General Design Considerations

It was pointed out previously that the casing itself represents only losses and does not add anything to the total energy developed by the pump. In designing pump casings it is therefore important to utilize all available means of minimizing casing losses. However, commercial considerations dictate some deviations from this approach, and experience

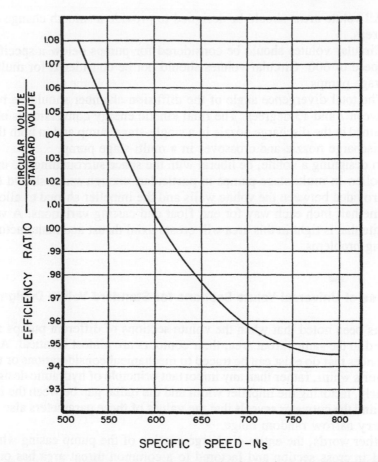

Figure 5-4. Efficiency comparison of circular and conventional volutes.

has shown that these do not have a significant effect on casing losses. The following design rules have shown themselves to be applicable to all casing designs:

1. Constant angles on the volute sidewalls should be used rather than different angles at each volute section. Experience has shown that these two approaches give as good results and the use of constant wall angles reduces pattern costs and saves manufacturing time.

2. The volute space on both sides of the impeller shrouds should be symmetrical.

3. All volute areas should be designed to provide a smooth change of areas.
4. Circular volutes should be considered for pumps below a specific speed of 600. Circular volutes should not be considered for multi-stage pumps.
5. The total divergence angle of the diffusion chamber should be between 7 and 13 degrees. The final kinetic energy conversion is obtained in the discharge nozzle in a single-stage pump and in both the discharge nozzle and crossover in a multi-stage pump.
6. In designing a volute, be liberal with the space surrounding the impeller. In multi-stage pumps in particular, enough space should be provided between the volute walls and the impeller shroud to allow one-half inch each way for end float and casting variations. A volute that is tight in this area will create axial thrust and manufacturing problems.

The Use of Universal Volute Sections for Standard Volute Designs

It has been noted that when the volute sections of different pumps are factored to the same throat area; their contours are almost identical. Any differences that do exist can be traced to mechanical considerations or the designer's whim, rather than any important principle of hydraulic design. Similarly, factoring the impeller width and the radial gap between the impeller and the cutwater reveals that the values of these parameters also lie in a very narrow random range.

In other words, the entire discharge portion of the pump casing when viewed in cross section and factored to a common throat area has only minor variations throughout the entire specific speed spectrum. This fact enables us to eliminate the usual trial-and-error method of designing volute sections while still consistently producing casings to a high standard of hydraulic design. To facilitate this process we have prepared a set of "universal" volute drawings on which the typical volute sections described above have been laid out for a 10 sq in. throat area. Once the designer has chosen his throat area, he can quickly produce the required volute sections by factoring the sections shown for the "universal" volute. Sections for a single-volute pump are shown in Figures 5-5 and 5-6, and sections for a double volute pump are shown in Figure 5-7.

The Design of Rectangular Double Volutes

For low capacity (500–600 GPM) or low to medium specific speed ($N_s < 1,100$) pumps, a rectangular volute design should be considered.

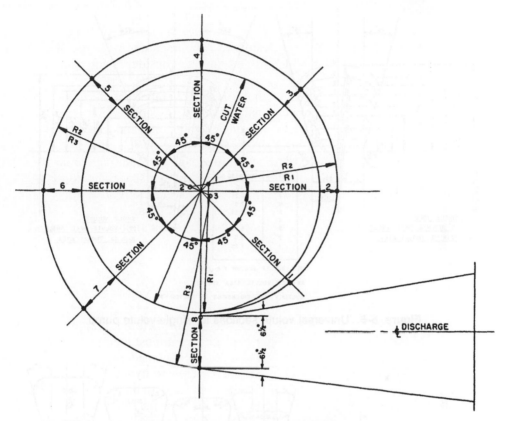

Figure 5-5. Typical single-volute layout.

The universal volute sections for such a design are shown in Figure 5-8. A rectangular volute casing requires the same throat area as a standard volute casing and should be laid out according to the principle of constant velocity.

Rectangular volutes are widely used in small single-stage and multistage pumps. The benefits of the rectangular volute are strictly economical. The simple volute section yields a considerable cost savings due to reduced pattern costs and production time. Over the range of specific speeds where it is used the hydraulic losses are negligible.

The Design of Circular Volutes

The details of a typical circular volute casing design are shown in Figure 5-9. The ratio between the impeller diameter, D_2, and the volute di-

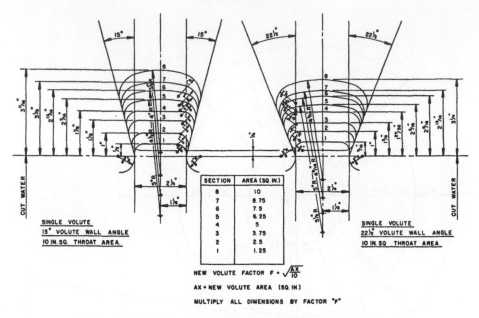

NEW VOLUTE FACTOR F = √(AX/10)

AX = NEW VOLUTE AREA (SQ. IN.)

MULTIPLY ALL DIMENSIONS BY FACTOR "F"

Figure 5-6. Universal volute sections for single-volute pump.

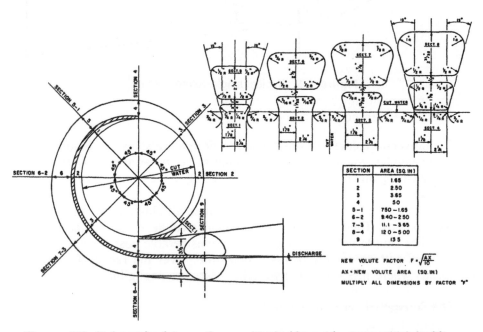

NEW VOLUTE FACTOR F = √(AX/10)

AX = NEW VOLUTE AREA (SQ. IN)

MULTIPLY ALL DIMENSIONS BY FACTOR "F"

Figure 5-7. Universal volute sections and typical layout for trapezoidal double-volute pump.

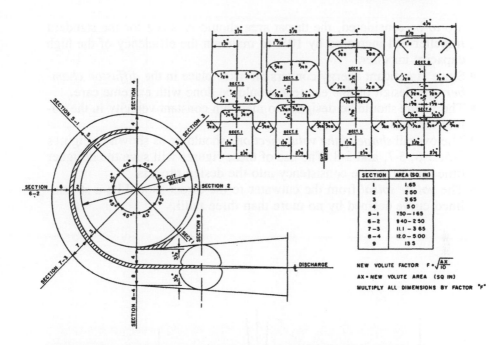

Figure 5-8. Universal volute sections and typical layout for rectangular double volute pump.

ameter, D_3, is quite important and should not be less than 1.15 or more than 1.2. The volute width, t, should be chosen to accommodate the widest (maximum flow) impeller that will be used in the casing. The capacity at BEP can be controlled by the choice of the volute diameter, D_4. Generally, the best results are obtained by selecting the volute width and diameter for each flow requirement. To minimize liquid recirculation in the volute, a cutwater tongue should be added. Tests have shown the addition of a cutwater tongue can reduce radial loads by up to 20%.

General Considerations in Casing Design

There are several considerations in the casing design process that apply to all volute types. These are as follows:

• The most important variable in casing design is the *throat area*. This area together with the impeller geometry at the periphery establishes the pump capacity at the best efficiency point. The throat area should be sized to accommodate the capacity at which the utmost efficiency is required, using Figure 3-8. Where several impellers in the same casing

are to be considered, the throat area should be sized for the standard impeller and increased by 10% to maintain the efficiency of the high capacity impeller.
- Since significant energy conversion takes place in the *diffusion chamber,* the design of this element should be done with extreme care.
- The *volute* should be designed to maintain constant velocity in the volute sections.
- The overall *shape of the volute sections* should be as shown in Figures 5-5, 5-6, 5-7, and 5-8. The use of these figures will save the designer time and introduce consistency into the design process.
- The *volute spiral* from the cutwater to the throat should be a streamlined curve defined by no more than three radii.

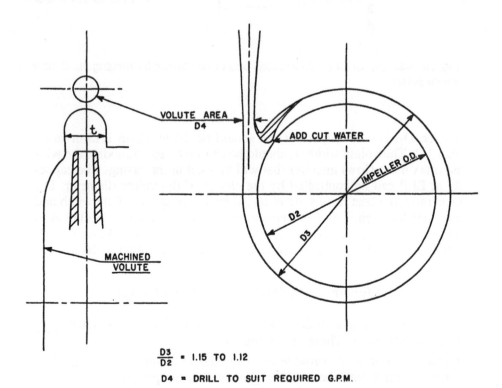

$$\frac{D3}{D2} = 1.15 \text{ TO } 1.12$$

D4 = DRILL TO SUIT REQUIRED G.P.M.

t = TO SUIT IMPELLER WIDTH

Figure 5-9. Typical layout for circular volute pump.

Manufacturing Considerations

Casings, particularly of the double-volute design, are very difficult to cast. In small- and medium-sized pumps the volute areas are small and the liquid passages are long, requiring long unsupported cores. In volute-type multi-stage pumps the problem is more pronounced since there are several complicated cores in a single casing.

Casing Surface Finish

To minimize friction losses in the casing, the liquid passages should be as clean as possible. Since cleaning pump casings is both difficult and time consuming, an extreme effort to produce smooth liquid passages should be made at the foundry. The use of special sand for cores, ceramic cores, or any other means of producing a smooth casting should be standard foundry practice for producing casings.

Particularly with multi-stage pumps, however, even the best foundry efforts should be supplemented by some hand polishing at points of high liquid velocities such as the volute surfaces surrounding the impeller and the area around the volute throat. Both of these areas are generally accessible for hand polishing. In addition, both cutwater tongues should be sharpened and made equidistant from the horizontal centerline of each stage. The same distance should be maintained for each stage in a multi-stage pump.

Casing Shrinkage

Dimensional irregularities in pump casings due to shrinkage variations or core shifts are quite common. Shrinkage variations can even occur in castings made of the same material and using the same pattern. The acceptance or rejection of these defects should be based upon engineering judgments. However, knowing that shrinkage and core shifts are quite common, the designer should allow sufficient space for rotating element end float. The allowance for total end float should be a minimum of one-half inch.

Conclusion

Although it is often claimed that casings are very efficient, this is misleading, since the hydraulic and friction losses that occur in the casing can only reduce the total pump output and never add to it. It is the designer's responsibility to do his utmost to minimize these losses.

Notation

P   Radial force (lbs)
H   Impeller head (ft)
D_2 Impeller diameter (in.)
B_2 Total impeller width including shrouds at D_2 (in.)
K   Experimental constant
sp gr   Specific gravity

Reference

Stepanoff, A. J., *Centrifugal and Axial Flow Pumps,* 2nd edition, John Wiley and Sons, Inc., New York, 1957, pp. 111–123.

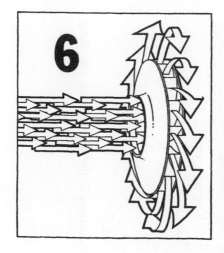

Design of Multi-Stage Casing

Multi-stage casing discussed in this chapter will be related to one type of pump, namely, a horizontally split, opposed impellers, double-volute design, similar to one shown in Figure 6-1. This type of pump offers the following features:

- Pump casings, when properly bolted, are suitable for high working pressures. Many pumps in service are operating at 3,000 to 4,000 psi discharge pressure.
- The pumps are axially balanced allowing the use of standard ball bearings for many services. Only when shaft diameter is too big or rotating speed too high, will Kingsbury-type bearings be required.
- When NPSHA is low, double-suction first-stage impellers are available as a part of standard design.
- A completely assembled rotating element can be checked for shaft and ring run-out over the full length of the shaft.
- The most modern designs have split impeller hub case rings, horizontally split, allowing the assembled rotating element to be dynamically balanced.
- The pumps are easily designed for high pressure and speeds above 7,000 rpm.
- Removal of the top half of the pump casing exposes the complete rotating element for inspection or repair.

The impeller design for the multi-stage pump is the same as that for a one-stage unit, as described in Chapter 3. The double-volute design is also the same as that for a one-stage pump as described in Chapter 5.

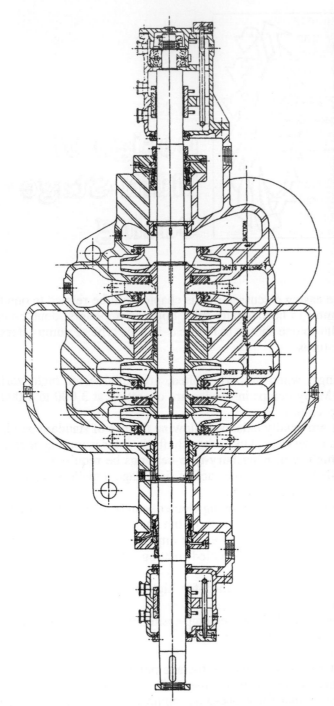

Figure 6-1. Horizontally split opposed impellers double-volute multi-stage pump (courtesy BW/IP International, Inc. Pump Division, manufacturer of Byron Jackson/United™ Pumps).

However, in a multi-stage casing, the liquid from one stage to the next stage must be transferred by means of a crossover passage. The term "crossover" refers to the channel leading from the volute throat of one stage to the suction of the next. Crossovers leading from one stage to the next are normally referred to as "short" crossovers and are similar to return channels in diffuser pumps. These are normally designed in right hand or left hand configurations, depending upon the stage arrangement. Crossovers that lead from one end of the pump to the other or from the center of the pump to the end are normally referred to as "long" crossovers.

The stage arrangements used by various pump manufacturers are shown schematically in Figure 6-2. Arrangement 1 minimizes the number of separate patterns required and results in a minimum capital investment and low manufacturing costs. However, with this arrangement a balancing drum is required to reduce axial thrust. Arrangement 2 is used on barrel pumps with horizontally split inner volute casings. Arrangement 3 is the most popular arrangement for horizontally split multi-stage pumps and is used by many manufacturers. Finally, with Arrangement 4 the series stages have double volutes while the two center stages have staggered volutes. This design achieves a balanced radial load and an efficient final discharge while requiring only one "long" crossover, thereby reducing pattern costs and casing weight.

General Considerations in Crossover Design

The principal functions of a crossover are as follows:

- To convert the velocity head at the volute throat into pressure as soon as possible, thereby minimizing the overall pressure losses in the crossover.
- To turn the flow 180° from the exit of one stage into the suction of the next.
- To deliver a uniformly distributed flow to the eye of the succeeding impeller.
- To accomplish all these functions with minimum losses at minimum cost.

Velocity cannot be efficiently converted into pressure if diffusion and turning are attempted simultaneously, since turning will produce higher velocities at the outer walls adversely affecting the diffusion process. Furthermore, a crossover channel that runs diagonally from the volute

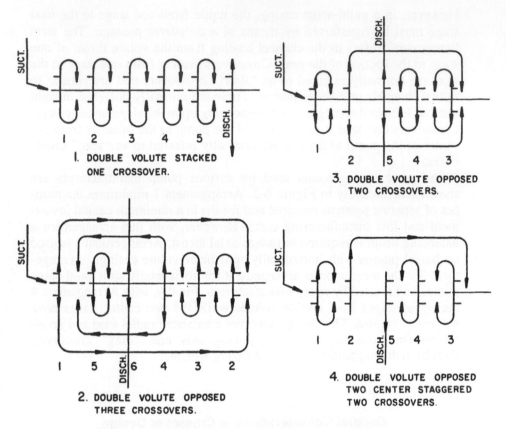

1. DOUBLE VOLUTE STACKED
ONE CROSSOVER.

2. DOUBLE VOLUTE OPPOSED
THREE CROSSOVERS.

3. DOUBLE VOLUTE OPPOSED
TWO CROSSOVERS.

4. DOUBLE VOLUTE OPPOSED
TWO CENTER STAGGERED
TWO CROSSOVERS.

Figure 6-2. Multi-stage pump stage arrangements.

throat to the suction of the next stage imparts a spiral motion to the flow resulting in prerotation and hydraulic losses. For these reasons, the multi-stage pumps of 25 or more years ago were designed with high looping crossovers. To achieve radial balance these crossovers were in both the top and bottom casing halves. This design, referred to as the "pretzel" casing, was very costly, difficult to cast, and limited to a maximum of eight stages.

These problems prompted a study to evaluate the performance of various crossover shapes. A 4-in. pump delivering 1,200 GPM at 3,550 RPM was selected as a model and the three crossover configurations shown in Figure 6-3 were tested. For these tests the pump hydraulic passages were highly polished (60–80 micro-inches), ring clearances were minimized and component crossover parts were carefully matched using a template. Configuration 1 was designed with a total divergence angle of

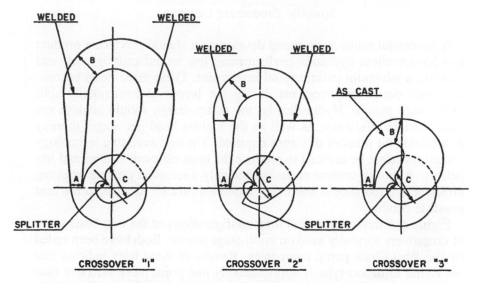

Figure 6-3. Configurations evaluated during crossover performance study.

7° in the passage between the volute throat and the entrance to the "U" bend. From this point the area was held constant to the impeller eye. To prevent prerotation a splitter was added to the suction channel. Crossovers 2 and 3 were designed maintaining the same areas at sections A, B, and C with the same divergence angle but progressively reducing the radial extent of the crossover. The "U" bend on Crossovers 1 and 2 were cast separately from the casing and highly polished before welding. Crossover 3 was cast as a single piece, and the "U" bend polished only in the accessible areas.

The results of testing all three crossover configurations are shown on Figure 6-4. The tests indicated that Crossover 1 yielded a peak efficiency four points higher than Crossover 3. Subsequent testing of commercial units, however, indicated the difference to be only two points. The difference in improvement was attributed to the poor quality of the commercial castings and the use of normal ring clearances. The two-point efficiency loss associated with Crossover 3 was deemed commercially acceptable and was incorporated in multi-stage pumps of up to fourteen stages by all the West Coast manufacturers. These pumps were suitable for higher pressures, easily adaptable to any number of stages, odd or even, and readily castable even in double-volute configurations.

Specific Crossover Designs

A successful multi-stage pump development should produce a product that has excellent hydraulic performance, low manufacturing cost, and requires a minimum initial capital investment. These three items become the basic design requirements during the layout of horizontally split multi-stage pumps. Hydraulically, the pump design should achieve the best possible efficiency, as well as the highest head per stage, thereby minimizing the number of stages required. The best available technology should therefore be utilized to produce the most efficient volutes and impellers. Although crossover design has only a secondary effect on pump efficiency, it too should use every available "trick" to achieve the best possible results.

Figure 6-5 shows short and long configurations of the two basic types of crossovers normally used on multi-stage pumps. Both have been tested by the West Coast pump companies. Results of these tests indicate that the radial diffusion type is approximately one point more efficient than the diagonal diffusion type.

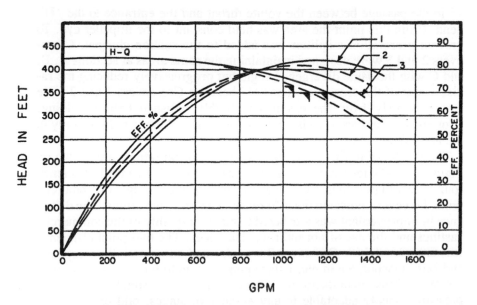

Figure 6-4. Results of crossover performance study.

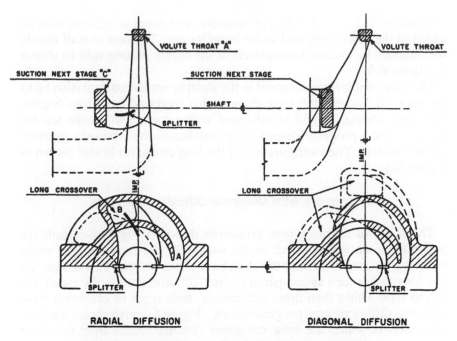

Figure 6-5. Radial and diagonal diffusion crossovers.

Crossovers with Radial Diffusion Sections

The radial diffusion type crossover shown in Figure 6-5 has a diffusion section that follows the volute periphery along the impeller centerline, diffusing with a total divergence angle of 7° up to the point where area at "B" is four times the volute throat area. This point should be reached before the "U" bend to the suction channel. The suction channel should be sized to accommodate the largest capacity impeller that will be used in the pump. The area of the suction return channel should be consistent immediately after the "U" bend. Some designers prefer to decelerate slightly at the impeller eye; however, recent tests indicate that better efficiency is obtained if the liquid is accelerated as it approaches the impeller eye.

Tests on this type of crossover indicate a total head loss equal to 86% of the inlet velocity head. The addition of a welded splitter in the "U" bend will reduce this loss to 65%. However, the cost of adding this splitter is generally prohibitive, and it is not generally used.

From a theoretical viewpoint, crossover channels should have a circular cross section to minimize friction losses. However, all the designs on the market today have rectangular shapes for practical reasons.

To attain the best hydraulic performance, anti-rotation splitters must be added to the suction channel at the impeller eye. The best overall results are obtained by placing two splitters at the casing parting split as shown in Figure 6-5.

The long crossover is identical to the short or series configuration up to the area "B," where the long channel that traverses the pump begins. This long channel should be designed with a "window" at the top for cleaning and a properly shaped plate matched to the crossover opening before welding. The configuration of the long crossover is also shown in Figure 6-5.

Crossovers with Diagonal Diffusion Sections

The diagonal diffusion type crossover shown in Figure 6-5 leads the liquid from the volute throat to the suction of the next impeller while traveling diagonally around the periphery of the volute. This design has one long radius turn as compared to the "U" bend used in the radial diffusion type. Other than these differences, both types of crossover have the same diffusion, area progression etc. Figure 6-5 also shows the configuration in which the long crossover channel climbs over the short crossover.

Even though this crossover has only a single long-radius turn, it is not as efficient as the radial type. This can be attributed to the diagonally located channel, which imparts a spiral motion to the fluid leaving the volute throat, resulting in hydraulic losses larger than those in the "U" bend.

Mechanical Suggestions

In previous chapters, we have described design procedures for centrifugal pump impellers and volutes applicable to one-stage or multi-stage units and in this chapter crossovers for multi-stage pumps only. However, hydraulic considerations alone for multi-stage pumps are not sufficient to complete a final unit. Mechanical details must be considered. This refers to patterns, foundry methods, mechanical bolting, and quality controls.

Patterns

Multi-stage casings have quite complex shapes of liquid passages, including crossovers, crossunders, double volutes at each stage, etc. For this reason, pattern equipment must be of high quality sectional design to allow for variations in number of stages. Normal practice is to make the first pattern a four stage of hard wood with each stage being a separate section. For additional stages, the sections are duplicated in plastic material. For each stage combination, plastic sections should be assembled on their own mounting boards. This arrangement will allow several pumps of different stages to be produced at the same time.

Foundries

The core assembly for multi-stage pumps is very complex as shown in Figure 6-6. It shows a 12-stage 4-in. pump with a single-suction first-stage impeller.

In order for the rotating element to fit into the pump casing, each volute core must be assembled perpendicular to the shaft centerline. To assure perpendicularity, a special gauge should be made for this purpose.

It is also vitally important to cast casing with wet area surfaces as smooth as possible. For this reason, the casing cores should be made from "green sand" or ceramic materials.

The major hydraulic loss in multi-stage pumps is friction loss. To minimize this, the as-cast-surface roughness of the internal passages should be a minimum of 125 micro inches. The smoother the wet areas, the less the cost of hand polishing or grinding will be.

Figure 6-6. Core assembly.

Casing Mismatch

On horizontally split pump casings mismatching between the upper and lower casing halves is quite common. This mismatch is normally corrected by hand filing using a template (Figure 6-7). The same template is used by the machine shop to define bolt locations and also by the foundry. It is quite important that the fluid passages in both halves of the casing match in both the horizontal and vertical planes since any mismatching will adversely affect pump performance.

Check for Volute Interference

After the pump casing is completely finished to satisfy quality specification, the assembled rotating element should be installed in the casing (as shown in Figure 6-8) to check for interference against volute walls. Also check the element for total end float, which should be no less than one-half inch total or one-quarter inch on each end.

General Design Suggestions

- It is important to have the volute symmetrical about shaft centerline.
- A ¹/₁₆-in. deep relief should be machined around each stud at the split. At each stud at assembly the metal will rise. The relief will allow the gasket to lay flat, reduce gasket area, and increase gasket unit pressure.
- It is very important to install splitters at each impeller suction entrance. These can be installed recessed at the casing split or cast on casing hub rings.
- Horizontally split centrifugal pumps are designed for relatively high pressures; 3,000 psi hydrotest is very much standard by many manufacturers. Pressure up to 6,000 psi hydro can be obtained by special design of bolting and case.

It is essential (to prevent stage-to-stage bypassing) to have all bolts located as close as possible to the open area. The high pressure differential in opposed impeller design multi-stage pumps is between two center stages and at the high pressure end. Particular attention to bolting should be paid in these two areas. It is sometimes advantageous to provide elevated bosses to bring bolting closer to open area.

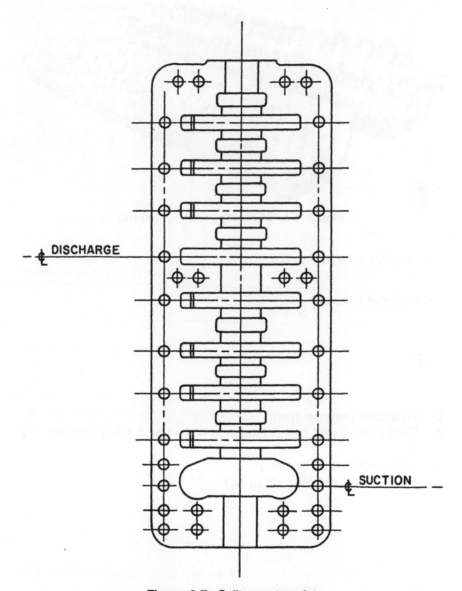

Figure 6-7. Split case template.

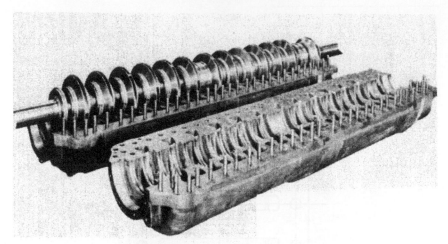

Figure 6-8. Multi-stage rotating element.

As a guide to selecting bolt size use:

$$\text{Number of bolts} = \frac{P \times A}{S \times A_R}$$

Notation

P   Hydrotest pressure (psi)
A   Total area at split minus total area of bolt hole relief diameter (sq in.)
S   Allowable bolt stress (psi)
A_R Root area of bolt thread (sq in.)

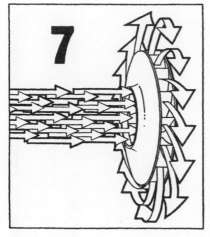

Double-Suction Pumps and Side-Suction Design

7

The double-suction single-stage pump is, perhaps, the most widely used pump throughout the industrial world. Applications range from light duty building trade pumps to heavy duty pipeline injection pumps (Figure 7-1).

The double-suction is a very simple machine whose initial cost is relatively low. Above 700–1,000 GPM, efficiency is high and required NPSH is low. All modern double-suction pumps are designed with double-volute casing to maintain hydraulic radial balance over the full range of the head-capacity curve. Having a double-suction impeller, the pumps are theoretically in axial balance. Double-suction pumps often have to operate under suction lift, run wide open in a system without a discharge valve, or satisfy a variable capacity requirement. These pumps may be quite large, pump high capacities, and handle pumpage with gas or entrained air. In spite of all this, they are expected to operate without noise or cavitation. Suction passage design should therefore be based on the best available technical "know-how," and liberties should not be taken during the design process.

Double-Suction Pump Design

Pump Casing

The double-suction, double-volute casing is designed in identical manner to the single-suction pump, as described in Chapter 5. The specific speed, N_s, of a double-suction pump is identical to the single-suction unit. *Do not divide* calculated pump specific speed by the square root of

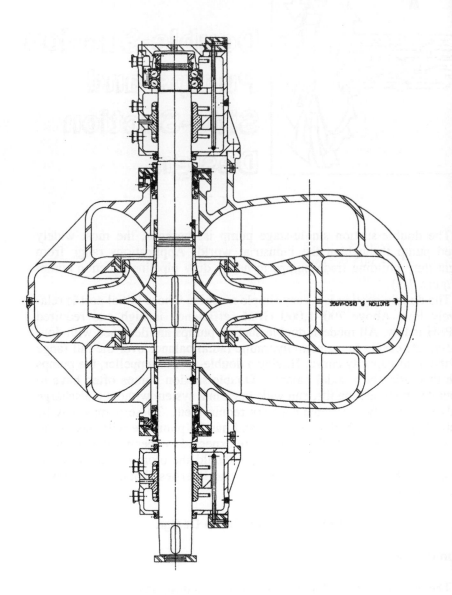

Figure 7-1. Single-stage, double-suction, pipeline pump (courtesy BW/IP International, Inc. Pump Division, manufacturer of Byron Jackson/United™ Pumps).

two. Experience shows that this procedure will give misleading design factors and unfavorable test results.

Design of the suction approach to the pump impeller for double-suction pumps will differ from the single-suction design. This will be covered in detail in the following paragraphs.

Double-Suction Impeller

The method for calculating impeller diameter, impeller width, number of vanes, and vane angularity is identical to the procedure for the single-suction impeller described in Chapter 3. The method for impeller layout will also follow Chapter 3, with a double-suction impeller being considered two single-suction impellers back to back. With double entry the eye area is greater and the inlet velocity lower, thus reducing NPSHR.

Side Suction and Suction Nozzle Layout

The importance of hydraulic excellence in the design of liquid passage areas from suction nozzle to the impeller eye or eyes is quite often minimized or unfavorably adjusted for economic reasons. Experience shows that this approach leads to many field NPSH problems. The current trend in industry is one of reducing NPSHA; therefore, it is essential for optimum NPSHR that the design of the suction approach to the impeller eye be carefully controlled.

We know from experience that in the design of the side suction inlet a certain amount of prerotation of the incoming liquid is desirable. To obtain this condition, the baffle (or splitter) is provided. This splitter is rotated 30° to 45° from suction centerline in the direction of pump rotation. The splitter will locate the radial section of zero flow, and the areas will progressively increase in both directions away from it.

The following drawings and information must be available to design a side suction.

1. Volute layout.
2. Impeller layout.
3. Shaft or sleeve diameter at the impeller.
4. Suction nozzle size.

Layout of the laterally displaced side suction should be done in two parts:

1. Sketch an approximate end view and profile (Figures 7-2 and 7-3) using the following guidelines:

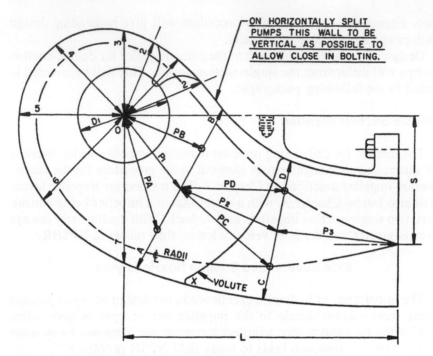

Figure 7-2. Double-suction layout—end view.

Table 7-1
Linear Dimensions of Suction Sections

Section	Approx. Dimension
1	$D_1 \times 0.84$
2	$D_1 \times 0.90$
3	$D_1 \times 0.95$
4	$D_1 \times 1.06$
5	$D_1 \times 1.17$
6	$D_1 \times 1.30$
7	$D_1 \times 1.65$
A-B	$D_1 \times 1.8$ to 2

Table 7-2
Areas of Suction Sections

Section	Area
6	.5 (Area at A-B)
5	.375 (Area at A-B)
4	.25 (Area at A-B)
3	.125 (Area at A-B)

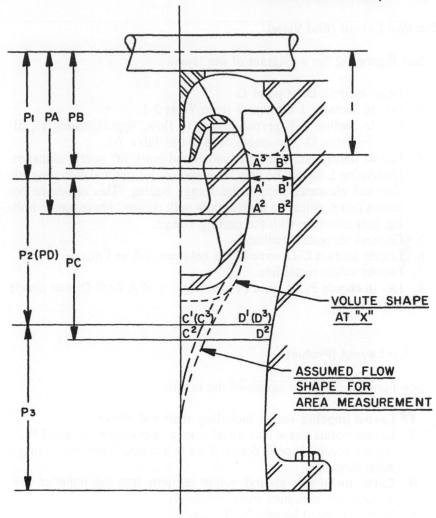

Figure 7-3. Double-suction layout—profile.

a. For linear dimensions of Sections 1 through 7, and A-B, use Table 7-1.

b. Area progression from nozzle to impeller should follow Figure 7-4. The range suggested permits impellers of different suction specific speeds to be used with a common suction.

c. Areas of Sections 3 through 6 measured normal to the flow are shown in Table 7-2.

2. Make final layout after checking all areas and dimensional location of suction nozzle.

Suction Layout (End View)

See Figure 7-2 for a diagram of the layout.

1. Draw a circle D_1 at point O.
2. Locate Sections 1 through 7 from Table 7-1.
3. Locate Section A-B perpendicular to flow, approximately D_1 dimension from O. For length of A-B use Table 7-1.
4. Locate nozzle dimensions L and S and mark off nozzle diameter. Dimension L should be only long enough for gradual area progression and clearance behind the flange bolting. This clearance becomes more critical on horizontally split pumps, where nozzle bolting may interfere with the parting flange.
5. Connect all points freehand.
6. Locate Section C-D somewhere between A-B and nozzle.
7. Layout volute metal line.
8. Lay in chords P_1, P_2, and P_3 to mid point of A-B, C-D, and nozzle respectively.

Suction Layout (Profile)

See Figure 7-3 for a diagram of the profile.

1. Layout impeller shape, including shaft and sleeve.
2. Layout volute shape and metal line at intersection of chord P_2.
3. Layout volute shape in dotted lines at location X to observe maximum blockage.
4. Curve metal line around volute sections into the impeller eye, maintaining minimum metal thickness.
5. Mark off chord lengths P_1, P_2, and P_3.
6. Locate nozzle diameter.
7. Approximate width of A' - B' and C' - D' to satisfy area progression from Figure 7-4.
8. Connect outer wall points freehand with ample curvature into impeller eye.
9. Lay in Sections 1 through 7. In this example only Section 1 is shown.
10. Develop Sections A-B and C-D as surfaces of revolution (Figure 7-5). Sections can be divided into any number of increments (e.g., PA and PB). Transfer these chords from the end view to the profile, measure dimensions P' - A' P' - B' and transfer to section layout.

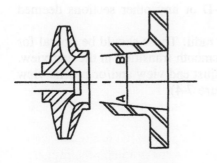

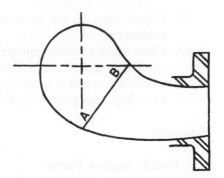

END TYPE SUCTION   ELBOW TYPE SUCTION

LOCATION	AREA
IMPELLER EYE	100
SECTION A – B	120 TO 140
SUCTION FLANGE	132 TO 169

Figure 7-4. Suction area progression.

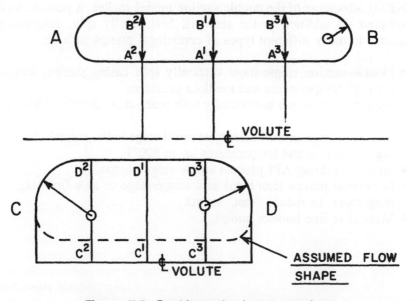

Figure 7-5. Double-suction layout—sections.

11. Repeat Step 10 for Sections C-D or any other sections deemed necessary.
12. Close sections with appropriate radii. These should be liberal for castability and should follow a smooth transition in the end view.
13. Check areas and if necessary adjust end view and/or profile view to satisfy area progression (Figure 7-4).

Example

Double Suction Pump:

- Eye diameter D_1 = 5.5 in.
- Shaft diameter = 3 in.
- Total eye area = 32 in.^{2}
- Area at A-B = 39 in.^{2}
- Suction nozzle = 8 in. Area = 50 in.^{2}
- Area section #6 = 19.5 in.^{2}
- Area section #5 = 14.6 in.^{2}
- Area section #4 = 9.75 in.^{2}
- Area section #3 = 4.9 in.^{2}

The double-suction casing in combination with a double-suction impeller has an inherent NPSH advantage over a single-suction combination. Reduction of the required NPSH is normally 40% to 50%. This NPSH advantage of the double suction model makes it possible to be adopted (in addition to the standard horizontally split, single-stage pump) to many different types of centrifugal pumps such as:

- Double-suction single-stage vertically split casing pumps, suitable for high temperatures and medium pressures.
- As a first stage in a horizontally split multi-stage double-volute-type pump.
- As a first stage in the barrel-type multi-stage units, suitable for very high pressures and temperatures up to 800°F.
- To the overhung API process single-stage pumps.
- In vertical pumps (can type) as a single-stage or as a first-stage arrangement, to reduce "can" length.
- Vertical in-line booster pumps.

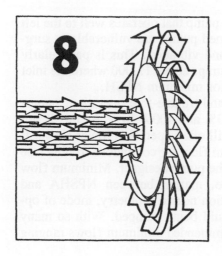

8

NPSH

The expressions NPSHR and NPSHA are accepted abbreviations for net positive suction head required and net positive suction head available. Probably more has been written on NPSH than any other subject involving pumps. With so much literature available, one might assume that NPSH and its relationship to cavitation is well understood. Nothing could be further from the truth. To this day, NPSH is still misunderstood, misused, and misapplied, resulting in either costly over-design of new systems or unreliable operation of existing pump installations. Avoiding these problems requires accurate prediction of NPSHR, supply of sufficient NPSHA, and attention to suction piping approach. NPSHR can be considered the suction pressure required by the pump for safe, reliable operation.

Establishing NPSHA

Establishing NPSHA, which is the head available characteristic of the system that provides flow of liquid to the pump, is the responsibility of the system designer. As NPSHR increases with pump capacity, normal practice is to establish NPSHA at the operating condition, then add a reasonable margin to accommodate any anticipated increase in pumping capacity. All too often, future operating problems begin here. It is not unusual for the ultimate user to add some anticipated increase in capacity, then for insurance the contractor designing the system adds even more. When the pump designer finally gets the data sheet, he designs the impeller inlet and suction nozzle geometry for operating capacity, plus user margin, plus contractor margin. If these margins are not carefully controlled, the result can be an over-sized pump that operates well to the left of BEP (Figure 8-1). Such over-designed pumps are vulnerable to surging, recirculation, cavitation, noise, and vibration. This is particularly true with high-suction specific speed pumps above 11,000 where the inlet geometry has already been extended for minimum NPSH.

For minimum energy consumption and trouble-free operation, pumps should ideally be operated between 80% and 100% BEP. As this is not always possible or practical, pumps will often operate at lower flows. It is therefore important that the minimum flow for continuous trouble free operation be carefully considered by the pump designer. Minimum flow is influenced by physical pump size, margin between NPSHA and NPSHR, impeller inlet geometry, suction nozzle geometry, mode of operation, and last but not least, the liquid being pumped. With so many variables it is not unusual to find recommended minimum flows ranging from 10% to 60% BEP.

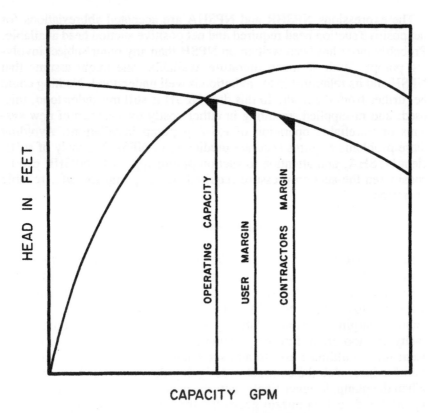

Figure 8-1. Margins of safety result in oversized pump.

Predicting NPSHR

The other side of the coin necessary for reliable operation is of course accurate prediction of NPSHR by the pump designer.

In considering NPSHR, it is necessary to understand that a centrifugal pump is designed as a hydraulic machine to move liquids. Any amount of entrained air or gas present will cause a deterioration in pump performance. Various tests substantiate the claim that a volume of only one percent air or gas will cause a loss of head and efficiency. As liquid travels from suction nozzle to impeller eye, it will experience pressure losses caused by friction, acceleration, and shock at blade entry. If the summation of these losses permits vaporization of the liquid, vapor bubbles will form in the impeller eye, travel through the impeller, and upon reaching a high pressure region, collapse. This collapse or implosion of the vapor bubbles is classic cavitation, which can lead to impairment of performance and impeller damage (Figures 8-2 and 8-3). Thus, predicting

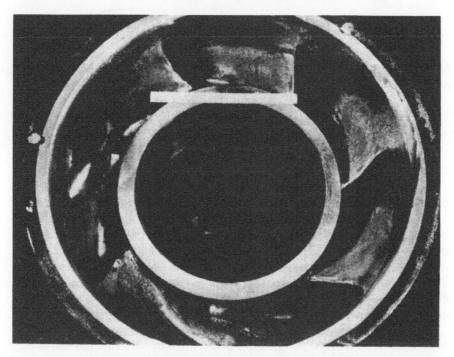

Figure 8-2. Cavitation damage—looking into impeller eye.

Figure 8-3. Cavitation damage—at leading edge of one vane.

NPSHR is in fact predicting the losses in the critical area between suction nozzle and the leading edge of the first-stage impeller blades (Figure 8-4).

Moderate Speed Pumps

One method used successfully for many years by pump designers will predict NPSHR with reasonable accuracy when the pump liquid is water. The inlet velocity of the liquid entering the impeller eye, C_{M1}, and the peripheral velocity of the impeller blade, U_t, are calculated and the ratio used to predict NPSHR (Figure 3-6). The chart is valid only for flows between 50% and 100% BEP at maximum impeller diameter. For capacities above 100% BEP, apply the factors indicated on the chart. Development of this chart is a result of acquisition of many years of pump test data. Values read from the chart will approximate the NPSHR established during pump testing using 3% head loss as criteria. When velocities exceed those shown, the chart should not be extrapolated.

The most commonly used method in determining the cavitation characteristics of a centrifugal pump is to cause a breakdown in the normal head capacity curve. This is done by holding the speed and suction pressure

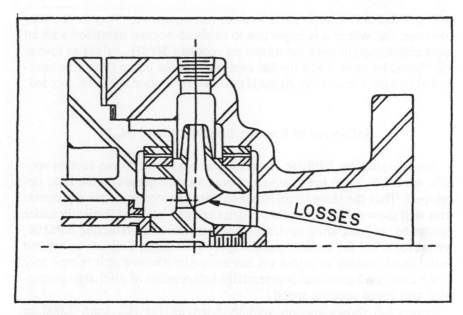

Figure 8-4. Pressure loss between suction nozzle and leading edge of impeller vane.

constant and varying the capacity, or by holding the speed constant and reducing the suction pressure at various capacities. Either of these methods will produce a breakdown in the head characteristic as shown in Figure 8-5, indicating a condition under which the performance of the pump may be impaired.

Accurately determining the inception of cavitation requires extreme control of the test and involves sophisticated instrumentation. The Hydraulic Institute has permitted a drop in head of 3% to be accepted as

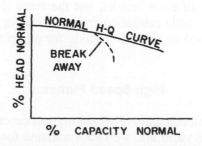

Figure 8-5. Loss in head during cavitation test.

evidence that cavitation is present under test conditions. They do caution, however, that where it is important to establish normal operation with an appreciable margin over the minimum required NPSH, values as low as 1% should be used. They further caution, that the pump should be operated above the break-away sigma if noise and vibration are to be avoided.

Influence of Suction Specific Speed (N_{ss})

When evaluating NPSHR one should always refer to the suction specific speed (N_{ss}) and not to sigma, whose value depends on the head developed. Thus the same impeller when tested before and after a diameter trim will show different values of sigma yet will behave identically under cavitation. In comparing cavitation performance and predicting NPSHR, we prefer to use the suction specific speed parameter. This can be applied in the same manner as sigma yet has more significance as it relates only to the inlet conditions and is essentially independent of discharge geometries and pump specific speed.

Figure 8-6 shows suction velocity triangles for N_{SS} from 7,000 to 16,000. It graphically shows that as suction specific speed increases, normal vane entrance angle becomes flatter, C_{M1} relative to shaft becomes smaller, and peripheral velocity at impeller eye becomes greater. The ratio of C_{M1} to U_t for higher suction specific speed, is very small and chances for cavitation are greatly increased. This cavitation would appear for the following reason: To move liquid from one point to another, a change of pressure gradient must take place. But as C_{M1} (velocity) is reduced, velocity head, $V^2/2g$, is also reduced. When it becomes lower than the head required to overcome impeller entrance losses, the liquid will backflow creating cavitation and pump damage.

Figure 8-7 represents a test of a four-inch pump with eight different suction specific speed impellers. Best efficiency point of all impellers is the same. Impeller profile is also the same, but impeller eye geometry is different for each suction specific speed. Some tests on different types of pumps might show different results, but the trend should be the same. This test shows the stable cavitation-free window for seven suction specific speeds and could be used as a guide for pump selection.

High Speed Pumps

The prediction methods and NPSH testing just described under moderate speed pumps are valid and well substantiated for low- to moderate-speed pumps. It is our opinion, however, that neither method will ensure

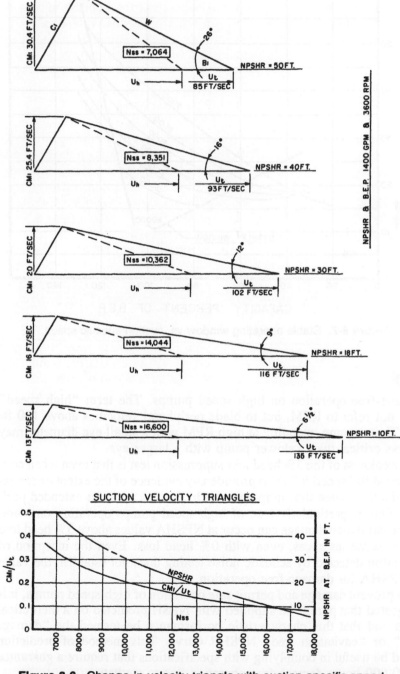

SUCTION VELOCITY TRIANGLES

Figure 8-6. Change in velocity triangle with suction specific speed.

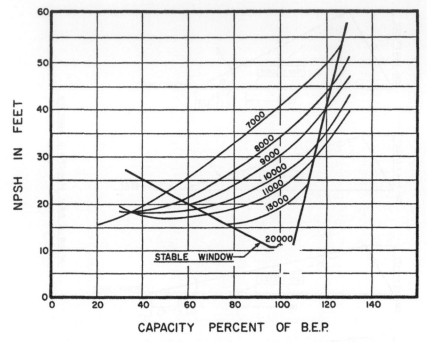

Figure 8-7. Stable operating window vs. suction specific speed.

damage-free operation on high speed pumps. The term "high speed" does not refer to RPM, but to blade peripheral velocities above 160 ft/ sec. Thus a pump operating at high RPM with a small eye diameter may be less critical than a slower pump with a larger eye.

A weakness of the 3% head loss suppression test is that even when conducted at full speed it fails to provide any evidence of the extent of cavitation damage when the pump operates in this 3% zone for extended periods. This is particularly true of high-speed pumps operating on water where cavitation damage can occur at NPSHA values above 3% head loss and in some instances, even with 0% head loss. Even the inception of cavitation detected by acoustic noise testing does not establish the value of NPSHA for damage-free operation.

To prevent damage and permit safe operation of high-speed pumps, it is suggested that pumps be supplied with NPSH predicted on a theoretical basis and that this characteristic performance be termed the "damage free" or "cavitation free" NPSHR curve. This method of prediction could be useful in complying with specifications that require a guarantee of 40,000 hours damage-free operation due to cavitation.

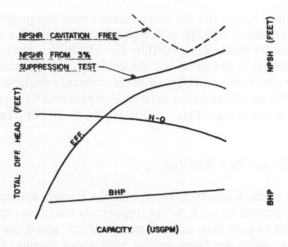

Figure 8-8. Performance curve showing NPSHR cavitation free and NPSHR 3% head loss.

"Cavitation-free NPSHR" cannot be demonstrated by a suppression test, as no head loss will be evident. Therefore, to satisfy the normal requirement for testing on the pump manufacturer's test stand, it is suggested that two NPSHR curves be offered with the pump quotation. These would be "cavitation-free NPSHR" and the conventional 3% head loss NPSHR (Figure 8-8).

Cavitation-Free NPSHR

As described earlier, cavitation is the formation of vapor-filled cavities in the pumped liquid resulting from a sufficient reduction of the liquid pressure to vaporize a proportion of the liquid. To prevent this vaporization and the damage associated with it, the pump designer must first consider the head losses in the most critical area, which is between the inlet nozzle of the pump and the leading edges of the first-stage impeller blades. This head loss is a result of the following factors:

1. Head loss due to friction.
2. Head drop due to fluid acceleration, which is the energy required to accelerate the flow from the suction nozzle to the impeller eye.
3. Head shock loss due to blade entry, which is the localized drop at the blade leading edge and is a function of the angle of attack and blade entry shape.

Both the friction losses and the acceleration losses are proportional to the square of the liquid velocity as it approaches the impeller eye with a constant of proportionality designated by K_1. The losses due to blade entry are proportional to the square of the velocity of flow relative to the blade leading edge with a constant of proportionality designated by K_2. To prevent cavitation, these losses must be compensated by supplying adequate NPSH to the pump. This "cavitation-free" NPSHR can then be expressed as:

$$NPSHR = K_1 C_{M1}^2/2g + K_2 w^2/2g \qquad (8\text{-}1)$$

where the first term, $K_1 C_{M1}^2/2g$, represents the friction and acceleration losses, and the second term, $K_2 w^2/2g$, represents the blade entry losses. From this, it can be seen that, in small pumps of low speed, the first term is predominant, while for large and/or high speed pumps, the second term is the controlling factor and the first term is of secondary importance. This explains why it is often possible to reduce NPSHR on moderate speed pumps by changing to a larger eye impeller. As cavitation is most likely to occur in the region where the relative velocity, w, is highest, the calculation is based only on the maximum diameter of the blade tip, D_t, at the impeller entry.

The incidence angle, α, that influences K_2 is the difference between the inlet blade angle, B_1, and the flow angle, θ (Figure 8-9). B_1 is determined from C_{M1} multiplied by factor R_1, which allows for the effects of recirculated flow, Q_L, and nonuniform velocity distribution. As the leakage, Q_L, does not remain constant due to internal erosion, and as many engineers differ in their selection of R_1, it is seldom if ever that α equals zero. Leakage Q_L through impeller wear ring clearances and balance

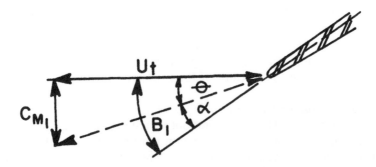

Figure 8-9. Blade and flow angle at impeller inlet.

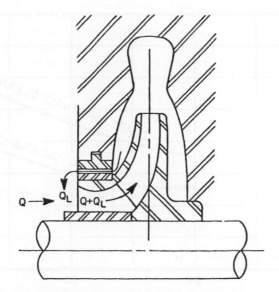

Figure 8-10. Leakage across wear ring back to impeller eye.

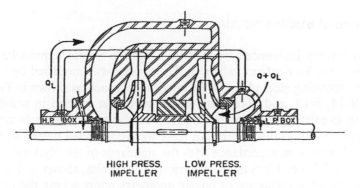

Figure 8-11. Leakage across high pressure bushing back to impeller eye.

lines in a multi-stage pump (Figures 8-10 and 8-11) should be added to the flow Q entering the impeller eye. This leakage will vary with the head developed and therefore has more influence on α during low flow operation. Pumps of low specific speed where ring leakage can be a significant percentage of pump flow will show increased NPSHR with increased ring clearance. One example is shown in Figure 8-12.

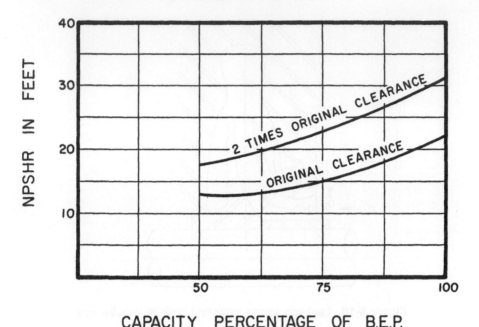

CAPACITY PERCENTAGE OF B.E.P.

Figure 8-12. Influence of wear ring clearance on NPSHR.

Influence of Suction Nozzle

K_1 is largely influenced by the pump suction nozzle approaching the impeller eye. To confirm values of K_1, a pump was modified by installing area reducing steel plates in the suction approach as shown in Figures 8-13, 8-14, 8-15, and 8-16. The plates were first formed in wood conforming to existing suction nozzle core box shape, then manufactured in steel and welded in place. NPSHR tests were conducted before and after the modification in accordance with the standards of the Hydraulic Institute, using 3% head loss as the criterion. The results, shown in Figure 8-17, illustrate the influence of nozzle geometry approaching the impeller and lead to the observations now described.

The test pump was of the double-inlet split-case type. This complex suction passage from nozzle to impeller eye, makes it difficult to determine analytically the expected losses. It is documented that a design of this type, which changes the direction of flow at least four times and moreover splits the flow into two separate channels, can have appreciable influence on the energy loss—sometimes as high as 2% to 3% of the effective head of one impeller. The improvements experienced in the test are difficult to confirm analytically. However, it is obvious that the reduced energy loss could not result from velocity change alone.

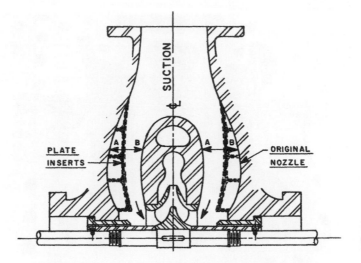

Figure 8-13. Plate inserts in double-suction nozzle.

Figure 8-14. Wooden templates for plate inserts.

Figure 8-15. Plate inserts installed in upper half case.

Figure 8-16. Plate inserts installed in lower half case.

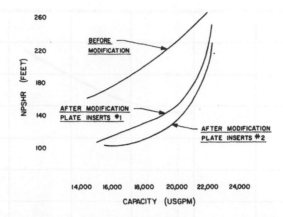

Figure 8-17. Influence of plate inserts on NPSHR.

It can be assumed that this improvement is influenced by reduced separation, improved flow stability and streamlining, a progressive increase in velocity, and the resulting reduced turbulence. It must also be accepted that any disturbance in the approach to the impeller can cause unequal distribution of flow rates into the two impeller eyes at different locations. These diversions from the correct angle of attack at the leading edge of the blades produce a corresponding head loss.

The test confirms that the cavitation characteristics of a good impeller design can be impaired by poor suction nozzle design. This is particularly true with double-entry impeller pumps where the complex nozzle geometry can adversely affect K_1. A well designed suction nozzle has a gradual decrease in area from nozzle to impeller, allowing a progressive increase in velocity. Area distribution guidelines are shown in Figure 7-4. The range suggested permits impellers of different suction specific speeds to be used with a common suction.

Using the actual suction nozzle area ratio at Section A-B gives a reasonable means of estimating K_1 (Figure 8-18). For the same incidence angle α, K_2 has a higher value at capacities above design. For estimating K_2 use Figure 8-19.

Influence of Liquid

The boiling of the liquid in the process of cavitation is a thermal process and is dependent on the liquid properties, pressure, temperature, latent heat of vaporization, and specific heat. To make this boiling possible, the latent heat of vaporization must be derived from the liquid flow.

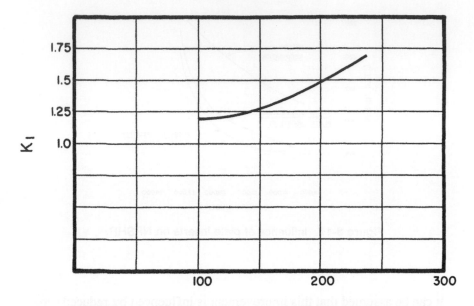

SUCTION AREA AT A-B (PERCENTAGE OF Ae)

Figure 8-18. Estimating K_1.

The extent of cavitation damage depends on the proportion of vapor re-leased, the rapidity of liberation, and the vapor specific volume. Taking this into consideration the cavitation face calculation can be corrected by applying a gas-to-liquid ratio factor C_b (Figure 8-20). On this basis, cold water must be considered the most damaging of the commonly pumped liquids. Similarly, this difference in behavior applies to water at different temperatures. A review of the properties of water and its vapor at several temperatures shows the specific volume of vapor decreases rapidly as pressure and temperature increase. This difference in behavior under cavitating conditions makes cold water more damaging than hot.

The problems associated with cold water are substantiated by operating experience in the field, where pumps handling certain hydrocarbon fluids or water at temperatures significantly higher than room temperature will operate satisfactorily with a lower NPSHA than would be required for cold water.

$$\text{NPSHR} = [K_1C_{M1}^2/2g + K_2w^2/2g]C_b$$

$$= [(K_1 + K_2)C_{M1}^2/2g + K_2U^2/2g]C_b \tag{8-2}$$

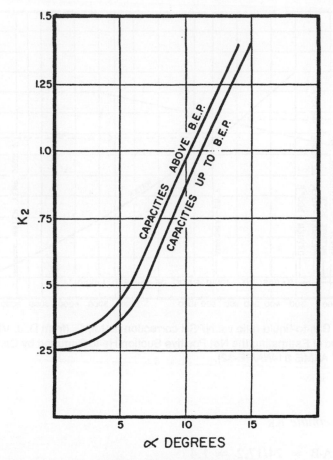

Figure 8-19. Estimating K_2.

Example

Ignoring internal leakage back to the impeller, calculate NPSHR cavitation free for the pump now described.

- Capacity—1,800 GPM
- Product—Water
- Temperature—70°F
- Speed—8,100 RPM
- Impeller eye area—17.2 sq in.
- Eye diameter—5 in.
- Inlet blade angle—15°
- Suction area at A-B—24 sq in.

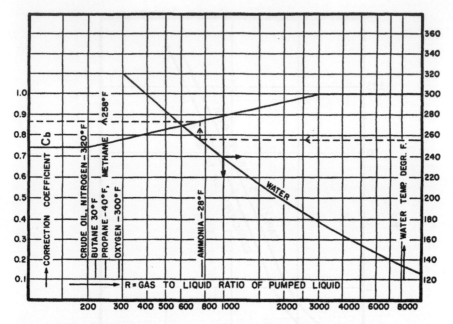

Figure 8-20. Gas-to-liquid ratio vs. NPSH correction factor, C_b (from D. J. Vlaming, "A Method of Estimating the Net Positive Suction Head Required by Centrifugal Pumps," ASME 81-WA/FE-32).

Step 1: Determine K_1.

Ratio at A-B $= 24/17.2 = 1.4$

From Figure 8-18:

$K_1 = 1.25$

Step 2: Calculate θ.

From Figure 8-9:

$$
\begin{aligned}
\tan \theta &= C_{M1}/U_t \\
&= .321 \ Q/Ae \div D_t N/229 \\
&= .321 \times 1,800/17.2 \div 5 \times 8,100/229 \\
&= 33.6/176.8 = 0.19
\end{aligned}
$$

$\theta = 10.75°$

Step 3: Determine K_2.

From Figure 8-9:

$$\alpha = B_1 - \theta$$
$$= 15 - 10.75$$
$$= 4.25°$$

From Figure 8-19, as 1,800 GPM is less than BEP:

$$K_2 = .32$$

Step 4: Calculate NPSHR from Equation 8-2.

$$NPSHR = [(1.25 + .32)33.6^2/64.4 + .32 \times 176.8^2/64.4]C_b$$

From Figure 8-20,

$$C_b = 1.0$$
$$NPSHR = 27.5 + 155.3$$
$$= 183 \text{ ft}$$

Figures 8-21, 8-22, and 8-23 show examples of calculating NPSHA.

Suction Piping

As described earlier, NPSHR is influenced by suction nozzle design. Similarly, poor suction piping can adversely affect NPSHR and pump performance. Double-suction first-stage impellers are particularly vulnerable to a nonuniform approach of the liquid. If elbows are located close to the pump, they should be oriented to provide equal distribution of flow into both eyes of the impeller. An elbow parallel to the pump shaft directly before the pump is conducive to spiral flow and unbalanced flow distribution into the two eyes. Unequal flow distribution can result in excessive vibration, high axial thrust loads, noise, and cavitation.

Effect of Viscosity

Although the influence of viscosity is predictable on other hydraulic characteristics, particularly head, capacity, and efficiency, little general information is available to indicate the effect on NPSHR. From experience we know that up to 2,000 SSU we are safe in using water NPSHR

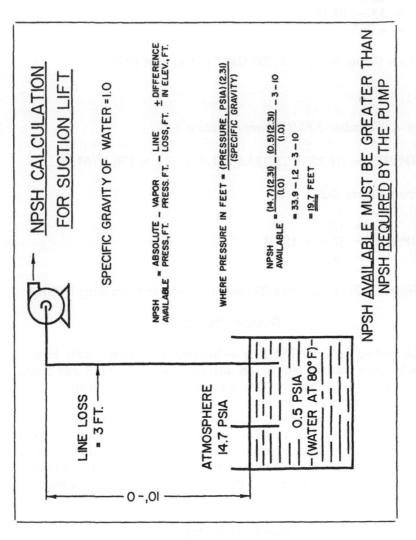

Figure 8-21. Calculating NPSHA for suction lift.

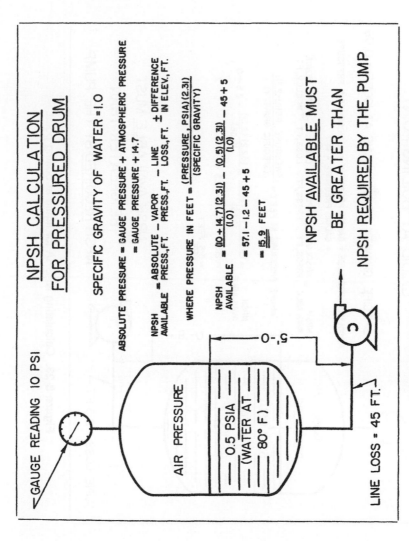

Figure 8-22. Calculating NPSHA for pressure drum.

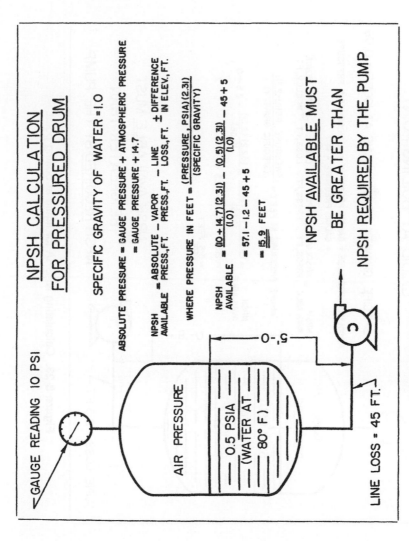

NPSH CALCULATION
FOR PRESSURED DRUM

SPECIFIC GRAVITY OF WATER = 1.0

ABSOLUTE PRESSURE = GAUGE PRESSURE + ATMOSPHERIC PRESSURE
= GAUGE PRESSURE + 14.7

NPSH   = ABSOLUTE − VAPOR − LINE   ± DIFFERENCE
AVAILABLE   PRESS.,FT. PRESS.,FT. LOSS.,FT. IN ELEV., FT.

WHERE PRESSURE IN FEET = (PRESSURE, PSIA)(2.31)
                          (SPECIFIC GRAVITY)

NPSH   = (10+14.7)(2.31) − (0.5)(2.31) − 45 + 5
AVAILABLE   (1.0)   (1.0)

          = 57.1 − 1.2 − 45 + 5

          = 15.9 FEET

NPSH AVAILABLE MUST
BE GREATER THAN
NPSH REQUIRED BY THE PUMP

GAUGE READING 10 PSI

AIR PRESSURE

0.5 PSIA
(WATER AT
80° F)

5'-0

C

LINE LOSS = 45 FT.

Figure 8-22. Calculating NPSHA for pressure drum.

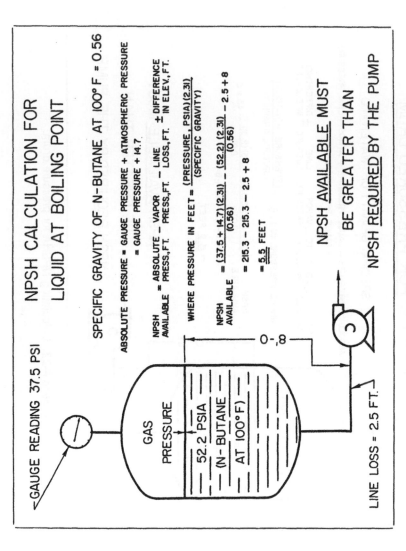

Figure 8-23. Calculating NPSHA for liquid at boiling point.

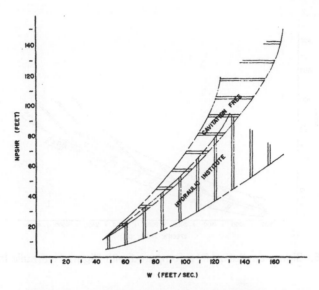

Figure 8-24. NPSHR cavitation free compared with NPSHR Hydraulic Institute.

values. Above 2,000 SSU, it is necessary to use the pump designer's judgment or experimentally determine the cavitation characteristics.

Figures 8-24 through 8-26 compare our cavitation-free NPSHR with NPSHR as permitted by the Hydraulic Institute. The Hydraulic Institute values shown were established by test and cavitation-free values calculated. Of 98 pump tests reviewed, 60 were selected for compilation of data. Criterion for selection was a witnessed suppression test conducted in accordance with the standards of the Hydraulic Institute. To reduce scatter, test data were confined to capacities limiting α to a maximum of $6\frac{1}{2}°$. The pumps used were limited to a maximum impeller blade peripheral velocity of 120 ft/sec. At higher velocities, the difference between cavitation-free and Hydraulic Institute will be more evident. Suction specific speed values lower than those shown in Figure 8-26 can be expected. General details of the pumps are:

- Number of pumps—50
- Number of tests—60
- Pump size (discharge nozzle)—1 in. to 30 in.
- Impeller diameter—7 in. to 37 in.
- Number of stages—One
- Impeller entry—Single suction: 26 pumps. Double suction: 24 pumps.
- Specific speed range—500 to 4,750

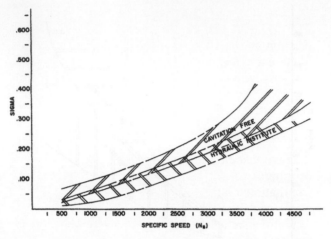

Figure 8-25. Sigma cavitation free compared with Sigma Hydraulic Institute.

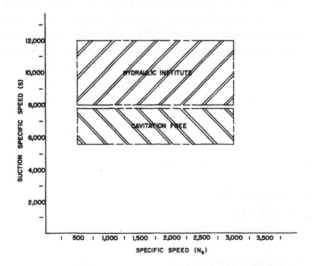

Figure 8-26. Suction specific speed cavitation free compared with suction specific speed Hydraulic Institute.

Notation

K_1 Friction and acceleration loss coefficient

K_2 Blade entry loss coefficient

NPSHR   Net positive suction head required (based on test or by calculation)

NPSHA   Net positive suction head available on site

C_{M1} Average meridional velocity at blade inlet (ft/sec) = .321 Q/Ae

w   Relative velocity of flow (ft/sec)

U_t Peripheral velocity of impeller blade (ft/sec) = $D_t N/229$

g   Gravitational acceleration (ft/sec^2)

θ Angle of flow approaching blade

α Angle of incidence

σ Sigma $= \dfrac{NPSH}{H}$

H   Head generated per impeller (ft)

B_1 Blade angle at outer radius of impeller eye

C_b Gas-to-liquid ratio factor

Q   Pump capacity (GPM)

Q_L Recirculated leakage entering impeller (GPM)

R_1 Factor in determining B_1

D_t Diameter at blade tip (in.)

N   Speed (RPM)

N_{SS} Suction specific speed $= \dfrac{N\,(Q)^{0.5}}{NPSH^{0.75}}$

        (for double suction impellers, use $(Q/2)^{0.5}$)

BEP   Best efficiency point on performance curve

N_s Specific speed $= \dfrac{N\,(Q)^{0.5}}{H^{0.75}}$

Ae   Impeller eye area at blade entry (sq in.)

U_h Peripheral velocity of impeller blade at hub (ft/sec)

C_1 Absolute velocity of flow (ft/sec)

References

Hydraulic Institute Standards Of Centrifugal Rotary And Reciprocating Pumps, 14th edition.

Lobanoff, V., "What Is This NPSH, *Oil Gas Journal*, February 24, 1958.

Ross, R. R., "Theoretical Prediction of Net Positive Suction Head Required (NPSHR) for Cavitation Free Operation of Centrifugal Pumps," United Centrifugal Pumps.

Part 2

Applications

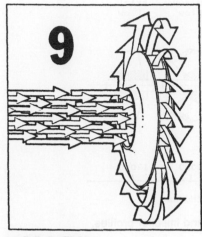

Vertical Pumps

by **Erik B. Fiske**
BW/IP International, Inc.
Pump Division

This chapter discusses radially split bowl pumps that are typically mounted vertically. In older literature these pumps are often, but improperly, referred to as vertical turbine pumps. This pump type is unique in that designs with optimum efficiency can be obtained over the full specific speed range, normally with values from 1,500 to 15,000. In the upper specific speed range, the pumps are referred to as axial flow or propeller pumps. The impeller profile changes with the specific speed as shown in Figure 9-1.

The hydraulic performance parameters, including efficiency, compare favorably with centrifugal pumps of the volute and diffuser type. However, except for highly specialized designs, vertical pumps are seldom used for high speed applications above 3,600 rpm.

Vertical pumps can be designed mechanically for virtually any application and are the only suitable configurations for certain applications such as well pumping. They are commonly used for handling cryogenic liquids in the minus 200°F to minus 300°F range as well as for pumping molten metals above 1000°F. The radially split bowl design lends itself to safe, confined gasketing. For high pressure applications, typically above 1,000 psi discharge pressure, an outer pressure casing can be employed, similar to that which is used for double-case, horizontal pumps. Except for conventional well pumps, the mechanical design for the majority of vertical pumps is customized in accordance with the application requirements. This requires close cooperation between the pump manufacturer and the architect/engineers responsible for the pump mounting structure and system piping.

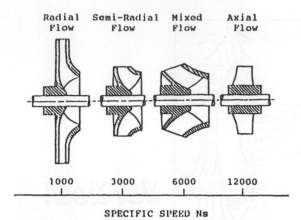

Figure 9-1. Specific speed and impeller profiles.

Configurations

There are three primary types of vertical pump configurations that are used for a broad range of applications.

Well Pumps

Designed to be installed in cased wells, these pumps consist of a multi-stage pumping element or bowl assembly installed at sufficient depth below the dynamic water level (the water level when the pump is operating) and with sufficient NPSH to preclude cavitation. The subject of NPSH is dealt with in detail Chapter 8. The bowl assembly, as illustrated in Figure 9-2, consists of a series of impellers mounted on a common shaft, and located inside diffuser bowls. The number of stages is determined by the height to which the liquid must be raised to the surface plus the design pressure required at the surface. The bowl assembly is suspended from a segmented column pipe that directs the flow to the surface where the column pipe is attached to a discharge head. The column also houses the lineshaft with bearings for transmitting the torque from the driver to the bowl assembly. The discharge head, in addition to providing the required connection to the customer's piping, also serves as the base for the driver. The driver can either be a direct electric motor drive, typically of hollow shaft construction, see Figure 9-3, or a right angle gear drive powered by a horizontal engine or turbine. The discharge head must be supported on a foundation adequate to carry the water-filled weight of the pumping unit plus the driver weight. However, the hydraulic thrust developed by the pump impellers is not transmitted to the foundation.

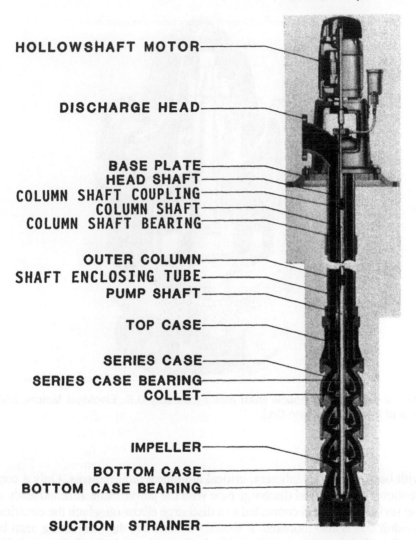

HOLLOWSHAFT MOTOR

DISCHARGE HEAD

BASE PLATE
HEAD SHAFT
COLUMN SHAFT COUPLING
COLUMN SHAFT
COLUMN SHAFT BEARING

OUTER COLUMN
SHAFT ENCLOSING TUBE
PUMP SHAFT

TOP CASE

SERIES CASE
SERIES CASE BEARING
COLLET

IMPELLER
BOTTOM CASE
BOTTOM CASE BEARING

SUCTION STRAINER

Figure 9-2. Well pump with hollow shaft, electric motor (courtesy BW/IP International, Inc. Pump Division, manufacturer of Byron Jackson/United™ Pumps).

An alternate well pump design uses a submersible electric motor drive, which is close coupled to the pump as shown in Figure 9-4. The motor can either be of the "wet winding" or "dry winding" design. For large motors, 250 HP and up, the preferred construction is the dry winding with the motor sealed and oil filled [2]. The motor is typically mounted below the pump, so there will be continuous flow of liquid around the outside of the motor for cooling. The submersible configuration eliminates the need for a lineshaft

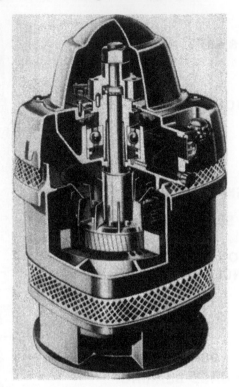

Figure 9-3. Vertical hollow shaft motor (courtesy U.S. Electrical Motors, Division of Emerson Electric Co.).

with bearings and its inherent, critical alignment requirements. Only a conventional, taper thread discharge pipe with the power cable attached leads to the surface. Here it is connected to a discharge elbow on which the electrical conduit box is also mounted. It should be noted that the well casing must be sized so that there is room alongside the bowl assembly for the power cable and a protective guard.

Wet Pit Pumps

This pump configuration, illustrated in Figure 9-5, can be either of the single-stage or multi-stage design, depending on the application requirements and covers the complete range of specific speeds. Installed in a pit or inlet structure, the water surface on the suction side of the pump is *free* and subject to atmospheric pressure. The available NPSH for a pump in an open system of this type is therefore equal to the atmospheric pres-

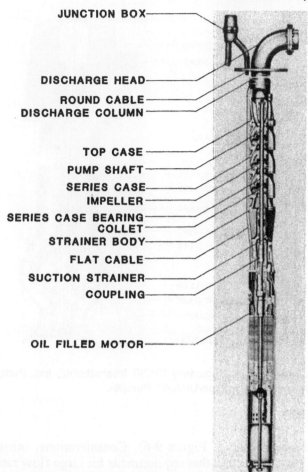

JUNCTION BOX

DISCHARGE HEAD
ROUND CABLE
DISCHARGE COLUMN

TOP CASE
PUMP SHAFT
SERIES CASE
IMPELLER
SERIES CASE BEARING
COLLET
STRAINER BODY
FLAT CABLE
SUCTION STRAINER
COUPLING

OIL FILLED MOTOR

Figure 9-4. Submersible well pump (courtesy BW/IP International, Inc. Pump Division, manufacturer of Byron Jackson/United™ Pumps).

sure, plus the static liquid level above the first-stage impeller, less correction for the liquid vapor pressure at the pumping temperature.

Because the cost of a pit or intake structure is high and dependent on the depth of the structure, the submergence is typically kept to a minimum in line with sound design practices. As a result, the maximum pump speed is limited by the NPSH available and the required flow rate. Single-stage pumps can usually be furnished for pumping heads up to 200 feet, but multi-stage pumps are required for higher heads. The bowl diameters of well pumps and their corresponding flow rates are restricted. However, wet pit pumps can be furnished in any size and therefore for

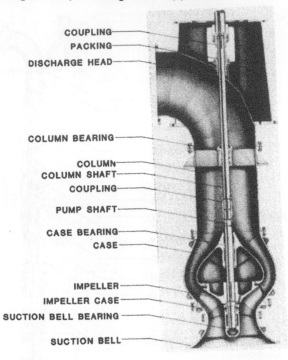

COUPLING
PACKING
DISCHARGE HEAD

COLUMN BEARING

COLUMN
COLUMN SHAFT
COUPLING

PUMP SHAFT

CASE BEARING
CASE

IMPELLER
IMPELLER CASE
SUCTION BELL BEARING

SUCTION BELL

Figure 9-5. Wet pit pump (courtesy BW/IP International, Inc. Pump Division, manufacturer of Byron Jackson/United™ Pumps).

any desired flow rate (see Figure 9-6). Considerations, other than the pump itself, usually dictate that requirements for large flow rates be split between two or more pumps operating in parallel. The pump setting (the axial length of the bowl assembly plus the length of the discharge column from which the bowl assembly is suspended) is normally less than 100 feet. The column houses the lineshaft, which is connected to the driver shaft with a rigid coupling in the discharge head (see Figure 9-5). The discharge head also houses a shaft sealing device. The driver, which is supported on the top of the discharge head, is generally provided with a thrust bearing of adequate size to carry the weight of the motor rotor and pump rotating element plus the hydraulic axial thrust developed by the pump. When the driver is not designed to carry the total axial thrust from the pump, a thrust bearing assembly must be provided in the discharge head above the shaft sealing device. A flexible type coupling must then be provided between the pump and the driver.

While the discharge elbow is normally located in the head above the pump mounting floor, it may be advantageous for certain applications to

Figure 9-6. Installation of 100,000 GPM wet pit pump (courtesy BW/IP International, Inc. Pump Division, manufacturer of Byron Jackson/United™ Pumps).

locate the elbow on the discharge column below the mounting elevation. For this type of configuration, care must be taken by the piping designer to make sure that any horizontal expansion of the discharge pipe is contained and will not force the pump column with the lineshaft out of alignment. Furthermore, the column must be free to elongate axially at the discharge elbow when filled with water. This can be accomplished by locating a flexible coupling in the discharge pipe near the discharge elbow.

Wet pit pumps are usually driven by a vertical, solid shaft induction or synchronous motor as depicted in Figure 9-7. However, horizontal drivers that transmit torque through a right angle gear mounted on the discharge head may also be used. For applications where a wide variation in flow and/or head requirements exist, a variable speed driver may be used. Eddy current clutches or variable frequency electric drives are often used, the latter being the more efficient of the two.

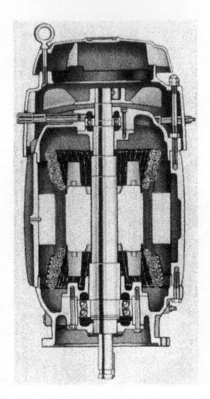

Figure 9-7. Vertical solid shaft motor (courtesy U.S. Electrical Motors, Division of Emerson Electric Co.).

Barrel-Mounted or Can-Mounted Pumps

This pump type, as depicted in Figure 9-8, is mounted in a suction barrel or can that is filled with liquid from the suction source. In this type of closed system, the NPSH available is not related to the atmospheric pressure. It is a function of the absolute pressure (above absolute vacuum) at the centerline of the suction flange, plus the liquid head from the center line of the suction to the first-stage impeller, less barrel losses, and less the vapor pressure of the pumped liquid. The available NPSH can therefore be increased by lowering the elevation of the first-stage impeller and extending the suction barrel until the available NPSH meets or exceeds the required NPSH. This feature is in fact frequently the reason for selecting a vertical, barrel-mounted pump. To achieve the same result with a horizontal pump would require lowering the entire unit at great ex-

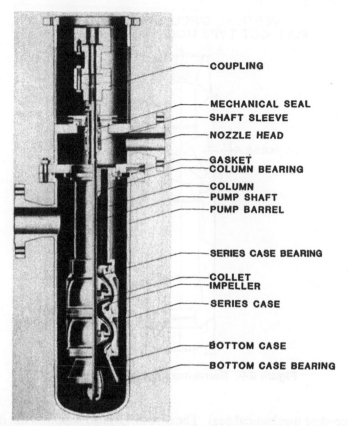

COUPLING

MECHANICAL SEAL
SHAFT SLEEVE
NOZZLE HEAD

GASKET
COLUMN BEARING

COLUMN
PUMP SHAFT
PUMP BARREL

SERIES CASE BEARING

COLLET
IMPELLER

SERIES CASE

BOTTOM CASE
BOTTOM CASE BEARING

Figure 9-8. Barrel-mounted pump (courtesy BW/IP International, Inc. Pump Division, manufacturer of Byron Jackson/United™ Pumps).

pense. Often a special first-stage impeller with superior NPSH characteristics is furnished. Otherwise, the bowl assembly is of a multi-stage design with identical impellers of the radial or semi-radial flow type. The bowl assembly is either directly suspended from the discharge head or connected to the head with a discharge spool, the length of which is determined by the NPSH required.

The configuration shown in Figure 9-9 is of the pull-out type that permits removal of the pump without disturbing either the discharge or suction nozzle connections. Pump alignment is, within reason, not affected by any nozzle forces imposed. The suction nozzle can be located either in the suction barrel or in the discharge head, in line with the discharge nozzle. The latter "in-line" construction is commonly used for booster applications in pipelines. The shaft sealing device in the discharge head is usu-

VERTICAL CIRCULATING PUMP
PULL-OUT TYPE MOUNTED IN BARREL

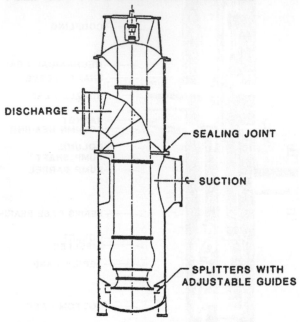

Figure 9-9. Barrel-mounted pull-out pump.

ally a face-type mechanical seal. The discharge head supports the driver, which should preferably be of the solid shaft design, either as a direct electric motor drive or a horizontal driver through a right angle gear. The unit is typically supported under the top flange of the suction barrel and bolted to an adequate foundation with a desired mass of five times the total unit weight. If desired, the entire barrel can be embedded in concrete or thermally insulated for high or low temperature applications.

Applications

Well Pumps

The most common applications for well pumps are:

- Water well, or bore hole installations, using either a surface-mounted driver or a close-coupled submersible motor.
- Incline-mounted water pumps installed on lake or river banks and driven by a conventional electric motor or a close-coupled submersible motor.

- Loading pumps in underground caverns used for storing petroleum products. The pump is mounted in a caisson.
- Dewatering pumps in mines. The pump is mounted in a mine shaft.

Water Well Pumps

This is the most common application and covers a broad range of services such as municipal water supply, irrigation service, and industrial service water. For settings down to 400 feet, a line shaft construction with a surface mounted driver is normally used. It should be noted that line shaft construction requires that the well be straight, so that the column shaft bearings can be kept in alignment. The well must be checked for this purpose prior to pump installation. For settings beyond 400 feet, and where electric power is available, the close-coupled submersible unit is usually the most cost-effective and also the most reliable. The close-coupled design often permits running at higher rotative speeds. With elimination of line shafting and corresponding bearings, well straightness is not as important. However, the well should be "caged" (checked with a dummy pump/motor assembly) prior to pump installation to make sure the unit will not bind in "dog legs."

Incline-Mounted Pumps

The cost of excavating and providing an adequate intake and mounting structure for vertical pumps on lakes and river banks can be substantial, particularly where large fluctuations in water level require high structures. A less costly installation may be achieved by mounting vertical pumps on an incline on a lake or river bank, as shown in Figure 9-10. The pump is mounted inside a pipe or in a trough, permanently anchored on piers along the bank with the bowl assembly at a sufficient depth to provide adequate submergence. The pump can either be of the line shaft type with an electric motor drive or close coupled to a submersible motor, in which case both the pump and motor are mounted in the pipe or on the trough.

Cavern Pumps

For ecological as well as safety and economic reasons, petroleum products, ranging from propane to crude oil, are often stored in natural or manmade caverns rather than in large surface tanks [7]. Well type pumps, most commonly driven by submersible motors, are used for unloading the cavern before the product is further transported by pipeline or ship. The pumps are usually mounted in a caisson, which is sealed at the top and terminates near

Figure 9-10. Incline-mounted line shaft pump.

the bottom of the cavern, where the pumps take suction. During maintenance, water is let into the cavern, and with the petroleum floating on top of the water, the bottom of the caisson is sealed off with a water "plug," preventing undesirable gases from escaping. The water level in the cavern can be maintained with separate pumps.

Mine Dewatering Pumps

Vertical well pumps are often preferred for mine dewatering. All sensitive electrical equipment, including control panels, can be located well above levels where accidental flooding might occur or where explosive gasses may be present. This includes both conventional motor driven pumps and submersible motor pumps. The pump may be installed in an open mine shaft, or a separate well may be sunk for the purpose of dewatering. The mechanical construction of the pump is similar to a water well pump.

Wet Pit Pumps

The most common applications for wet pit pumps are:

- Water supply pumps for municipalities and industry. The pumps are mounted in intake structures on lakes or rivers.
- Condenser cooling water pumps for central power plants. The pumps take suction from a natural body of either fresh or salt water.
- Cooling tower pumps. Take suction from a cooling tower basin and circulate water through a closed system.
- Flood control pumps mounted at dams and in collection basins, often as part of large flood control systems.
- Transfer pumps for central irrigation districts and water treatment facilities.

Water Supply Pumps

This pump type is normally installed as multiple, parallel operating units in a simple intake structure or as a stand-alone pumping plant located on a reservoir, lake, or river and discharging into a pipeline or an open canal. Depending on the system requirements, multi-stage pumps or single-stage pumps of the desired specific speed are used. A combination of fixed and variable speed drivers may be desirable to obtain optimum system efficiency. While structural integrity and cost are critical items in design of intake structures, hydraulic considerations and protection of the pumping equipment are equally important. The structure should be physically located so that a minimum of debris and silt will be diverted toward it. Trash racks and rotating screens, which can routinely be cleaned, must be provided to keep foreign objects from entering the pumps. The intake structure must be designed for low approach velocity and with dividing walls forming individual bays as required [5]. The pumps should be located within the structure in such a fashion that uniform velocity distribution is provided at each pump suction bell. Obstructions, changes in flow direction, or velocity changes that may cause formation of vortices and air entrainment must be avoided. Flow patterns within the structure, when one or more pumps are idle, must also be considered. Pump settings, the distance from the mounting floor to the suction intake, typically vary from 15 feet to 80 feet. The discharge may be located above or below the mounting floor, depending on the system requirement. In either case, the discharge pipe should be anchored downstream from the pump discharge flange to prevent pump misalignment from pipe reaction forces. When a flexible discharge piping connector is used, tie bars must be provided across the connector to restrain the hydraulic separating forces and prevent pump misalignment. Figure 9-11

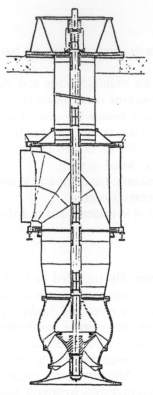

Figure 9-11. Water supply pump of pull-out design with below ground discharge.

shows a below ground discharge application with pull-out construction. This construction is particularly well suited for locations where the discharge is located below the water level because the nozzle can be permanently welded to the discharge piping system. Solid shaft electric motors are most commonly used as drivers.

Condenser Cooling Water Pumps

Installation considerations, configuration, and driver requirements for these pumps are in general the same as for water supply pumps. However, in addition, the following must be considered:

● When selecting lubrication method for the column bearings, neither oil nor grease can be used, because small amounts of hydrocarbon entering the condenser will impair the heat-transfer properties.

- Fresh water bearing lubrication is preferred, but salt water injection is acceptable as long as it is filtered and continuous. If a pump sits idle for extended periods without injection, accelerated corrosion may take place as well as the build-up of harmful crystals and marine life.
- Care must be taken in the selection of base materials for salt water. Consideration should be given to the average temperature of the water and the potential absence of oxygen.

Cooling Tower Pumps

These pumps typically have short settings, less than 20 feet, operate against a fixed head, and are connected to fixed speed drivers. The cooling tower basin and associated sump for pump installation are usually very limited in depth and area, requiring specific precautions to avoid vortices and ensure uniform velocity distribution [5]. Model structure testing is recommended where design margins are small. Water quality tends to become questionable in this type of closed system, warranting caution in material selection.

Flood Control Pumps

These are typically low lift, short setting pumps, but vary greatly in size depending on historic demands. A 2,000 to 3,000 gpm sump pump may be adequate for protecting a small area, while a large flood control district may require multiple pumps of several hundred-thousand gpm capacity each. Such large pumps can partly be formed in concrete at the site. This can be a major cost advantage, particularly if the water quality demands high alloy metals. Figure 9-12 shows a propeller pump where the suction bell, column, and discharge elbow have been formed in concrete.

Pump efficiency is not the primary consideration for flood control pumps because operating time and therefore power consumption is limited. Reliability is the main concern, and the pumps must be capable of handling large amounts of silt and sand. Serious flood conditions may be connected with loss of electric power, and flood control pumps are therefore often driven by diesel engines through right angle gear drives.

Transfer Pumps

This general category of pumps covers a wide variety of applications, from highly efficient central irrigation pumps and canal lift pumps, to simple, non-clog pumps in sewage and water treatment plants. Commonality in design is therefore minimal. When efficiency is one of the

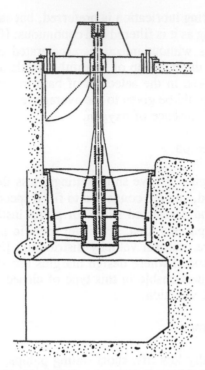

Figure 9-12. Propeller pump partly formed in concrete.

primary design criteria, the design is generally the same as for cooling tower pumps. On the other hand, non-clog pumps are designed for maximum reliability and availability. The impeller is typically of semi-open construction with two or three vanes and contoured to prevent adherence of stringy material. Where suspended solids are a problem, provision for clean water injection to the bearings can be provided. Handhole covers are provided at locations where buildup of solids will require removal.

Barrel-Mounted or Can-Mounted Pumps

The most common applications for barrel-mounted pumps are:

- Condensate and heater drain pumps for power plant service.
- Process pumps for products with limited NPSH available.
- Small boiler feed pumps for industrial applications.
- Cryogenic process and transfer pumps.
- Loading pumps on tank farms.
- Booster pumps for pipelines handling either water or petroleum products.

Condensate and Heater Drain Pumps

This pump type is typically installed as two 50% capacity pumps taking suction from a header connected to a condenser or heater for boiler feed water. The available NPSH is normally only two to four feet at the mounting floor, requiring additional NPSH to be built into the barrel. The multi-stage bowl assembly, typically in the 1,500 to 2,500 specific speed range, can be fitted with a special first-stage impeller to meet the required NPSH. Figure 9-13 shows a unit with a double suction first-stage impeller. The suction nozzle may be located either in the barrel or the discharge head, whichever the user prefers. The shaft seal in the discharge head is typically of the mechanical face type and must be water quenched because the seal is under vacuum when on standby. A continuous vent line must be provided from the top of suction side in the pump to the vapor phase in the suction tank (condenser). A minimum flow bypass line may be required at the discharge control valve if extended low flow operation cannot be avoided. Induction motor drive is the most common, but a variable speed drive offers advantages for peak loaded plants. Condensate pumps normally operate in the 130°F range, and cast iron bowls with bronze impellers and bearings are usually adequate. For applications where the peripheral vane velocity in the suction eye exceeds 80 feet per second, a stainless steel impeller should be used. Some users will not permit bronze materials in the system because it may contribute to corrosive attack on condenser tube welds. In these cases, all impellers should be furnished in martensitic steel. Heater drain pumps may operate up to 350°F and require impellers of martensitic steel and bearings of a carbon-graphite composite. Because flashing in the first stage cannot always be avoided in this service, injecting second-stage pump pressure into the suction case bearing is recommended.

Process Pumps

These pumps are of multi-stage construction, with a special first-stage impeller to meet the limited available NPSH. Handling liquids near their boiling point requires a continuous vent line from the pump suction side back to the suction source. The mechanical shaft seal can either be mounted internally in the discharge stream, flushed and cooled by the pumped liquid, or mounted in an external, water jacketed stuffing box for high temperature applications. Materials for the bowl assembly and fabricated components are selected to suit the liquids handled, including cavitation resistant material for the first-stage impeller, when applicable.

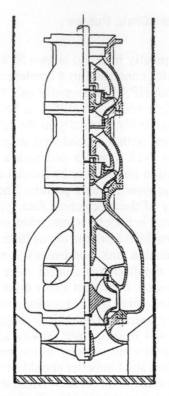

Figure 9-13. Bowl assembly with double suction first stage.

Small Boiler Feed Pumps

In design, these pumps are quite similar to heater drain pumps, although the NPSH margin is usually sufficient not to require injection to the suction case bearing. The mechanical seal should be located in a stuffing box, mounted externally on the discharge head. The minimum recommended material selection is cast iron bowls, martensitic steel impellers, and carbon/graphite bearings.

Cryogenic Pumps

Vertical barrel pumps are particularly well suited for cryogenic applications. Being vertically suspended, thermal contraction and expansion will not cause pump misalignment as long as reasonable precaution is taken in dealing with nozzle forces at the suction and discharge flanges. Because the motor is supported on top of the discharge head, it is automatically aligned to the pump. The external pump configuration is simple and easy to thermally insulate with jackets or in a "cold box." A thermal barrier, also known as a "warming box," with a throttle bushing and a double mechanical seal or a gas shaft seal, is located in the discharge head, just above the discharge nozzle. Here the cryogenic liquid is flashed and bled back to the suction source, while an inert gas blanket under the shaft seal prevents leakage to atmosphere. Depending on the liquid pumped, pump materials with adequate impact strength are bronzes, aluminum, and austenitic stainless steels. Materials with similar thermal coefficient of expansion must be used where tolerances are critical.

Loading Pumps

Normally installed immediately adjacent to large storage tanks used for loading product into pipelines or transport vessels, these pumps are mounted so that the storage tank can be emptied, even when the available NPSH becomes zero at the bottom of the tank. Adequate provisions must be made for venting the suction barrel and providing a minimum flow bypass when applicable.

Pipeline Booster Pumps

These pumps typically operate unattended and must be designed for reliability. The discharge and suction nozzle should both be located in the discharge head for simplicity in piping and valve placement. When handling liquids where leakage to the atmosphere is hazardous, a tandem or double mechanical shaft seal with a buffer fluid should be provided for the stuffing box. Pumps can be arranged to operate singly, in parallel, or in series.

Design Features

For comparison and evaluation of design features, the three basic assemblies of vertical pumps should be addressed, namely the bowl assembly, the column assembly, and the head assembly. (See Figure 9-14).

The Bowl Assembly

The simplest bowl assembly configuration consists of a straight shaft with taper collet mounted impellers, bowls that are joined together with straight threads and furnished with a shrink fitted bearing (Figure 9-2). The impellers can be either of the enclosed design, with both a front and a back shroud, or the semi-open design without a front shroud (see Figure 9-15). The bottom case bearing is normally permanently grease lu-

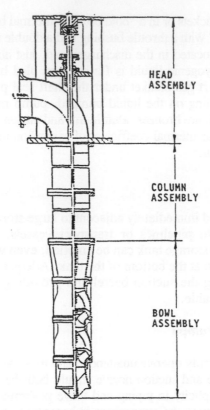

HEAD
ASSEMBLY

COLUMN
ASSEMBLY

BOWL
ASSEMBLY

Figure 9-14. Vertical pump assemblies (courtesy BW/IP International, Inc. Pump Division, manufacturer of Byron Jackson/United™ Pumps).

bricated and the other bearings lubricated by the pumped liquid. This design lends itself well to smaller pumps with up to 18-inch bowl diameter and 2-inch shaft diameter. However, the following limitations must be noted:

- Taper collet mounting of impellers, depending on shaft diameter and material combinations, is only recommended for handling liquids from 0°F to 200°F. The same temperature limitation applies to semi-open impellers.
- A grease lubricated bottom bearing is only recommended for ambient temperature water service; otherwise, lubrication with the pumped liquid should be used with filtration when required.

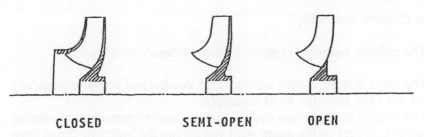

CLOSED   SEMI-OPEN   OPEN

Figure 9-15. Impeller configurations.

For larger pumps or when handling hot or cold liquids, the following design practices are recommended:

- The bowl joints should be flanged and bolted. Gaskets, when required, should be of the "O"-ring type, so that joints are made up metal to metal, and pump alignment therefore maintained.
- The impellers should be mounted on the shaft with key drives and secured axially with split rings and thrust collars.
- When selecting closed vs. semi-open impellers, the following must be noted:

  - Closed impellers should always be used for handling hot or cryogenic liquids.
  - Closed impellers exhibit lower downthrust when in the axially unbalanced configuration.
  - Closed impellers are easier to assemble for large pumps with more than three stages.
  - Semi-open impellers are more efficient due to elimination of disc friction from the front shroud.
  - Efficiency loss due to wear on the semi-open impeller vanes can be regained by adjusting the impeller setting at the adjustable pump to driver coupling, typically in increments of 0.016 inch.
  - When semi-open impellers are axially balanced to reduce axial thrust, downthrust can be maintained over the full operating range. This prevents shaft whip from upthrust with associated bearing wear.
  - Semi-open impellers can readily be hardsurfaced for erosion protection.
  - Semi-open impellers are less likely to seize when handling sand or foreign material.

The Column Assembly

The column assembly consists of three primary components:

- The outer column, which serves as the conduit and pressure boundary for the flow from the bowl assembly.
- The column shaft, or line shaft, which transmits torque from the driver to the impellers on the pump shaft and carries the hydraulic thrust from the bowl assembly to the thrust bearing in the head/driver assembly.
- The shaft enclosing tube, or inner column, which houses the column bearings, serves as a conduit for bearing lubrication, and protects the shafting. The liquid pumped determines whether or not a shaft enclosing tube is required.

Outer Column

The simplest outer column construction consists of pipe sections, normally of 10-feet length, with straight thread on both ends, and joined with pipe couplings. This design is commonly used for 12-inch column diameters or less. For handling relatively clean liquids, bearings of a rubber compound are located in housings with a three-legged or four-legged spider and a mounting ring, which is centered within the column coupling and clamped between the column pipe ends. Metal to metal contact provides an adequate liquid seal. This configuration is often referred to as *open lineshaft construction*.

For larger column sizes, or where corrosive or other properties of the pumped liquid make threaded joints undesirable, flanged column joints are used. Registered fits are used to provide alignment, with "O"-ring gaskets for sealing because they provide metal-to-metal flange face contact for alignment. Bearings housed in spiders can be clamped between column faces; however, superior alignment is provided with a design incorporating spiders welded into the outer column, with the flange register and spider bore machined in the same operation. When a shaft enclosing tube is required, the larger column sizes require a metal stabilizing spider clamped or welded at the column joint with a snug, machined fit around the enclosing tube. Again a tensioning device is required at the top end of the threaded enclosing tube.

Column Shaft

Shaft sections with three-inch shaft diameters or less are commonly joined by threaded couplings, which transmit both torque and axial thrust. For pumps with this construction, it is imperative that drivers be checked for correct rotation before being connected to the pump. Threaded couplings torqued in the reverse direction will unscrew, and the resulting jacking motion may cause serious damage. However, reverse rotation from backflow through the impellers will not cause the couplings to unscrew because the direction of shaft torque remains the same as for normal operation.

For column shafts four inches and larger in diameter, a keyed sleeve coupling should be used for transmitting torque. Axial thrust should be carried through split rings retained by thrust collars. Flanged bearings, both in the column and the bowl assembly, are recommended for bores four inches and larger.

Shaft Enclosing Tube

When abrasives or corrosive properties prohibit the pumped liquid from being used for flushing and lubricating the column bearings, the bearings should be fitted inside a shaft enclosing tube. The bearings, typically of bronze material, are threaded on the O.D. and serve as joiners for the five-feet long enclosing tube sections. Bearing alignment is provided by placing the enclosing tube assembly in tension through a threaded tensioning device located in the discharge head. The enclosing tube is stabilized within the outer column by random placement of hard rubber spiders, the hub of which fits tightly around the enclosing tube and the three legs fit tightly against the inside of the outer column. The desired bearing lubrication, which can be oil, grease, clean water, or any fluid compatible with the pumped liquid, is injected at the top end of the enclosing tube assembly.

To overcome the mounting number of assembly and handling problems that can occur with increase in column size, an enclosing tube integrally welded with ribs at the top and bottom of the outer column is a preferred design. Alignment is assured with simultaneous machining of the registered column fits, the inner column joints, with slip fit and "O"-rings, and the bearing seats. Furthermore, the need for a tensioning device in the discharge head is eliminated. The result is that both assembly and disassembly time for the pump is significantly reduced.

The Head Assembly

The discharge head is designed to serve the following functions:

• Support the suspended, liquid-filled weight of the pumping unit.
• Provide support for the driver.

- Incorporate a discharge nozzle to guide the flow from the outer column to the system pipe. For barrel-mounted pumps, a suction nozzle may also be located in the discharge head.

The discharge head must house a shaft sealing device suitable for the maximum pressure the pump can be subjected to. The sealing device is located in a stuffing box that can be placed either in the discharge stream for flushing or mounted externally for cooling and flushing. The actual sealing can be done with packing or a mechanical face seal. A pressure breakdown bushing, with bleed-back to pump suction, can also be included in the sealing device for high pressure applications.

The standard drive coupling for vertical pumps with solid shaft drivers is of a rigid design, capable of transmitting the maximum torque from the driver and the combined axial force from hydraulic thrust plus rotating element weight. The coupling typically incorporates a disc threaded on to the top end of the column shaft, clamped between the two coupling halves, that permits adjustment of the impeller setting within the bowls (see Figure 9-14). For drivers with limited thrust carrying capability, a thrust bearing must be incorporated into the discharge head design.

For pumps using hollow shaft drivers, torque is transmitted to the top column shaft or head shaft through a keyed clutch at the top of the motor, and impeller adjustment is made by a nut seated on top of the clutch (see Figure 9-3).

Except for the smaller well pumps and barrel-mounted pumps, most vertical pumps are of a structurally flexible design. This means that the structural, natural frequency of the first order is of the same magnitude as the operating speed. A careful analysis must therefore be made of the discharge head design in relation to its foundation and the connected driver and system piping to ensure that the combined natural frequency does not coincide with the pump operating speed. Similarly, deflection calculations for the unit must be made to ensure that pump alignment is not impaired when it is subjected to nozzle loads and the liquid filled weight.

Pump Vibration

The vibration pattern of a vertical pump is an inherent characteristic of its configuration, manufacture, and physical condition. Vibration results from factors such as rotating element unbalance, misalignment, looseness in the assembly, bent shafting, or bad driver bearings. Also, the operating parameters and the installation, including the rigidity of the supporting structure and attached piping, have a strong influence. The latter may cause vibration from sources such as hydraulic resonance in piping, turbulence at the pump intake, cavitation problems, low flow recirculation, and structural resonance in the pump/driver assembly.

The availability of data collectors and matching computer hardware and software has greatly facilitated collecting and analyzing vibration data. The establishment of vibration signatures is not only a means of verifying the satisfactory condition of a new installation, but can also serve as the basis for scheduling pump maintenance.

Measurement of axial and lateral vibration on the pump and driver is measured either as absolute movement on driver bearing housings or as relative movement between the shaft and the bearing housing or pump structure. Torsional vibration is seldom a problem in vertical pumps because the exciting force generated by the rotating impeller vanes passing the stationary bowl vanes is small. However, for applications where right angle gears and engines are used, exciting forces can be generated that may cause damaging torsional vibration. When these types of drivers are used, an analysis for torsional critical frequencies should be performed at the design stage. The computer models for performing these analyses are quite accurate and give good results. This subject is discussed in more detail in Chapter 18.

Figure 9-16 shows the desired locations for taking vibration measurements. The axial reading is taken as an absolute measurement directly on the motor thrust bearing housing. The lateral readings on the motor and discharge head are also taken as absolute measurements and should be taken in line with, and at right angle to, the discharge nozzle. Measurements on the shaft should be taken as relative measurements, 90° apart, just above the stuffing box.

Absolute vibration measurements are taken with velocity transducers or accelerometers. Accelerometers should either be permanently attached or attached with a magnetic base, while velocity transducers can be handheld. Velocity transducers and accelerometers are directional and must be installed with the base perpendicular to the desired direction of measurement. Relative vibration measurements are taken with proximity transducers. It should be noted that proximity transducers, due to their working principle, are sensitive to shaft material properties as well as surface finishes.

Because of the wide varieties and sizes of vertical pumps in use, the issue of acceptable vibration levels becomes rather complicated. However, both the Hydraulic Institute and the American Petroleum Institute have published acceptance criteria, specifically applicable to pumps, covering both overall vibration levels and filtered vibration, i.e., read-

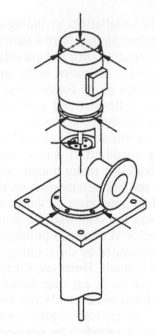

Figure 9-16. Locations for taking vibration measurements.

ings taken at discreet frequencies. Caution should be used in applying severity charts published for general machinery.

References

1. API Standard 610 *Centrifugal Pumps for General Refinery Service*, 7th Edition, American Petroleum Institute, Washington, D.C., 1990.
2. *Byron Jackson Pump Division Design Fundamentals in Submersible Motors*, Technical Report Vol. 3, No. 1, 1956.
3. deKovats, A. and Desmur, G., *Pumps, Fans and Compressors*, Glasgow, Blackie & Son Limited, 1978.
4. Fiske, E. B., "Fundamentals of Pump Design," *2nd International Pump Symposium*, Texas A&M University, April 1985.
5. Hydraulic Institute Standards 2.2, *Vertical Pumps—Application*, 15th Edition, Hydraulic Institute, Cleveland, OH, 1992.
6. Sanks, R. L., *Pumping Station Design*, Stoneham, MA, Butterworth Publishers, 1989.
7. Stamps, D. B., "Submersible Pumps for Underground Storage Caverns," *Byron Jackson Pump Division Publication No. 7520*, August 1984.

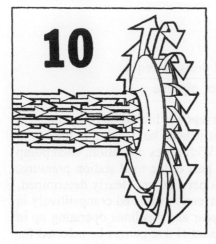

10

Pipeline, Waterflood, and CO$_2$ Pumps

Pipeline Pumps

Unlike most other pump applications, pipelines constantly have changes in throughput and product. This is particularly true in the transportation of crude oil and hydrocarbons. This variation in liquid characteristics, throughput, and pressure can result in a wide range of system head curves, requiring extreme flexibility in pump operation. Selecting pumps can be complicated, usually requiring multiple pumps installed in series at each station. Selection may involve parallel operation, variable speed, and/or modifications to meet future requirements.

As pumping requirements must match pipeline characteristics, the first step in pump selection is analysis of the hydraulic gradient, and profile. This defines the length and elevation change of the pipeline and is used to establish pipeline pressure, pipeline horsepower, number of stations, number of pumps, and appropriate mode of operation. Pipe size is determined from throughput requirements and optimum investment, plus operating cost economics. Calculation of friction loss and static head establishes the pressure required to move throughput. With pipeline throughput and pressure known, pump efficiency can be estimated and pipeline horsepower calculated. The number of pumps required is then estimated by selecting the preferred driver size. Number of stations is estimated by the safe working pressure (S.W.P.) of the pipeline and the pressure required to move throughput.

$$\text{Pipeline HP} = \frac{\text{Pressure req'd (PSI)} \times \text{Throughput (BPD)}}{58,700 \times \text{Estimated pump efficiency}}$$

$$\text{No. of pumps} = \frac{\text{Pipeline HP}}{\text{Driver HP}}$$

$$\text{Min. no. of stations} = \frac{\text{Throughput pressure req'd}}{\text{S.W.P. of pipeline}}$$

Variable capacity requirements or horsepower limitations may dictate a need for multiple pumps. In this event it must be decided if the pump should operate in series or in parallel. With series operation, each pump delivers full throughput and generates part of the total station pressure.

Once pump conditions and mode of operation are clearly determined, pumps can be selected. With pumps currently evaluated competitively in excess of $1,000 per horsepower per year and pipelines operating up to 40,000 horsepower per station, it is essential that pumps be selected for optimum efficiency. This requires an understanding of the losses that occur inside a pump. These are:

- Friction losses.
- Shock losses at inlet to the impeller.
- Shock losses leaving the impeller.
- Shock losses during the conversion of mechanical power to velocity energy then to potential energy.
- Mechanical losses.
- Leakage losses at impeller rings and interstage bushings.
- Disc friction losses at the impeller shrouds.

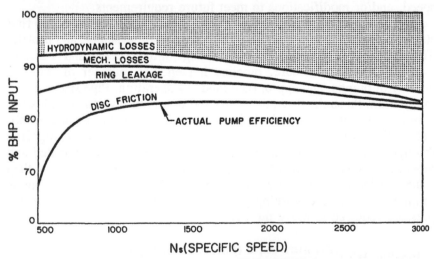

Figure 10-1. Typical analysis of pump losses.

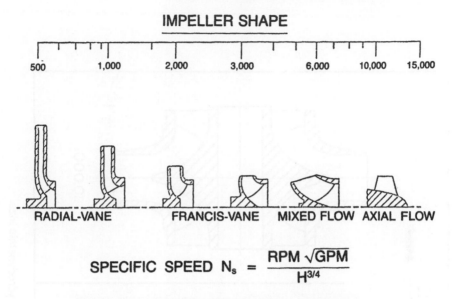

Figure 10-2. Specific speed describes impeller shape.

These losses can be generally classified as hydrodynamic, mechanical, ring leakage, and disc friction. Analysis of these losses for one specific performance at various speeds is shown in Figure 10-1. Pump efficiency is a result of the sum of these losses and is influenced by specific speed, which is basically a non-dimensional number. As discussed in Chapter 2, the physical meaning of specific speed has no practical value; however, it is an excellent means of modeling similar pumps and describes the shape of the impeller under discussion (Figure 10-2).

Pump efficiency is also influenced by hydrodynamic size (Figure 10-3). For any given speed, pump efficiency increases with size of pump or with hydrodynamic size (Figure 10-4). Through careful selection of pump speed and stage number, optimum specific speed and hydrodynamic size can be determined for maximum efficiency.

Condition Changes

Many pipeline conditions require low-capacity, low-pressure start-up with ultimate change over to high-capacity, high-pressure. By considering this requirement at the design stage, pumps can be built to accommodate the initial and ultimate conditions through field modifications. One method is to adjust the ratio of liquid velocity leaving the impeller to liq-

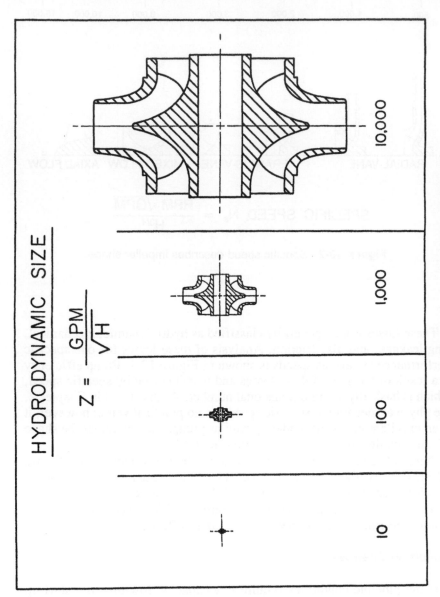

Figure 10-3. Pump size increases with hydrodynamic size.

PUMP   EFFICIENCY

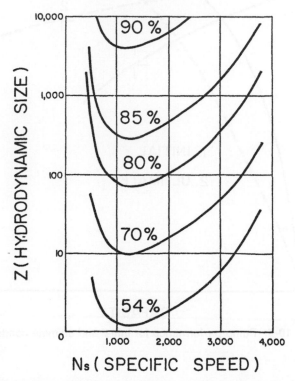

Figure 10-4. Efficiency increases with hydrodynamic size.

uid velocity entering the volute throat. By careful selection of this velocity ratio, pump throughput and optimum efficiency can be moved from
the initial to the ultimate condition (Figure 10-5).

The volute throat velocity can be adjusted by cutting back (chipping)
the stationary volute lip or lips to a predetermined dimension (Figure
10-6). This technique can be used on single- or double-volute pumps.
Similarly, the velocity leaving the impeller can be adjusted by removing
metal (underfiling) from the non-working side of the impeller blade (Figure 10-7). Through a combination of volute chipping, impeller underfiling, and ultimate installation of a high-capacity impeller, a wide variety
of pump conditions all at optimum efficiency becomes possible (Figure
10-8).

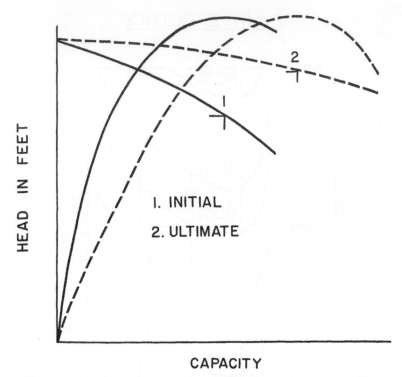

HEAD IN FEET

I. INITIAL

2. ULTIMATE

CAPACITY

Figure 10-5. Pump performance for initial and ultimate condition.

Another method of satisfying two capacities at optimum efficiency is the installation of volute inserts. These removable pieces reduce the internal fluid passages and physically convert a high-capacity pump to low capacity (Figures 10-9 and 10-10). Inserts of this type are ideal for low-capacity field testing or reduced throughput operation with improved efficiency (Figures 10-11 and 10-12).

Destaging

It is not unusual for pipelines to start up at low pressure then later change to high pressure at constant throughput. This change in pump head requirements can be suitably handled by selecting a multi-stage pump for the ultimate high-pressure condition. For the initial low-pressure condition, appropriate impellers are removed and the interstage chambers isolated by destaging tubes (Figure 10-13). When changing to high pressure the destaging tubes are removed and additional impellers installed (Figure 10-14).

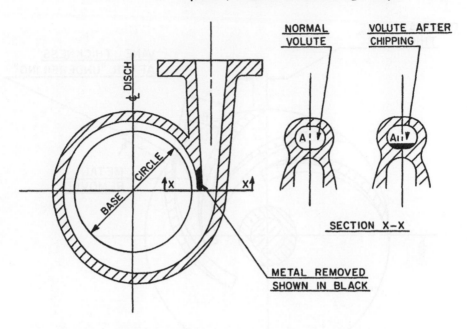

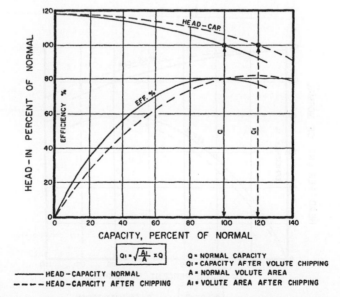

Figure 10-6. Changing pump performance by volute chipping.

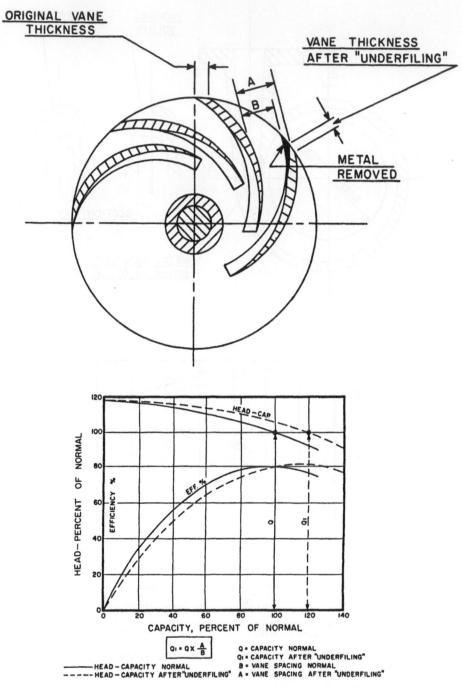

Figure 10-7. Changing pump performance by impeller underfiling.

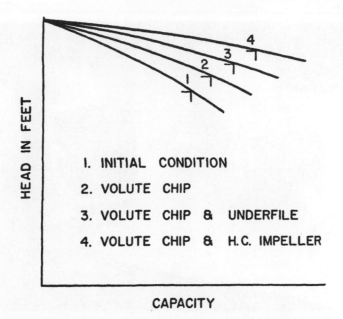

Figure 10-8. Performance range by volute chipping, impeller underfiling, and installation of high-capacity impeller.

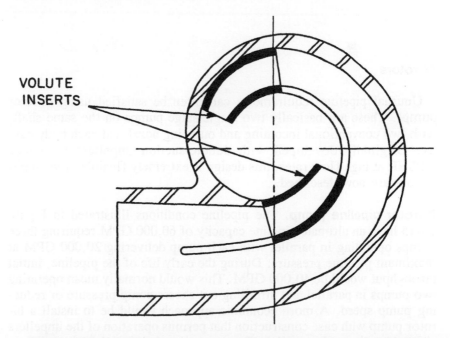

Figure 10-9. Volute inserts convert high-capacity pump to low capacity.

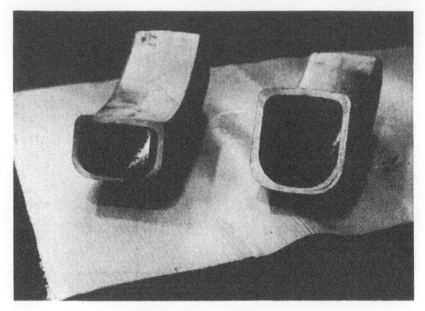

Figure 10-10. Volute inserts.

Bi-rotors

Unusual pipeline requirements can often be satisfied using bi-rotor pumps. These are basically two single-stage pumps on the same shaft, with two conventional incoming and outgoing nozzles at each body cavity. Originally developed with double-suction impellers to reduce NPSHR at high flow rates, this design is extremely flexible. Two applications are now described.

Bi-rotor pipeline pump. The pipeline conditions illustrated in Figure 10-15 have an ultimate pipeline capacity of 60,000 GPM requiring three pumps operating in parallel with each pump delivering 20,000 GPM at maximum pipeline pressure. During the early life of the pipeline, initial throughput would be 40,000 GPM. This would normally mean operating two pumps in parallel and throttling out excess pump pressure or reducing pump speed. A more economic approach would be to install a bi-rotor pump with case construction that permits operation of the impellers either in series or in parallel. The conventional, integral cast cross-over

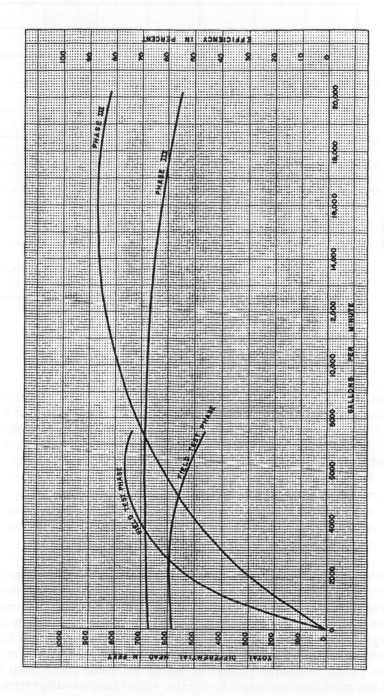

Figure 10-11. Performance change with volute inserts and low-capacity impeller.

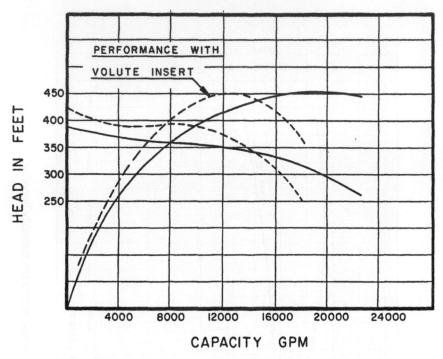

Figure 10-12. Performance change with volute inserts.

(Figure 10-16), which permits transfer of liquid from the first- to the second-stage impeller, is replaced by a bolted-on cross-over (Figure 10-17). With the cross-over bolted to the pump case, both impellers operate in series. With the cross-over removed, both impellers operate in parallel (Figures 10-18 and 10-19). With both impellers operating in parallel, the pump will deliver twice-normal pump capacity at half-normal pump pressure. In this configuration, initial throughput can be handled by one pump instead of two. For ultimate throughput, the cross-over is bolted in place and three pumps operate in parallel. Driver size is not affected, as the required horsepower is identical for either configuration.

Pulsation. The bolted-on cross-over has the additional benefit of permitting corrective action in the event of sympathetic acoustical frequency. All centrifugal pumps have a source of energy at blade passing frequency (Figure 10-20). Normal pressure pulsations generated by the pump can be magnified by system resonance when they are coincidental with fluid or mechanical natural frequencies within the system. These can be in the suction piping, discharge piping, or within the pump itself when the pump has more than one stage. Corrective action involves either relocat-

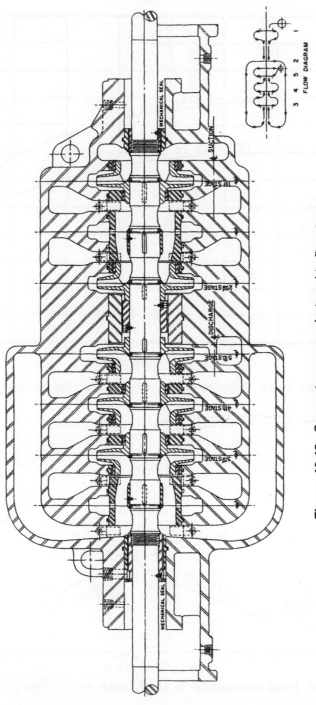

Figure 10-13. Seven-stage pump destaged to five stages.

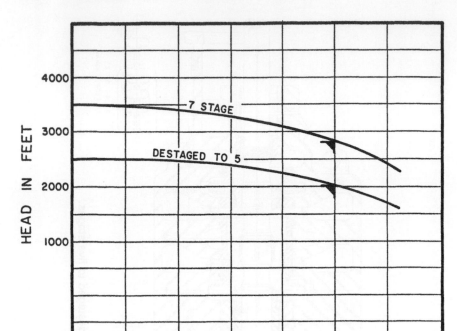

Figure 10-14. Pump performance before and after destaging.

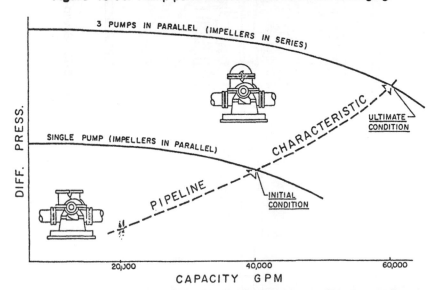

Figure 10-15. Pump configuration to satisfy initial and ultimate pipeline requirement.

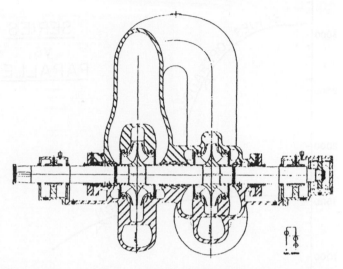

Figure 10-16. Two-stage pipeline pump with integral cast crossover.

Figure 10-17. Two stage pipeline pump with bolted on crossover.

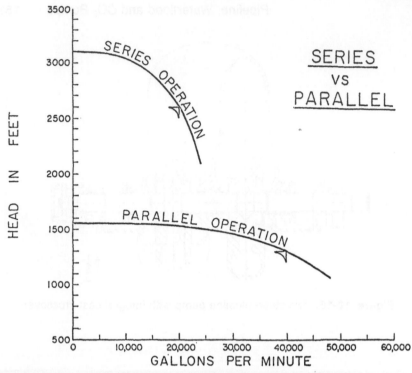

Figure 10-18. Performance of two-stage pipeline pump with bolted on cross-over (series operation) and without crossover (parallel operation).

Figure 10-19. Two-stage pipeline pump being tested with crossover removed and both impellers operating in parallel.

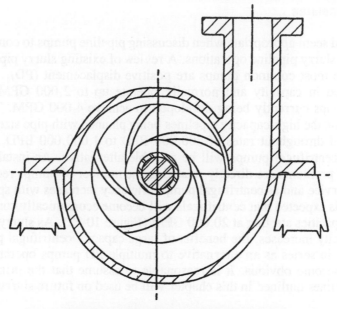

Figure 10-20. Blade passing frequency.

ing one or more of the sympathetic frequencies, or changing the pump generating frequency by a change of speed or blade number. With a bolted-on crossover a simpler solution is to add calculated pipe length between pump case and cross-over, which will change the wave length and relocate the acoustic frequency away from the generating frequency (Figure 10-21).

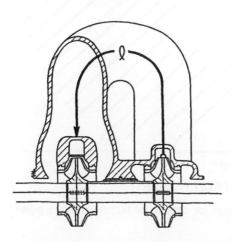

Figure 10-21. Acoustic frequency.

Slurry Pipelines

It would seem appropriate when discussing pipeline pumps to comment on future slurry pipeline operations. A review of existing slurry pipelines shows the most common pumps are positive displacement (PD). These are limited in capacity and normally operate up to 2,000 GPM with larger pumps currently being developed to achieve 4,000 GPM. To accommodate the high-capacity pipelines being planned with pipe size up to 42 in. and throughput rates of 500,000 BPD to 1,000,000 BPD, high-capacity centrifugal pumps will be a viable alternative. As installation costs and problems are directly related to the number of pumps required for the service and as centrifugal pump efficiency increases with specific speed, it is expected that centrifugals will become economically competitive at capacities starting at 20,000 GPM (Figure 10-22). As slurry pipeline capacity increases, the benefits of large-capacity centrifugal pumps operating in series as an alternative to multiple PD pumps operating in parallel become obvious. It is reasonable to assume that the principles and guidelines outlined in this chapter will be used on future slurry pipelines.

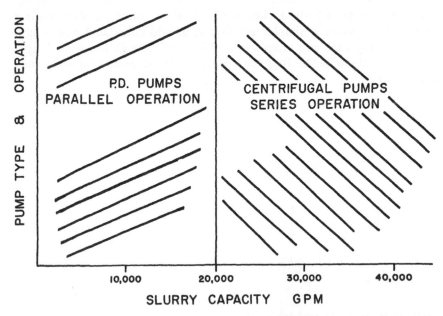

Figure 10-22. Suggested economic range for centrifugal pumps in slurry pipeline applications.

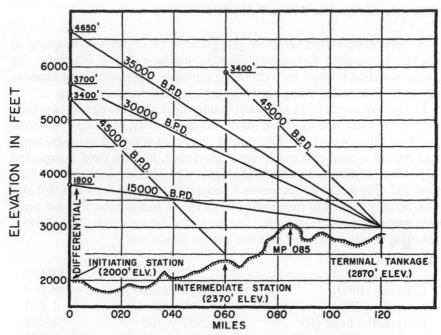

Figure 10-23. Hydraulic gradient and profile of 120-mile-long pipeline.

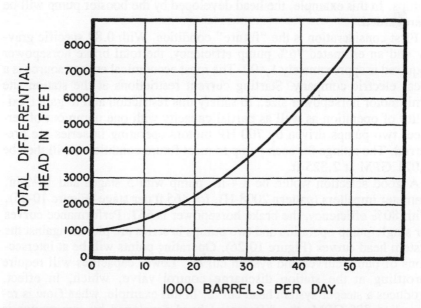

Figure 10-24. System head curve for 15,000 to 30,000 BPD production.

Example of Pipeline Pump Selection

As described earlier, pipeline pumps must be capable of adapting to change in pipeline throughput. The following exercise illustrates the pump selection process for a crude oil pipeline, where condition changes necessitate impeller changeouts, destaging, and volute chipping.

Pumps are required to transport crude oil from a developing oil field through a trunk line to a tank farm 120 miles away. Routing, pipe size, and pipe rating have been determined. A profile with hydraulic gradients and typical system curves has been developed, and the field is expected to produce 15,000 to 30,000 BPD (Figures 10-23 and 10-24). Note Milepost 085 (Figure 10-23) becomes a "control point." At least 1,200-ft station head is needed to overcome the elevation and to insure 50 psi positive pressure at this high point. The projected future rate is 35,000 BPD, resulting in the following pipeline design conditions:

	Initial	Future
Capacity (BPD)	15,000–30,000	35,000
Capacity (GPM)	437–875	1,021
Differential head (ft)	1,800–3,700	4,650
Differential head (psi)	694–1,246	1,792

A booster pump has been sized to provide adequate NPSH to the mainline pumps. In this example, the head developed by the booster pump will be disregarded.

First consideration is the "future" condition. With 0.89 specific gravity and an estimated 76% pump efficiency, the total brake horsepower required is approximately 1,400. The most economical energy source is a local electric company. Starting current restrictions at the station site limit motor horsepower size. To satisfy this restriction and to gain flexibility of operation as well as partial capacity with one pump out of service, two pumps driven by 700 HP motors operating in series are preferred. The ratings for each pump to meet future conditions, will then be 1,021 GPM at 2,325 ft.

A good selection would be a 4-in. pump with 5 stages and 10³/₈-in. diameter impellers (pattern 2008-H), for 465 ft per stage (Figure 10-25). With 80% efficiency, the brake horsepower is 667. Performance curves for single-pump operation and two pumps in series are plotted against the system head curves (Figure 10-26). Operating points will be at intersections of pump curves and system curves. Lower capacities will require throttling at the station discharge control valve, which, in effect, produces a steeper system head curve. In this example, when flow is reduced to 885 GPM, the differential head developed by two pumps is

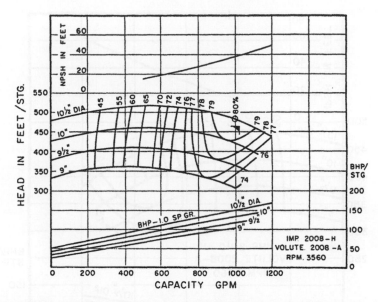

Figure 10-25. Performance for one stage of multi-stage pipeline pump selected for "future" condition.

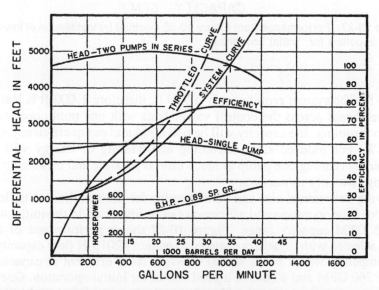

Figure 10-26. Performance of pump from Figure 10-25 for single pump operation and two pumps in series plotted against system head curves.

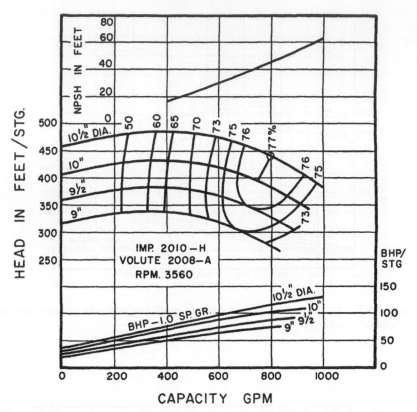

Figure 10-27. Performance from Figure 10-25 modified by installation of low-capacity impellers for "initial" condition.

4,850 ft. The system requires only 3,750 ft, therefore 1,000 ft is lost to friction (head) across the control valve. Note with one pump operating and no throttling, the capacity will be 630 GPM and pump efficiency will be 72%. For reduced rates, throttling and wasting of energy will be avoided by running one pump as much time as possible and making up by running two only as necessary.

Having determined size and configuration for the mainline units, let's consider how operation and efficiency can be improved in the initial 437 to 875 GPM capacity range. Figure 10-27 shows performance of the pump selected with impellers changed to pattern #2010-H (low capacity). This impeller, which peaks at 800 GPM, is more efficient at capacities below 760 GPM and would be a good choice for initial operation. One of the five-stage pumps, say the #1 unit, can be furnished destaged to four stages. By operating with four, five, or nine stages, various rates can be

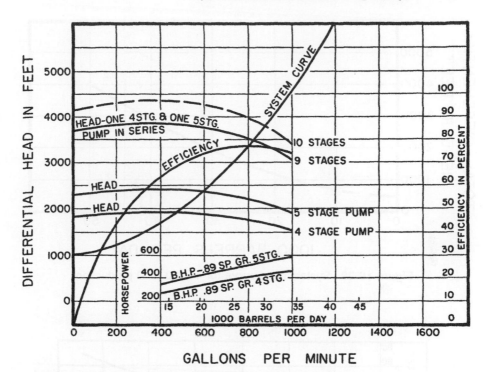

Figure 10-28. Performance of pump from Figure 10-27 in four, five, nine, and ten stages plotted against system head curve.

attained without throttling (Figure 10-28). When rates exceed 830 GPM, the #1 unit can be upstaged. At this point, throughput is approaching the 35,000 BPD future design rate and it is time to consider changing impellers to pattern #2008-H (high capacity).

Let's now assume after many years' operation there is a need to further increase capacity to 45,000 BPD (1,312 GPM). With 6,800 ft (2,620 psi) differential head, two stations are required to stay within the pipe pressure rating. The intermediate station is located near midpoint for hydraulic balance. Differential head required at each station is then half the total or 3,400 ft.

A new station system curve is developed (Figure 10-29). At 1,312 GPM and 3,400 ft with 76% efficiency, the station BHP would be 1,319. With two pumps in series, head required of each pump would be 1,700 feet, and 660 BHP per unit would be within horsepower rating of existing drivers. In this instance, volutes can be chipped to provide a throat area equivalent to volute pattern #2204-A. Without changing impellers, the higher capacity performance can be attained (Figure 10-30).

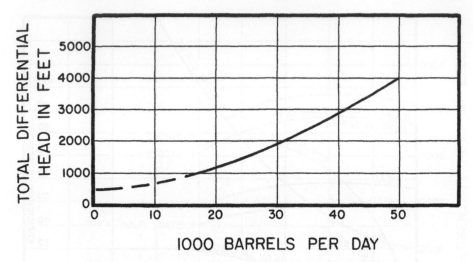

Figure 10-29. System head curve for 45,000 BPD production.

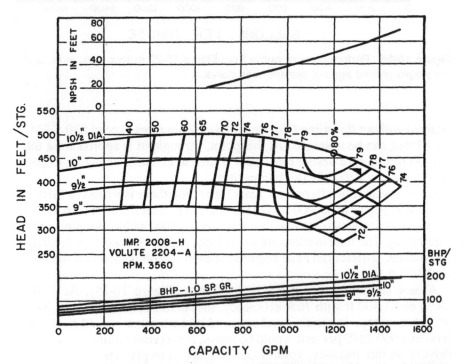

Figure 10-30. Performance from Figure 10-25 with volutes chipped to increase pipeline throughput to 45,000 BPD.

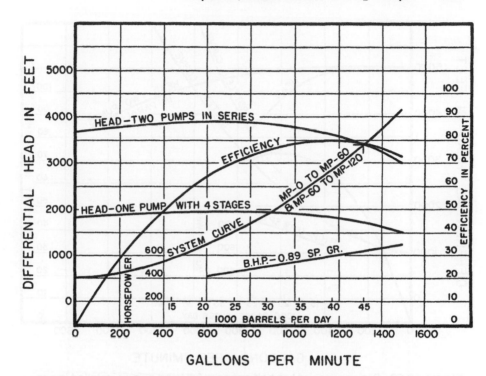

Figure 10-31. Performance of pump from Figure 10-30 in four-stage single-pump operation and two pumps in series plotted against system head curve.

To get 1,700 ft, we have two choices: The impellers of the 5-stage pumps can be trimmed to about a 9³/₄-in. diameter for 340 ft/stage. Alternatively, the pumps can be destaged to 4 stages and at 425 ft/stage will not require impeller trimming. Refer to iso-curve (Figure 10-30) and note how trimming affects efficiency at capacities approaching and beyond peak. At 1,319 GPM, destaging is obviously preferred. Figure 10-31 shows performance with one or two pumps operating.

Series vs. Parallel

A flat head-capacity curve is desirable for pipeline pumps installed in series. When station capacity is being controlled by throttling, less horsepower is lost than would be with a steep head-capacity curve. Capacity control of individual pumps does not present a problem since pumps operating in series will have the same flow rate.

Where pipeline pumps are to be installed in parallel, identical pumps with constantly rising head-capacity curves are usually called for. Load is

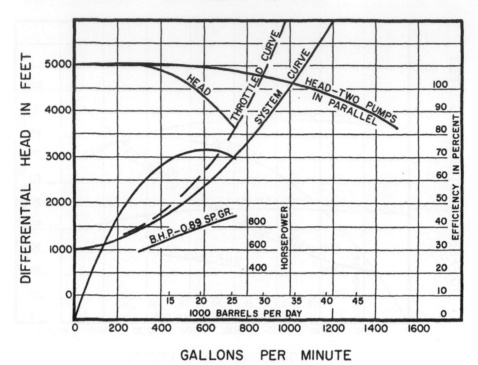

Figure 10-32. Performance of two half-capacity full head pumps installed in parallel plotted against system head curve.

shared equally, and there is less chance of a pump operating at less than minimum continuous stable flow. Figure 10-32 shows our system curve and two half-capacity full head pumps installed in parallel. In this particular system, parallel configuration is a poor choice. When only one pump is running, it is necessary to throttle to stay within operating range.

Parallel configuration should be considered in systems where a substantial portion of the head is static. When pump configuration is not clear cut, it is wise to plot each station curve along with pump curves and carefully analyze parallel versus series operation.

Waterflood Pumps

Many types of pumps are used to extend the life of declining oil fields by injecting displacing fluids at pressures ranging from 2,000 psi to 8,000 psi. In waterflooding applications (secondary recovery), centrifugals are preferred when injection capacities exceed 10,000 BPD. Typical

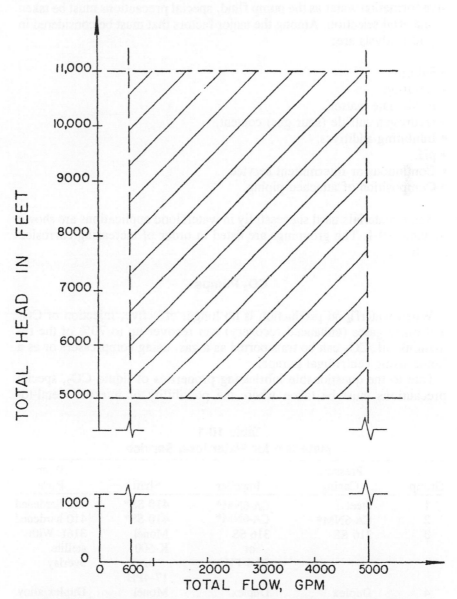

Figure 10-33. Performance coverage for typical waterflood applications.

pump conditions range in capacity up to 5,000 GPM and heads up to 11,000 feet (Figure 10-33). Pumps are normally multi-stage with horizontal-split or double-case construction. As many applications use corrosive formation water as the pump fluid, special precautions must be taken in material selection. Among the major factors that must be considered in liquid analysis are:

- Salinity.
- Aeration.
- Particulate matter.
- Hydrogen sulfide (sour gas) content.
- Inhibiting additives.
- pH.
- Continuous or intermittent service.
- Composition of attached piping.

Some materials used successfully in waterflood applications are shown in Table 10-1. The groupings are listed in order of increasing corrosion resistance.

CO₂ Pumps

When waterflood production is no longer effective, injection of CO_2 and other gases (enhanced recovery) can recover up to 70% of the remaining oil. CO_2 can be transported as a gas, using compressors or as a liquid using centrifugal pumps.

Due to the questionable lubricating properties of liquid CO_2, special precautions must be taken where the possibility of internal metal-to-

Table 10-1
Materials for Waterflood Service

Group	Pressure Casing	Impeller	Shaft	Wear Parts
1	Steel	CA-6NM*	410 SS*	410 hardened
2	CA-6NM*	CA-6NM*	410 SS*	410 hardened
3	316 SS	316 SS or 17-4 PH	Monel K-500 or 17-4PH	316L With stellite overlay
4	Duplex SS	Duplex SS	Monel K-500	Duplex alloy with stellite overlay

When H₂S is present or suspected, hardness shall not exceed R_c22 (240HB).

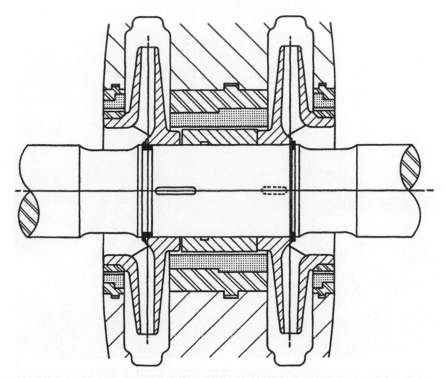

Figure 10-34. Bronze impregnated graphite inserts for dry running applications.

metal contact exists. One approach used successfully is to install bronze-impregnated graphite inserts into all stationary wear rings (Figure 10-34). Rings of this type have operated with no damage during dry running tests on multi-stage pumps where dynamic deflection permitted internal contact with rotating and stationary rings.

Mechanical Seals

A number of solutions in the sealing of CO_2 are offered by the various seal manufacturers who in their selection process must consider the poor lubrication quality, the possibility of icing at the faces, and the typical high-suction pressure. The CO_2 pipeline pump shown in Figure 10-35 had a pumping rate of 16,000 GPM and a maximum suction pressure of 1,800 psia. In this application, double seals were chosen using a compatible buffer fluid with adequate lubrication properties.

Figure 10-35. CO_2 pipeline pump (courtesy BW/IP Internal, Inc. Pump Division, manufacturer of Byron Jackson/United™ Pumps).

Horsepower Considerations

As liquid CO_2 is compressible, special consideration during the pump selection process must be given to horsepower requirements. Gas horsepower (GHP) and brake horsepower (BHP) should be calculated, and it is recommended that pump shaft and driver be sized to accommodate the larger of the two. Depending on the thermodynamic properties, GHP can be greater or less than BHP. To calculate GHP, it is necessary to predict the behavior of the liquid across the pump. This can be done either from interpolation of thermodynamic tables (assuming pure CO_2) or by computer calculations using the new equation of state for the actual composition (see Starling 1973). This method has been widely applied to predict the behavior of any mixture of hydrocarbons. While it is recommended that final pump design be based on computer calculations, preliminary pump selection can be based on thermodynamic tables. For a constant throughput in units of standard cubic feet per day (SCFD), the inlet and outlet pump capacity in GPM will change with change in specific volume (Figure 10-36).

Calculation Procedure

The following example assumes the use of appropriate Thermodynamic Tables.
Given

$P_1 = 1400$ psia
$P_2 = 1650$ psia
$T_1 = 80°F$
$F_R = 450 \times 10^6$ SCFD

Step 1. Find Inlet Conditions

A. From P_1, and T_1, find H_1, V_1, and S_1 from tables

$H_1 = -3858.81$ Btu/lb
$V_1 = 0.01925$ ft³/lb
$S_1 = 0.7877$ Btu/lb °R

B. Calculate γ_1, SG_1, Q_1, and H

$\gamma_1 = 1/V_1 = 1/0.01925 = 51.948$ lb/ft³
$SG_1 = \gamma_1/62.33 = 51.948/62.33 = .8334$
$Q_1 = F_R \times .116 \times 7.48/1440 \times \gamma_1$
$Q_1 = 450 \times 10^6 \times .116 \times 7.48/1440 \times 51.948 = 5220$ GPM
$H = (P_2 - P_1) 2.31/SG_1$
$H = (1650 - 1400) \times 2.31/.8334 = 693$ ft

C. Select Pump
Select pump for 5220 GPM and total pump head of 693 ft
Note, pump efficiency in this example is 85%.

D. Calculate BHP

$BHP = Q_1 \times H \times SG_1/3960 \times$ Pump Efficiency
$BHP = 5220 \times 693 \times .8334/3960 \times .85 = 896$

Step 2. Find Outlet Conditions

A. Assume constant entropy $S_1 = S_2$

B. From $P_2 = 1650$ psia and $S_2 = 0.7877$ Btu/lb °R interpolate tables to find H_2, V_2 and T_2

$H_2 = -3857.93$ Btu/lb
$V_2 = 0.0191665$ ft³/lb
$T_2 = 82.88$ °F

C. Calculate ΔH, and ΔH^1

$\Delta H = H_2 - H_1$
$\Delta H = (-3857.93) - (-3858.81)$
   $= 0.88$ Btu/lb
$\Delta H_1 = \Delta H / \text{Pump Efficiency}$
   $= .88/.85 = 1.0353$ Btu/lb

Step 3. Correct Outlet Conditions For Pump Efficiency

A. With P_2 remaining same calculate H_C

$H_C = H_1 + \Delta H^1$
   $= -3858.81 + 1.0353$
   $= -3857.77$ Btu/lb

B. From H_C and P_2 interpolate tables to find V_C, S_C, and T_C

$V_C = 0.019192$ ft³/lb
$S_C = 0.78798$ Btu/lb °R
$T_C = 83.078$ °F

C. Calculate γ_C and Q_2

$\gamma_C = 1/V_C = 1/0.019192 = 52.105$ lb/ft³
$Q_2 = F_R \times .116 \times 7.48/1440 \times \gamma_C$
   $= 450 \times 10^6 \times .116 \times 7.48/1440 \times 52.105$
   $= 5204$ GPM

Step 4.   Calculate GHP

$$GHP = \Delta H^1 \times M/2545$$

$$\text{where } M = Q_1 \times 500 \times SG_1$$
$$= 5220 \times 500 \times .8334$$
$$= 2,175,174 \text{ lbs/hr}$$

$$GHP = 1.0353 \times 2,175,174/2545$$
$$= 885$$

Step 5. Size Driver and Pump Shaft for 896 BHP

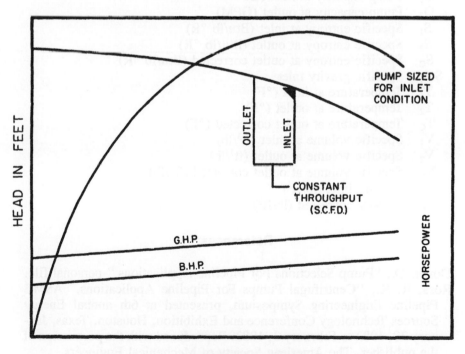

Figure 10-36. Performance curve for CO_2 pump.

Notations

BHP Brake horsepower
F_R Flow rate (SCFD)
f Blade passing frequency = N × vane no./60
f_1 Acoustic frequency = velocity of pulsation (ft/sec)/wave length l (ft)
GHP Gas horsepower
H Total pump head (ft)
H_1 Inlet enthalpy (Btu/lb)
H_2 Outlet enthalpy (Btu/lb)
H_C Outlet enthalpy corrected (Btu/lb)
ΔH Differential enthalpy (Btu/lb)
$ΔH^1$ Differential enthalpy corrected (Btu/lb)
M Mass flow rate (lbs/hr)
N Speed (RPM)
P_1 Inlet pressure (psi)
P_2 Outlet pressure (psi)
Q_1 Pump capacity at inlet (GPM)
Q_2 Pump capacity at outlet (GPM)
S_1 Specific entropy at inlet (Btu/lb °R)
S_2 Specific entropy at outlet (Btu/lb °R)
S_C Specific entropy at outlet corrected (Btu/lb °R)
SG_1 Specific gravity inlet
T_1 Temperature at inlet (°F)
T_2 Temperature at outlet (°F)
T_C Temperature at outlet corrected (°F)
V_1 Specific volume at inlet (ft³/lb)
V_2 Specific volume at outlet (ft³/lb)
V_C Specific volume at outlet corrected (ft³/lb)
$γ_1$ Density inlet (lb/ft³)
$γ_C$ Density corrected (lb/ft³)

References

Doner, O., "Pump Selections For Pipeline Applications," personal file.

Ross, R. R., "Centrifugal Pumps For Pipeline Applications," ASME Pipeline Engineering Symposium, presented at 6th annual Energy Sources Technology Conference and Exhibition, Houston, Texas, January 30, 1983 to February 3, 1983. Reproduced with the permission of the publisher, The American Society of Mechanical Engineers.

Starling, Kenneth E., *Fluid Thermodynamic Properties For Light Petroleum Systems*, Gulf Publishing Company, Houston, Texas, 1973.

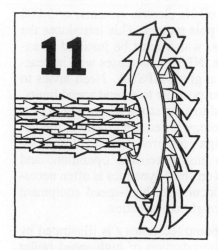

11

High Speed Pumps

By **Edward Gravelle**
Sundstrand Fluid Handling Division

The trend toward higher process pressures, which has developed over the past half century or so, has provided impetus to exploit the advantages of high speed to better provide high head capability in centrifugal pumps. High head centrifugal design may be provided by using high rotating speed, by series multi-staging, or by a combination of both.

The advantages of high-speed design are several. Fewer and smaller stages are required to meet a given head objective, and not infrequently, single-stage designs can provide capability that would otherwise require multi-staging. Smaller, more compact design tends toward shorter shaft spans that can result in lowered shaft deflection and improved shaft dynamics. Compactness, involving fewer and smaller components, is economical of materials, which becomes increasingly important when expensive materials are required for handling severe process fluids. Minimal spares inventory and relatively quick and easy maintenance are attributes of high speed, which are often very attractive to users, to whom pump availability is central to the viability of their businesses. Lightened pump weight can translate into smaller and less expensive mounting foundations.

Conversely, other considerations are involved in a movement toward higher speeds. About 95% of all pumps in industry are driven by electric motors in a world built around 50 and 60 Hz electric systems, or 3,000 and 3,600 RPM speed limits with two pole motors. This introduces the need for speed-increasing gear systems, which must be justified in exchange for high-speed pump advantages. NPSHR increases with increasing speed, placing limits on speed for a given NPSHA. Need arises to improve suction performance as much as possible to extend speed limits. Material capabilities must be recognized to keep stress levels within prudent design limits. Modern seal technology is required to meet the demands of combined high speeds and high pressures. Bearing design sophistication is frequently required to ensure reliable operation, and recognition of the influence of bearings on shaft dynamics is often necessary. Increased noise generation can occur with high-speed equipment due to high power densities and lightweight construction.

Industrial acceptance of high-speed pump technology is illustrated by Karassik, who has indicated that the introduction of high-speed boiler feed pumps in 1954 was followed by the steadily increasing use of these machines, culminating in total abandonment of the older 3,550 RPM equipment by 1971.

This chapter will deal primarily with an unconventional pump type particularly suited for operation at high to very high speeds to produce typically very high heads at low to moderate flow rates. Although this pump type has enjoyed wide acceptance in industry, comparatively little on this design has appeared in the literature.

An early commercial application of this design began in 1959 for an aircraft service. The pump was used in the Boeing 707 for takeoff thrust augmentation in jet engines, at a time when engine power was relatively low and the world had not caught up with the generally longer runstrip requirements for jet aviation. This pump rotated at 11,000 RPM and delivered 80 GPM of water to the combustors at 400 psi (well above combustor pressure to allow atomization), to increase the engine mass flow rate and increase thrust by 15% for takeoff. The unit weighed only 8½ pounds, including step-up gearing from the 6,500 RPM power takeoff pad to pump speed. Some 250 units were produced for this service.

In 1962, an industrial version of the new pump type rated to 100 HP and 6,000 feet of head in a single stage was introduced to the petrochemical industry. In the past two decades, this concept has grown into a family of products ranging from 1 to 2,500 HP, direct drive to 25,000 RPM and heads to 12,000 feet. Most commonly these products consist of a single high-speed stage, but as required, employ two or three stages to satisfy need for extreme heads or the combination of high head and low NPSHA. Over the past two decades many thousand machines of this type have been placed in service around the world.

History and Description of an Unconventional Pump Type

Developmental work on the pump type central to the discussions in this chapter was initiated in Germany prior to World War II to meet urgent wartime requirements and after 1947 continued in Britain. Need for a simple, lightweight, and easily manufactured pump suited to produce high heads at low flow rates existed in connection with aircraft and rocket propulsion systems. An unorthodox high-speed centrifugal pump concept resulted and was described by Barske in papers published in 1955 and 1960.

This pump is described as an open impeller type and is exemplified as highly unorthodox by Barske himself who states: "To a skilled designer the pump which forms the subject of this paper will, at first glance, appear most unfavorable and may well be regarded as an offense against present views of hydrodynamics." Reasons exist, however, to break with conventional design practice to meet objectives which would otherwise be difficult to achieve. Intentionally flaunting the rules, in fact, provides a pump design that can equal or exceed the performance of conventional pumps in the head-flow design range for which it is intended and for which it is best suited.

Typical Barske-type pump construction is illustrated by the sketch in Figure 11-1. The salient features of the design start with a simple open impeller, which rotates within a case bored concentrically with the impeller centerline. A single emission throat with a conical diffuser section is oriented tangentially to the case bore. Conical diffusers provide high recovery efficiency because of their minimal wetted area. A cone angle of $10°$ is commonly used, providing good recovery potential and reasonable cone length requirements.

Radical departure from conventional design practice exists in the exceptionally tall blade geometry used, with the impeller tip height, b_2, set equal to or moderately greater than the emission throat diameter, d_1. Blade angle, θ, is unimportant except that the flow area in the impeller eye must at least equal the area of the suction passage. Further obvious deviation from normal practice is the use of plain radial blades, with no attempt made to match inlet flow streamlines.

Performance trends of the Barske pump are generally as indicated in Figure 11-2A. The head at zero flow, or shutoff, is about equal to the design head, with a head peak a few percent higher than design in the neighborhood of half design flow. This curve shape is referred to as an unstable curve and is often viewed as undesirable, as described in Chapter 1. A stable curve is one in which the head rises continuously as flow is reduced from design to shutoff. Head drops rapidly for flows above design, and zero head or cutoff normally occurs around 130% of design

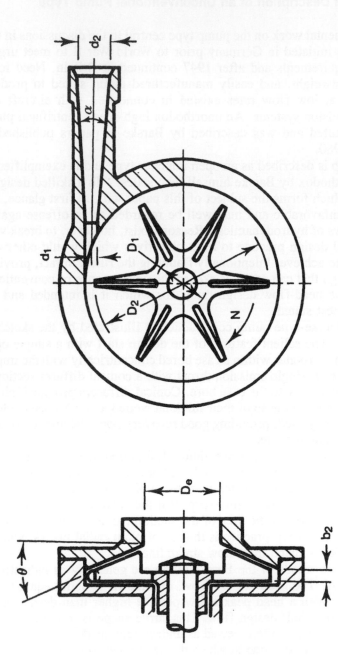

Figure 11-1. Barske open impeller centrifugal pump.

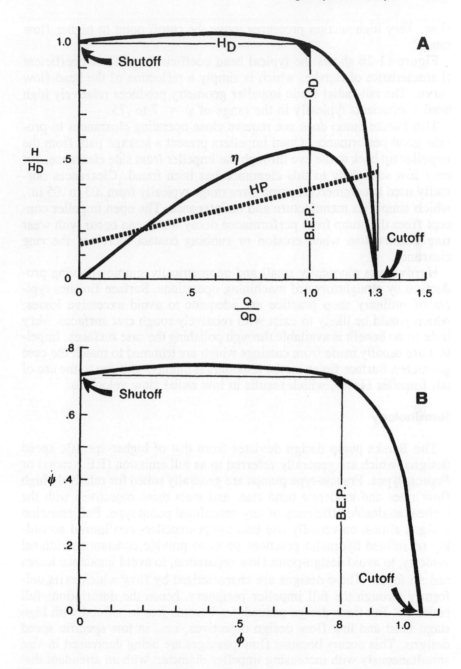

Figure 11-2. (A) Typical performance trends for Barske pump; (B) typical head coefficient vs. flow coefficient for Barske pump.

flow. Very high suction pressures move the cutoff point to higher flow rates.

Figure 11-2B shows the typical head coefficient and flow coefficient characteristics of Barske, which is simply a reflection of the head-flow curve. The tall radial blade impeller geometry produces relatively high head coefficients typically in the range of $\psi = .7$ to $.75$.

The Barske pump does not require close operating clearances to provide good performance. Open impellers present a leakage path from the impeller tip back to the eye through the impeller front side clearance, but only low sensitivity to this clearance has been found. Clearances normally used in commercial pump sizes range typically from .03 to .05 in., which simplifies manufacture and maintenance. The open impeller concept frees the pump from performance decay which can occur with wear ring construction when erosion or rubbing contact increases the ring clearance.

Hardware is physically small and geometrically simple allowing production by straightforward machining operations. Surface finishes typical of ordinary shop practice are adequate to avoid excessive losses, which would be likely to exist with relatively rough cast surfaces. Very little or no benefit is available through polishing the case surfaces. Impellers are usually made from castings which are trimmed to match the case geometry. Surface finish on the impellers is unimportant due to the use of tall impeller blades, which results in low radial flow velocities.

Terminology

The Barske pump design deviates from that of higher specific speed designs, which are generally referred to as full emission (F.E.) radial or Francis types. Francis-type pumps are generally suited for relatively high flow rates and moderate head rise, and meet these objectives with the highest attainable efficiency of any centrifugal pump type. Full emission designs almost universally use backswept impellers configured according to refined hydraulic practices so as to provide constant meridianal velocity, to avoid design-point flow separation, to avoid incidence losses and so forth. These designs are characterized by flow which exits uniformly through the full impeller periphery, hence the description: full emission. But these design procedures become less beneficial with high stage head and low flow design objectives, i.e., in low specific speed designs. This occurs because flow passages are being decreased in size simultaneously with increasing impeller diameter, with an attendant disproportionate increase in friction losses and lowered efficiency.

It has been established through experience that high-flow machines can be made to work relatively well at low flow rates by simply plugging some portion of the exit flow path; for example, plugging some of the diffuser passages in a vaned diffuser. This, of course, results in impeller passages that are oversized for the lower flow rates according to conventional design practice, but in fact can produce efficiencies superior to those attainable with the very narrow passages that would result in F.E. design procedures. The term partial emission (P.E.) arose to describe such pump geometry, apparently coined by Balje.

The Barske pump is correctly classified as a partial emission type, since the emission throat area is much smaller than the impeller emission area. More to the point, net through-flow in the Barske pump can occur only in a path extending generally from the inlet eye to the vicinity of the emission throat. This is true for the simple reason that the remainder of the case cavity is concentric with the impeller and is filled with incompressible fluid, precluding any possibility of a radial flow component. High circumferential fluid velocities exist in the forced vortex created by the impeller, which are superimposed on the through-flow stream extending from eye to throat. Through-flow is then in essence a fluid migration, where a given element of fluid makes a number of circuits within the forced vortex and moves to successively higher orbits in the eye-to-throat flow region.

Alternatively, the Barske pump can be referred to generically and geometrically as a concentric bowl P.E. pump or simply a concentric bowl pump. This is convenient for easier differentiation of the original pump type from its evolutionary offshoots to be described later.

Partial Emission Formulae

Use of tall, radial-bladed impellers in P.E. pumps results in flow conditions that must be described as disorderly. No attempt is made to match inlet geometry to the flow streamlines. Very low mean radial flow velocities combined with high tip speeds reduce the discharge vector diagram to essentially the tangential tip speed vector, u_2. Calculation procedures for P.E. pumps then are based on simple algebraic expressions involving impeller tip speed rather than on the vector diagrams used in F.E. design.

Barske starts with the assumption that the fluid within the case rotates as a solid body or forced vortex, and neglects the negligibly low radial component, resulting in a theoretical head of:

$$H' = \frac{u_2^2 - u_1^2}{2g} + \frac{u_2^2}{2g} \tag{11-1}$$

The first term represents the vortex or static head and the second term represents the velocity head or dynamic head. Even within the Barske paper, question arose as to whether inclusion of the $U_1^2/2g$ term in the static head expression recognizing the blade inlet diameter is appropriate. Low through flow and a strong forced vortex might well combine to extend rotation to the impeller centerline, i.e., might introduce prerotation of inlet flow. Measurements indicating that static pressure at the tip is close to $u_2^2/2g$ reinforce this view. Also, inlet prerotation is indicated by quite respectable suction performance despite radial blade inlet geometry. Thus, the theoretical head normally used in practice simplifies to

$$H' = \frac{u^2}{g} = \frac{u^2}{2g} + \frac{u^2}{2g} \qquad (11\text{-}2)$$

When the subscript is dropped, u is taken to indicate the impeller tip speed. Actual head for P.E. pumps is then stated as:

$$H = \psi \frac{u^2}{g} \qquad (11\text{-}3)$$

To further understand the workings of the P.E. pump, a somewhat simplistic exercise involving a mixture of theory and experience is put forth to establish how actual head is generated. As has been indicated, tall blade geometry produces a strong forced vortex resulting in a static head coefficient near unity, so the actual static head produced by the impeller is simply $u^2/2g$. Tests have shown that diffusion efficiency is nearly flat through much of the flow range, so we state that $\eta_d = .8$. Diffusion recovery potential is in accordance with the diffuser area ratio

$$\left[1 - \left(\frac{A_1}{A_2}\right)^2\right]$$

And finally, the P.E. flow coefficient is in the vicinity of $\phi = .8$. So actual head generation may be written

$$H = \frac{u^2}{2g} + \eta_d \frac{(\phi u)^2}{2g}\left[1 - \left(\frac{A_1}{A_2}\right)^2\right] \qquad (11\text{-}4)$$

Say, then, that the diffuser terminates in two throat diameters, i.e., has an area ratio of 4, so the actual head should be

$$H = \frac{u^2}{2g} + .8\frac{(.8u)^2}{2g}\left[1 - \left(\frac{1}{4}\right)^2\right]$$

$$H = .74 \frac{u^2}{g}$$

This is to say that the estimated head coefficient in this breakdown is ψ = .74, which falls within the ψ = .70 to .75 range typically occurring in test experience.

No pretense exists about the theoretical elegance of the actual head exercise, but it does provide some insight into the workings of the pump. It is clear that roughly 2/3 of the total head is provided by the forced vortex in the bowl and that the remaining 1/3 comes from diffusion recovery. Or, for example, assume that it would be possible to improve diffuser efficiency to 90% as is attainable in the relatively idealized case of a venturi meter, thus increasing the head coefficient to ψ = .77. Using this result, we go to the following expression relating head coefficient to efficiency:

$$\eta_2 = \eta_1 \frac{H_2}{H_1} = \eta_1 \frac{\psi_2}{\psi_1} \tag{11-5}$$

Then, assuming an original pump efficiency of 60%, the improved diffusion recovery would increase the pump efficiency to 62.4%. A dramatic (and probably unachievable) 10% improvement in diffusion recovery would dilute to 2.4% improvement in the overall pump efficiency. Similarly, truncation of the diffuser cone from an area ratio of 4 to an area ratio of 3 would reduce the overall efficiency only from 60% to 59%.

It is useful to express head in convenient terms. Non-homogeneous units are used in this chapter as is commonly done in everyday practice, so constants in the main result from unit conversions. Impeller tip speed is:

$$u_2 = \frac{\pi D_2 N}{720} \tag{11-6}$$

Specific Speed

To those familiar with algebra, but unfamiliar with pump technology, it would appear that specific speed, described in Chapter 2, can be altered by simply changing the rotational speed. Not so. To illustrate this, we note from the affinity laws (also described in Chapter 2) that flow is proportional to speed and head is proportional to the square of speed. Start with a given pump with a specific speed of:

$$N_s = \frac{NQ^{.5}}{H^{.75}}$$

Then, if the rotational speed is changed by some factor x, we have:

$$N_s = \frac{xN(xQ)^{.5}}{(x^2H)^{.75}} = \frac{NQ^{.5}}{H^{.75}}$$

So we see that specific speed is unchanged with change in rotational speed; head and flow change in such a way as to keep specific speed constant.

Mathematically, the specific speed expression is seen to vary from zero at zero flow or shutoff, to infinity at zero head or cutoff. By definition, specific speed has meaning only at the best efficiency point or flow rate at which maximum efficiency occurs. Then, the head expression can be expanded to read

$$H = \psi\frac{u^2}{g} = \psi\left(\frac{D_2N}{1,300}\right)^2 \tag{11-7}$$

Flow is proportional to the product of tip speed and discharge throat area according to the following expression

$$Q = \frac{720}{231}\phi u A_1 = \frac{\pi^2}{924}\phi D_2 N d_1^2 \tag{11-8}$$

In the flow expression, it is seen that the flow coefficient, ϕ, is simply the ratio of discharge throat velocity to impeller tip speed. It is at first surprising to note that $\phi = 1.3$ at cutoff indicates that throat velocity is outrunning the impeller tip speed by 30%. This is in reality the case, and is explained by conversion of the bowl static head into velocity head so that total head, less losses, appears as velocity head in the discharge throat.

Further perception of the meaning of specific speed for P.E. pumps is desirable. To provide this insight, we simply insert the head and flow expressions of Equations 11-7 and 11-8 involving impeller diameter, throat diameter and speed into the specific speed expression

$$N_s = \frac{NQ^{.5}}{H^{.75}}$$

which yields an alternate expression for specific speed:

$$N_s = 4,847\frac{\phi^{.5}d_1}{\psi^{.75}D_2} \tag{11-9}$$

This expression may be described as the geometric form of specific speed and shows first that specific speed is independent of rotational speed, and secondly that specific speed is basically defined by the ratio of emission throat diameter to impeller diameter, i.e., is related to the ratio of flow capacity to head capacity of the pump stage. Since the head and flow coefficients of the P.E. pump do not range broadly, the specific speed of a given P.E. pump geometry is expressed approximately as:

$$N_s \simeq 5,500 \frac{d_1}{D_2} \tag{11-10}$$

Accumulated experience reflecting the efficiency potential of well-designed pumps versus specific speed are shown in Chapter 2. Impeller geometry trends toward relatively large diameters and small flow passageways as specific speed decreases.

A first observation is that pumps with the highest efficiency potential have a specific speed in the neighborhood of 2,000, and that efficiency starts to drop substantially for specific speeds below 1,000. The fundamental reason for lowered efficiency potential at low specific speed lies in the disproportionate losses incurred in low specific speed design, particularly disk friction and flow losses. Disc friction, neglecting a modest Reynold's number modifier, is well known to vary as the cube of speed and the fifth power of diameter. Pump power is proportional to the product of pump head and flow or the cube of impeller diameter and speed, $(DN)^3$. Then, without pretense of mathematical completeness, the impact of disk friction on efficiency can be expressed as follows:

$$\eta = \frac{\text{Output}}{\text{Input}} = \frac{\text{Output}}{\text{Output} + \text{Loss}} \propto \frac{D^3 N^3}{D^3 N^3 + D^5 N^3}$$

This expression illustrates the disproportionate influence of disk friction on efficiency for low specific speed pumps which tend toward large diameter impellers. Further, for a given head objective, design choices are such that the product of DN is a constant, so indicating the general advantage inherent in selection of high speed in return for a smaller impeller diameter.

A second observation is at first disappointing in that a family of dimensional parametric curves indicative of pump size appear on the otherwise dimensionless $N_s - \eta$ plot. Small pumps are always less efficient than hydraulically similar large pumps. The prime reason for this scale or size effect is mostly easily explained by a pipe flow analogy: skin friction arises from the inner circumference of the pipe and so is proportional to the diameter, while flow or throughput is proportional to the cross-sectional area and so is proportional to the square of the diameter. Small pipes thus experience relatively higher flow loss than do large pipes. In fact, pipe friction data provide excellent corollary with the $N_s-\eta$ data for pumps in that pipe diameters commonly appear as parameters on a dimensionless plot of friction factor versus Reynolds Number.

With specific speed as well as head and flow expressions having been defined for P.E. pumps, convenient expressions for impeller and throat size may be derived

$$D_2 = \frac{1,300}{N}\left(\frac{H}{\psi}\right)^{.5} = \frac{1,300}{N_s H^{.25}}\left(\frac{Q}{\psi}\right)^{.5} \qquad (11\text{-}12)$$

$$d_1 = .268\left(\frac{Q}{\phi}\right)^{.5}\left(\frac{\psi}{H}\right)^{.25} \qquad (11\text{-}13)$$

Power for any pump is

$$HP = \frac{HQ(SG)}{3,956\eta} = \frac{pQ}{1,714\eta} \qquad (11\text{-}14)$$

The concentric bowl pump has been unjustly criticized as having only low efficiency potential, probably because this pump type is frequently designed for very low specific speed where only low efficiency potential exists. Barske states that efficiency was of secondary importance in his development efforts, yet reports an efficiency island of 57% in the vicinity of $H = 1,000$, $Q = 40$, $N = 28,000$ ($N_s = 1,000$). This is seen to be representative of good pump performance as indicated by the general pump population data discussed in Chapter 2.

Because partial emission pumps range so widely in speed, it is sensible to use impeller diameters for scale or size parameters on $N_s-\eta$ maps, rather than the flow parameters widely used for the higher specific speed types. Direct comparison of P.E. and F.E. efficiency potentials from these data is a little elusive since these maps define explicitly only two of the four parameters involved in the specific speed expression. But by making the quite reasonable assumption that the low specific speed data collected by Karassik derived from pumps at 3,600 RPM, direct comparison can be made as shown in Figure 11-3. The dotted curves reflect the Karassik data and the solid curves represent P.E. performance at 3,600 RPM. Distinct P.E. efficiency superiority is seen to exist at low specific speeds and low to medium flow rates.

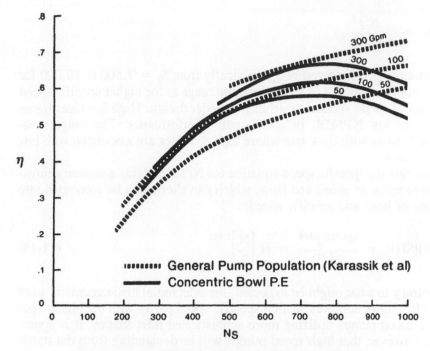

Figure 11-3. Partial emission efficiency comparison.

Concentric bowl pump peak efficiency occurs at about $N_s = 800$, and declines in efficiency at higher specific speeds as shown. This characteristic of P.E. pumps exists in general because with high specific speeds the inlet and discharge passages enlarge toward overlap causing declining head coefficients and efficiency reduction. This characteristic establishes a boundary region delineating the suitability of P.E. and F.E. pump types.

Although very low specific speed design must inherently entail efficiency sacrifice, such design can have overall attraction. Efficiency is not the sole consideration in pump selection, and can be overridden by factors such as simplicity, low initial cost, and quick, easy maintenance. These alternative considerations tend toward dominance at modest power levels and for intermittent or low-usage services.

Suction Specific Speed

Another dimensionless parameter highly important in pump design is known as suction specific speed involving a parametric group nearly identical to the pump specific speed expression. This subject is discussed in detail in Chapters 2 and 8.

$$S_s = \frac{NQ^{.5}}{NPSHR^{.75}}$$

Suction specific speed ranges typically from $S_s = 7,500$ to $10,000$ for P.E. pumps, which is in about the same range as for higher specific speed pumps of single suction, overhung impeller design. High S_s values translate into low NPSHR, or good suction performance. The range mentioned varies with flow rate where high S_s values are associated with low flow rates.

Solving the specific speed equation for NPSHR yields a suction expression in terms of speed and flow, which can alternately be converted into terms of head and specific speed:

$$NPSHR = \frac{N^{1.333}Q^{.666}}{S_s^{1.333}} = H \left[\frac{N_s}{S_s}\right]^{1.333} \tag{11-15}$$

Contrary to what might be expected, the inlet radial blade geometry used in P.E. pumps achieves suction specific speed parity with the higher specific speed pumps utilizing more sophisticated inlet shapes. It is apparent, however, that high speed pumps will be demanding from the standpoint of NPSHR. The bracketed term in the latter expression is known as the Thoma cavitation parameter, usually designated by sigma:

$$\sigma = \frac{NPSHR}{H} = \left[\frac{N_s}{S_s}\right]^{1.333} \tag{11-16}$$

The Thoma parameter, then, states NPSHR as a fraction of pump head and is a function of the ratio of specific speed to suction specific speed. Low specific speed thus offsets to a degree the higher NPSHR associated with high speeds.

The inlet eye size in the prior expressions is assumed to be generously sized, as is generally done so that only small NPSHR impact exists. The NPSHR expression expanded to include inlet eye size effect becomes:

$$NPSHR = \left[\frac{NQ^{.5}}{S_s}\right]^{1.333} + \frac{Q^2}{386D_e^4} \tag{11-17}$$

Inlet eye size has been found to have an influence on the efficiency potential of the pump, which as we have just seen, affects NPSHR. Availability of efficiency advantage via eye sizing then in reality hinges on NPSHA in the application. More will be said on this subject in the section "Partial Emission Design Evolution."

As an aside, pump users should be aware that overly conservative statements of NPSHA in an application can work to their disadvantage. The pump manufacturer must meet the stated NPSHA, so understated suction conditions can force the design toward lower speed or more and larger stages, which can result in an efficiency penalty or higher initial cost.

Inducers

Need to improve suction performance becomes quickly apparent in the move toward exploitation of high speed advantage. Inducer development began more than 50 years ago to provide this improvement. An inducer is basically a high specific speed, axial flow, pumping device roughly in the range of $N_s = 4,000$ to $9,000$ that is series mounted preceding a radial stage to provide overall system suction advantage. Inducers are characterized by relatively few blades, shallow inlet blade angles, and generally sophisticated hydraulic design.

The inducer must put up enough head to satisfy the needs of the radial impeller stage but in itself has a suction level requirement that establishes a new lower NPSHR for the system. Inducers are an important element in high speed pump design, and so have been and continue to be the subject of considerable interest and developmental work. Inducer design should be such that maximum suction performance is achieved, and such that cavitation erosion in the inducer itself is avoided in long-term operation.

Inducer performance is generally taken as the suction specific speed which corresponds to 3% pump head depression as NPSH is decreased. Theory exists establishing optimum suction performance in an expression known as the Brumfield criterion. A form of the Brumfield criterion developed in a comprehensive document on inducer design developed by NASA is as follows:

$$S_s = 3,574 \frac{(1 - 2\phi^2)^{.75}}{\phi} \tag{11-18}$$

Where ϕ is the inlet flow coefficient or the ratio of meridianal flow velocity to inducer tip speed:

$$\phi = \frac{93.62}{D_i^3} \left(\frac{Q}{N}\right) = \tan \beta \tag{11-19}$$

A plot of the Brumfield criterion is shown in Figure 11-4. It should be emphasized that this expression is theoretical but tempered by practical

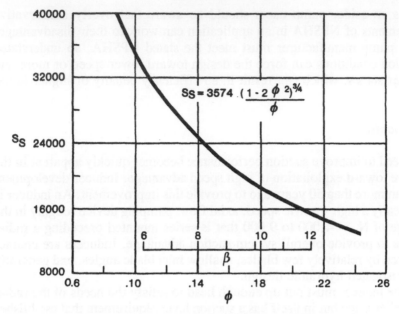

Figure 11-4. Brumfield performance criterion.

design considerations and that many details of inducer design are not addressed. The angle β is the fluid angle at inlet and differs from the blade angle, β_1, by a positive incidence angle, α. Prerotation is assumed to be zero, and other considerations such as blade shape, blockage, hub geometry, and leakage are simply ignored. The intent here is primarily to show the fundamental influence of the inlet blade angle on suction performance potential.

Optimal inducer design is distinctly a high-tech endeavor which must conform to hard-earned design guidelines and hydraulic disciplines. A well-designed inducer should possess a "sharp" breakdown characteristic as illustrated by the solid curve in Figure 11-5, rather than the "gentle" curve shown in broken line. The NPSHR disadvantage with gentle breakdown is evidenced by the NPSHR differential which exists at the 3% head depression level. It should not go unnoticed that the 3% head depression level refers to the inducer-pump combination, so the level of the 3% line on the inducer headrise curve will vary according to the head of the pump to which the inducer is coupled. Lower head units will suffer an NPSHR disadvantage with a gentle breakdown inducer compared to high head units equipped with the same inducer.

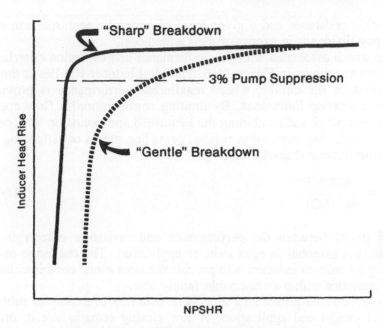

Figure 11-5. Inducer characteristics curves.

Commercial inducers have tended to result in performance suction specific speeds on the order of S_s = 20,000 to 24,000. This level of performance entails some compromise from the standpoint of consideration of inducer cavitation but provides respectable inducer design, though perhaps not ultimate design. These suction specific speeds generally produce dramatic improvement in suction performance, frequently providing up to 80% NPSHR reduction over uninduced pumps.

Freedom from long-term cavitation erosion to the inducer itself is provided by observing experimentally established cavitation limitations in the inducer design. The cavitation limitation is related to tip speed, fluid specific gravity, and the inducer material as follows:

$$S_s = \frac{K}{u_i(SG)} = \frac{720K}{\pi D_i N(SG)} \tag{11-20}$$

where K is an experimentally established constant which varies with the inducer hydraulics and material of construction.

Commonly used 316 stainless steel is considered to possess "good" resistance to cavitation erosion, but exploitation of materials with superior cavitation resistance and a given level of hydraulic sophistication may well pay dividends in excess of their added cost.

The trends associated with the performance and cavitation criteria for inducers are shown in Figure 11-6. The ideal inducer size lies at the intersection of the curves, where maximized performance is provided within cavitation limitations. By limiting consideration to flow coefficients of $\phi < .2$ and combining the Brumfield and cavitation limit criteria, the following expression results, providing means of estimating the optimum inducer diameter

$$D_i = \frac{(6KQ)^{.25}}{N^{.50}(SG)} \tag{11-21}$$

Exact parity between the performance and cavitation criteria is, of course, not essential in each inducer application. The challenge in designing a family of inducers is to provide the most useful combinations of characteristics within a reasonable family size.

This section has presented a superficial overview of a complex subject: inducer design and application. A few closing remarks are in order. Brumfield does not represent an ideal inducer such as with paper thin

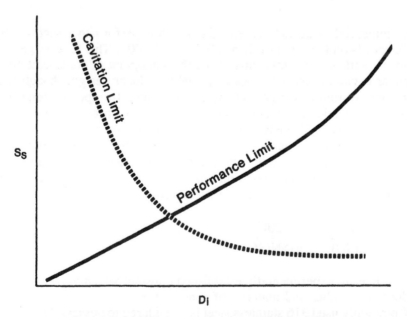

Figure 11-6. Trends of performance vs. inducer diameter.

Figure 11-7. Inducer family for partial emission pumps (courtesy Sundstrand).

blades, razor sharp edges, etc., but rather a useful "optimum" relating reasonable suction performance expectation to the inlet flow coefficient. The Brumfield suction performance can be exceeded with well-designed inducers. Tradeoff or compromise is not always required in performance and cavitation considerations, because modest speed, small inducers, or low specific gravity fluids can result in operating regimes far removed from cavitation concern. Substantial effort has been devoted to inducer development and will undoubtedly continue in the future, since inducers are a key element in extending the frontiers of high-speed pump technology.

A family of inducers for P.E. pumps is shown in Figure 11-7 which range from 1.25 to 3.5 inches in diameter and provide coverage from Q/N = .0005 to .1.

Partial Emission Design Evolution

Beyond the very substantial NPSHR improvement provided by inducers, improvement of the concentric bowl pump has been pursued in other areas including efficiency, curve shape, and noise reduction. Positive results have been achieved in all three of these areas as summarized below.

Areas exist where the concentric bowl pump is wanting by a few efficiency points to be more fully competitive with other pump types. A clue

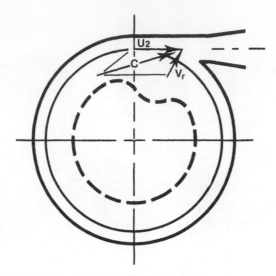

Figure 11-8. Diagram of concentric bowl static pressures and flows.

to a basic hydraulic "fault" in the concentric bowl pump exists in the eye-to-throat flow path described in the section "Terminology." The sketch in Figure 11-8 illustrates a concentric bowl pump, where the dashed line represents a polar plot of static pressure within the bowl. The static pressure is depressed in the vicinity of pump discharge, and this depression increases with increasing flow rate. This indicates unfavorable exchange of static head for velocity head in the area of discharge, which must be reconverted to static head in the diffuser. Furthermore, fluid approaches the discharge throat in the direction indicated by vector c, the vector sum of u_2 and v_r, detracting from the diffuser recovery potential.

These adversities can be eliminated by abandoning the concentric bowl geometry in favor of a volute collector geometry. This modification has been shown to improve efficiency by about 6 points with specific speeds in the range of N_s = 800 to 1,000, but this advantage fades to parity with the concentric bowl configuration at specific speeds of about N_s = 300 to 400.

Radial side load results from standing pressure variations around the impeller periphery. These hydraulically imposed radial loads are proportional to the product of pump head times the projected area of the impeller, and must be reckoned with from the standpoint of bearing loads. Side load trends vary dramatically with the pump design geometry as indicated in Figure 11-9, where magnitudes are shown in the upper plot and vector direction trends are shown by the polar plots within the lower fig-

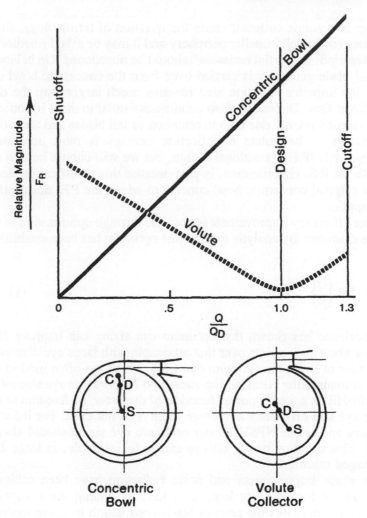

Figure 11-9. Radial load trends.

ures along with the letters S, D, and C, indicating shutoff, design, and cutoff flow rates. The concentric bowl side load increases continuously with flow and is always oriented in the general direction of the discharge throat. The volute radial load virtually vanishes at design flow, but undergoes about a 180° reversal in direction over the full flow range. These considerations are significant when fluid film bearings are used and the lube feed spreader groove location must not encroach into the bearing load zones.

Change to a volute collector raises the question of terminology, since fluid exists via the full impeller periphery and it may be asked whether or not the description "partial emission" should be abandoned. On balance, tall radial blade geometry is carried over from the concentric bowl design, so the impeller emission area remains much larger than the discharge throat area. Disorderly flow conditions similar to those in concentric bowl design prevail due both to retention of tall blades and the radial inlet geometry. The volute modification perhaps is most accurately viewed as a P.E./F.E. transitional design, but we will choose here to remain with the P.E. classification, in part because this modification stems from the original concentric bowl concept to which the P.E. designation clearly applies.

Further efficiency improvement is possible through optimization of the inlet eye diameter. By analytic means, this optimum has been established as:

$$D_e = 5.1 \left(\frac{Q}{N}\right)^{.333} \tag{11-22}$$

Test experience has shown that optimum eye sizing can improve efficiency by about four points over that attainable with large eye diameters on the order of twice the optimum diameter, which are often used in the interest of minimizing NPSHR. Equation 11-17 shows that eye size influences NPSHR as a fourth power function of diameter, so freedom to exploit the eye size efficiency advantage often does not exist. For inducerless design and ample NPSHA, near-optimum eye sizing should always be used. This situation nearly always exists, for example, in stage 2 of series-staged machines.

Curve shape improvement and noise reduction have been achieved through use of high-solidity impellers, i.e., by adding more impeller blades, which increases the ratio of blade cord length to blade spacing. High-solidity impellers tend toward minimizing flow stratification and blade loading because the total power is divided between a larger blade complement. To obtain benefit with high solidity impellers, it is necessary that all blades penetrate equally into the impeller eye. Use of splitter geometry or blades alternating in length as is often done in turbomachines to avoid eye crowding results in dominance of the larger blades and provides no advantage in P.E. pumps.

A rising-to-shutoff or stable curve shape can be provided with high solidity radial-bladed impellers as shown by the solid curve in Figure 11-10. This is contrary to the generally held view that backswept blade design is essential to providing a stable curve. The reason that improved curve shape results with high-solidity impellers is believed to lie with im-

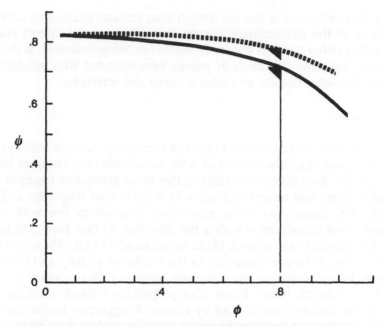

Figure 11-10. Curve shape variation with impeller eye size.

proved pitot recovery at low flow rates. Referring back to the concentric bowl characteristics shown in Figure 11-2, it is seen that shutoff head is about the same as design head. The ideal velocity head is $u^2/2g$, and some portion of this head may be converted to static head by either diffusion or pitot recovery. As flow is reduced, diffusion recovery potential decreases while pitot recovery potential simultaneously increases. The shutoff head coefficient would be expected to be only $\psi = .5$ without pitot recovery. A stable curve shape evidently results from improved pitot recovery provided by high solidity impellers.

As would be expected, increased impeller solidity results in an increased head coefficient. In "Partial Emission Formulae," the impeller blade inner diameter, D_1, was said to have small effect on pump performance, so was neglected in an illustrative exercise. But the impeller eye choice does in fact affect curve shape and efficiency with a volute collector, so presenting a designer's choice. A rising curve characteristic is achieved by setting D_1 appreciably larger than the inducer diameter. Choice of small D_1, or deeper blade penetration into the eye, provides higher design head and moderately higher efficiency, but results in a relatively flat curve shape.

It is interesting to note that the shutoff head remains unchanged with D_1 variation, so the characteristic curves "hinge" about the shutoff point. Although a rising curve is generally viewed as being desirable, it should be noted that many thousands of pumps have operated with satisfaction through the years despite an unstable curve characteristic.

Pump Noise

Noise generation becomes a subject of increasing concern with the increased power densities associated with sometimes very compact high-speed pumps. Investigation has shown that noise generation traces to hydraulic origins, and more specifically is at blade pass frequency in P.E. pumps. The underlying noise generating mechanism lies with pronounced flow stratification within the impeller, so that flow jets issue from the pressure side of each blade at the impeller exit. These jets impinge on the geometric anomalies in the discharge vicinity and produce pressure pulses which set the container walls in motion, which in turn broadcast airborne noise. Noise also propagates through metallic and fluid paths making noise control by means of lagging or enclosures extremely unattractive because the entire pumping system must be treated, including the pump, driver, base, and piping system.

High-solidity impellers have provided noise reduction typically on the order of 10 dbA, which viewed in different contexts translates into 90% sound power reduction but is perceived by the human ear as being half as loud. Roughly 5 dbA additional noise reduction is available by increasing the case size relative to the impeller size with virtually no sacrifice in efficiency. Industrial and governmental noise standards can always be met or exceeded with high-solidity impellers, whereas their low-solidity counterparts are sometimes marginal in this regard.

It should be noted that proper volute sizing is a prerequisite to achieving either the efficiency gains available with volute collectors or the noise and curve shape advantages associated with high-solidity impellers. Experience has taught that the cross-sectional area swept by the volute should be about 15% to 20% greater than the discharge throat area.

The photograph of Figure 11-11 compares low- and high-solidity partial emission pump impellers equipped with inducers. Advantages associated with the high-solidity impellers will result in this type supplanting in large measure their low solidity counterparts.

Design Configuration Options

High-speed partial emission pumps are well suited to provide high to very high heads. Inducers have augmented suction performance so that a

Figure 11-11. Comparison of high and low solidity impellers with inducers (courtesy Sundstrand).

majority of requirements can be satisfied by machines with single-stage simplicity. More difficult requirements can be met by staging arrangements that provide extremely high heads or very low suction requirements or a combination of both.

Samples of a family of pump designs that have been evolved to provide wide coverage and flexibility are illustrated in the collage shown in Figure 11-12 A-E, briefly described as follows:

A. Single-stage or two-stage HP to 1,500 and 2,500. Single-stage to H = 6,000, Q = 400. Series staging to H = 12,000, Q = 400. Parallel staging to H = 6000, Q = 800.
B. Three-stage same as Pump A with boost stage to provide extreme heads combined with low NPSHR.
C. Two-stage, two-speed. HP to 400 and 750. To H = 6000 with low NPSHR or H = 12,000 ft with ample NPSHR. Q to 400.
D. In-line vertical. HP to 50, 200 and 400. H = 6000 and Q = 400. Direct drive versions available to 75 HP.
E. Integral flange motor. HP from 1 to 200 in 3 size versions. H to 3500, Q to 400. Frame mounts optional.

As is readily apparent, a great deal of design capability and flexibility is available in this family of machines. The suction constraints associated

A

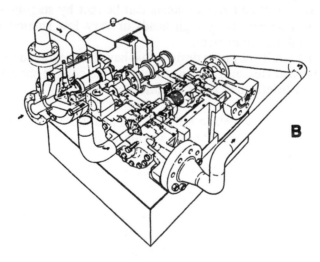

B

Figure 11-12. Examples of partial emission high speed pumps: (A) single- or two-stage pump; (B) three-stage pump; (C) two-stage, two-speed pump; (D) in-line vertical pump; (E) integral flange motor pump.

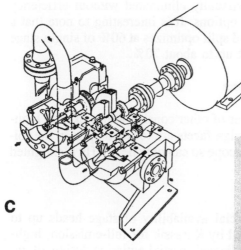

C

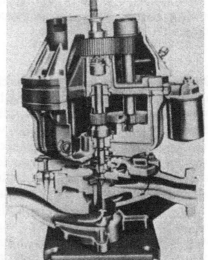

D

E

with high-speed pumps can be virtually eliminated without efficiency sacrifice via inducer and staging options. It is interesting to note that a two-stage machine with equal head split optimizes at 60% of single-stage speed, which can reduce NPSHR up to about 70%.

Other High Speed Considerations

Aside from hydraulics, a number of other considerations exist in high-speed design. Several of these design facets are touched upon in the following sections. Each is broad in scope so can only be briefly highlighted here.

Stress and Deflection. Commercial availability of stage heads up to about 3,000 feet has been indicated by Karassik for full-emission, high-speed pump types. As has been indicated, partial-emission design allows heads to 6,000 feet per stage even with relatively low-strength 316 stainless steel material, this potential accruing from rugged blade impeller design. Simple impeller geometry allows easy extension of this head limit, if such need arises, through use of high strength-to-weight materials such as 17-4 PH stainless steel or titanium alloys.

Size reduction associated with high-speed design is dramatically illustrated in Figure 11-13 showing high- and low-speed multi-stage rotors with equivalent pumping capabilities. Fewer high-speed rotors and the exponential relationships of span and shaft diameter combine to allow geometries with lower shaft deflection in the high-speed design.

Gears. Speed-increasing gearboxes are generally required for pump drive speeds above electric motor limits. Integral gear systems are frequently used with single overhung impeller design, where the gearbox high-speed shaft doubles to support the pump impeller. Multi-stage machines are usually designed with straddle-mounted rotors and free-standing gearboxes, so requiring strict attention to alignment and high-speed coupling design.

The single most important attribute of high-speed, high-pitch line-velocity gearing is precision. Hardened and ground gearing is attractive because modern gear grinding equipment provides very high precision capability, along with substantial size reduction over soft gearing. Hardened gearing does not undergo geometric change during break-in, so the required gear and mounting precision must exist at assembly.

Industrial gearing is generally designed in accordance with American Gear Manufacturers Association (AGMA) specifications in which complete design guidelines are presented. Gears corresponding to AGMA

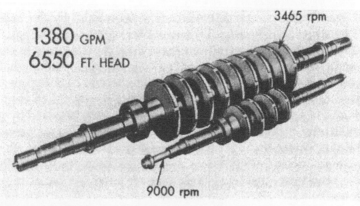

Figure 11-13. Multi-stage high and low speed rotor comparison (courtesy Worthington Division, McGraw Edison Company).

precision class 10 to 12 are commonly used in moderate- to high-pitch line-velocity gearing. The American Petroleum Institute (API) also publishes gearing specifications that are derived from AGMA, but demand more design conservatism. AGMA ratings compare to API ratings roughly in a ratio of 5:3.

Gear design considers ratings from two standpoints: strength and endurance. Strength rating is based upon evaluation of the gear tooth as a cantilever beam, and dominates in lower-speed, high-torque situations. Control of the case depth is important in hardened gear design to avoid through-hardening, or brittle teeth. Endurance rating evaluates gear design from the standpoint of wear resistance and becomes increasingly dominant with increasing pitch line-velocities. The lesser of the strength and endurance ratings at a given operating condition establishes the gearing rating. Best balance between strength and endurance results from coarse tooth selections for the lower operating speeds to fine-tooth selection in the high-speed ranges.

A tendency has existed to select spur gears for moderate power transmission and helicals for the higher power ranges. Spur gear geometry forces design with a contact ratio between 1 and 2, that is to say that the load is alternately carried by a single tooth or shared by a pair of teeth. Rating, then, is based upon single-tooth contact. Helical gears provide smooth meshing and continuous multi-tooth contact, and so in theory provide substantial increased capacity within a given envelope. This helical advantage, however, is highly dependent on gear precision, and is usually assumed to provide added design margin rather than increased capacity rating.

Helical gears introduce thrust which must be reacted by the bearing system. Thrust can be eliminated by use of herringbone gear design, but this option loses some packaging attraction with ground gears in that a wide central groove must be provided between the gear working halves for grinding wheel runout. Piecing helical gears back to back to provide a herringbone design detracts from already stringent precision requirements. Helical gears are generally (but not universally) believed by gear authorities to offer the advantage of lower noise, but here again precision looms more important than the spur-versus-helical choice per se. In any event, hydraulic noise in high-speed pumps has been found usually to overwhelm gear noise, divorcing noise consideration as a factor in gear-type selection.

For P.E. pumps, where speed is always tailored to a given application, desired speed is provided by simple gear size selection of standardized gears to fit within standardized gearboxes. Rarely do power and speed combine to require the maximum gearbox rating, so most often an added design margin exists at the rated power in a given application.

Bearings. Ball bearings have evolved to a high state of perfection and are attractive from the standpoints of low friction loss and modest lube system demands. Roller bearings have higher capacity than ball bearings, but are not well suited for high speeds due to a tendency of the rollers to skew in operation.

In general industrial equipment, the API guidelines are sometimes viewed as being unnecessarily conservative, and are modified in the interest of simplicity and low cost. Life projections should be tempered by full realization that only contact stress is considered, and that the quality of lubrication and many other practical aspects of bearing application are not addressed. In any event, high-speed/high-power design imposes bearing demands which soon outrun any realistic expectations of design adequacy with rolling contact bearings.

Hydrodynamic bearings possess capability to operate for indefinite periods at high load levels and high speed. This bearing type is self-acting with a film of lubricant separating the bearing elements in steady-state operation, precluding metallic contact and thus providing zero wear. The term "thick film" is used to describe these bearings, but this description must be taken in context since "thick" usually implies film heights of only a few ten-thousandths of an inch. Metallic contact cannot be tolerated in high-speed bearings, so the need for precise alignment is obvious. Materials selection is important largely because boundary lubrication or rubbing contact exists during start-stop cycles where full fluid-film separation cannot be achieved.

Plain journal bearings have excellent capacity and are nearly always suitable at pump speeds. For extreme speeds and powers, tilting pad journal bearings are sometimes used to take advantage of their excellent stability characteristics. Plain thrust bearings are inexpensive, and are generally used up to their load limits of 100 to 150 psi. More severe thrust loads require use of tilting pad thrust bearings with about 500 psi unit load capacity.

Bearings are obviously important elements in high-speed design, but the temptation to oversize bearings in the interest of unwarranted design conservatism should be suppressed because hydrodynamic bearing parasitic losses are not negligible.

Lube Systems. In small units equipped with ball bearings, lubrication needs can often be met with simple splash systems. Higher power units equipped with hydrodynamic bearings generally require a pressure lube system, including a lube pump, over-pressure relief valve, filter, and heat exchanger.

Free-standing lube pumps are sometimes used, but a pump driven from the gearbox input shaft is preferable, because lubricant is supplied during coastdown from high speed in the event of a power failure. Auxiliary lube pumps are sometimes required when start-up demands are severe. An example of this is an application with very high suction pressure acting over the shaft seal area producing high thrust at start-up. The thrust bearing must be copiously lubricated at start-up in order to survive the short-term boundary lubrication conditions existing until sufficient speed is achieved to provide lift-off to full film separation. Large machines and machines with very high stand-by suction pressure are often equipped with auxiliary lube pumps to provide full lubricant flow and start-up.

Shaft Dynamics. Shaft dynamics is a rather complex discipline that has evolved substantially over the years to ever higher levels of sophistication along with other engineering sciences. The advent of modern computer technology has raised analytic prowess to heights which would be otherwise impractical if not impossible.

The dynamic behavior of a shaft is strongly influenced by the characteristics of the bearings upon which it is invariably mounted; the important bearing characteristics being the spring rate and the damping coefficient. Rolling contact bearings have high, but finite, spring rates as opposed to relatively low spring rates in fluid-film bearings. Critical shaft speeds decrease with decreasing bearing spring rates. The high spring rates of rolling contact bearings usually vary over only a narrow range, so past experimental spring rate information generally suffices in shaft dynamics analyses. For hydrodynamic bearings, the relationship between load and film height are well established, and the spring rate is calculated by taking the first derivative of the W/h relationship, dW/dh. It must be recognized that the hydrodynamic spring rate will vary with load, so the ranges associated with gear loading and impeller hydraulic loads must be considered in detailed analyses.

All this appears a bit intimidating at first glance, but in practice shaft dynamics has generally shown itself to present neither incessant nor insurmountable problems. It should be pointed out that critical speed operation is not always destructive. Cases exist where a shaft can be made to run continuously at its critical for long periods, but needless to say, comfortable margin should always be provided between critical and design speed. With a new product family, thorough critical speed analysis is used in the design phase, and the analysis is confirmed by test experience in the hardware phase. Need for a full-blown analysis for each minor pump variation is alleviated. Normally, each production pump is tested at full rated capacity so any dynamic distress can be detected and corrected prior to shipment. Single-speed machines offer advantage in this regard since change of shaft stiffness, bearing stiffness, rotating mass, or a combination of all three can often cure a problem with modest hardware alteration.

Field problems with shaft dynamics are by far the exception rather than the rule. Such exception has an increased chance of occurring when full field operating conditions cannot be duplicated in the manufacturer's test facility. For example, water is the universally used test fluid, so pumps designed for low gravity fluids must often be operated at off-design conditions to simulate their full-power or full-speed operating characteristics. Or, a pump can interact differently with a user's system or foundation than it does in the laboratory. The computer has proven to be an invaluable aid in such occasional situations when a problem occurs.

Instrumentation is readily available to continuously monitor machine health if so desired. Noncontacting probes can directly observe high-speed shaft motion and can be arranged to provide display, alarm, or shut-down in the event of trouble. But this option is generally reserved for large and costly equipment. It is probably safe to say that this instrumentation is seldom opted for in machines under a few hundred horsepower. Experience has shown that the reliability and endurance of high-speed machines can be assumed to match that of their lower-speed counterparts, so similar ground rules on protective instrumentation should apply.

References

Balje, Dr. O. E., *Turbomachines; A Guide to Design Selection and Theory*, John Wiley and Sons, 1981.

Barske, U. M., and Dr. Ing. "Development of Some Unconventional Centrifugal Pumps" Proceedings of the Institution of Mechanical Engineers (Britain), Vol. 174 No. 11, 1960.

Barske, U. M., and Dr. Ing. "Formulas and Diagrams for the Calculation of Open Impeller Centrifugal Pumps" Royal Aircraft Establishment, Farnborough; Rocket Propulsion Department, Wescott; Technical Note RPD 127: 155.

Karassik, Igor J., and Hirschfeld, Fritz, "The Centrifugal Pump of Tomorrow" *Mechanical Engineering* (1982).

Karassik, Igor J., "The Centrifugal Pump-Out of the Past-Into the Future" Presented at the Second European Fluid Machinery Congress, The Hague, 1984.

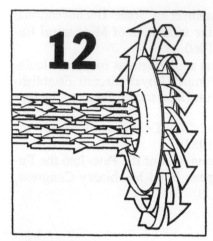

12

Double-Case Pumps

by **Erik B. Fiske**
BW/IP International, Inc.
Pump Division

Double-case pumps are also known as double-casing, barrel-case, or barrel-type pumps. They are used for temperatures and pressures above the range of single-case horizontally split, or diffuser-type multi-stage pumps, and pressures above the capabilities of radially split process pumps. The following are typical temperature and pressure limitations for centrifugal pumps. Horizontally split multi-stage pumps, such as those illustrated in Figures 6-1 and 10-13 are limited to operating temperatures of 400°F [1]. They can be designed for discharge pressures up to 4,000 psi, but often are limited to lower pressures by user specifications. Radially split process pumps are suitable for temperatures up to 800°F. Radially split double-suction, single-stage process pumps cover operating conditions up to 500 psi discharge pressure. Two-stage process pumps cover conditions to 1,000 psi. Double-case pumps are typically applied for operating conditions above these limits.

Configurations

Pump Casing

Double-case pumps got their name from being constructed with two cases: an inner case assembly that contains the complex shape of the stationary hydraulic passages and an outer case (barrel) that acts as the pressure boundary for the pumped fluid. The inner case assembly is subjected to external differential pressure, so any bolting required to hold it together is minimal. The outer barrel is designed as an unfired pressure vessel, and can be constructed to the requirements of well established industry codes.

Volute Casing with Opposed Impellers

Figure 12-1 illustrates the volute-type opposed-impeller configuration. The inner case assembly is horizontally split and consists of two identical halves, cast from the same pattern. The double-volute construction provides radial hydraulic balance. The opposed impeller arrangement minimizes resultant axial thrust, providing inherent axial thrust balance.

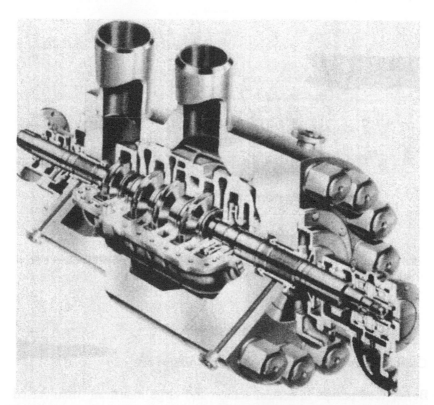

Figure 12-1. Volute-type opposed-impeller double-case pump (courtesy BW/IP International, Inc. Pump Division, manufacturer of Byron Jackson/United™ Pumps).

Diffuser Casings with Balance Drum

Diffuser-type pumps have all impellers facing in one direction, resulting in high axial thrust forces. Figure 12-2 shows a diffuser-type pump with a balance drum to carry the axial thrust forces. The inner case assembly is vertically split, and the symmetry of the diffusers provides radial hydraulic balance.

Diffuser Casings with Balance Disk

Figure 12-3 shows the diffuser-type configuration with a balance disk that carries the axial forces. Except for the axial balancing device, the construction is similar to the previous diffuser-type design. Balance disk construction is used for clean services such as boiler feed because of its ability to completely balance axial thrust at all operating flow rates.

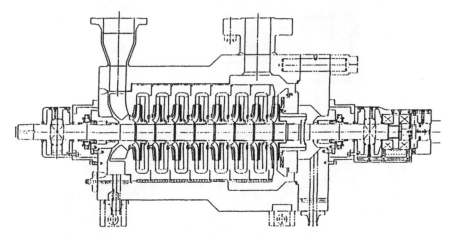

Figure 12-2. Diffuser-type in-line impeller, double-case pump with balance drum (courtesy Dresser Pump Division, Dresser Industries, Inc.).

Applications

Common applications for double-case pumps are:

- Boiler feed pumps in central station and large industrial fossil-fueled power plants.
- High pressure and/or high temperature pumps in oil refineries or chemical plants.

Figure 12-3. Diffuser-type in-line impeller, double-case pump with balance disk (courtesy Ingersoll-Rand Company).

- High pressure oil field water injection and offshore hydrocarbon condensate reinjection pumps.
- Pipeline pumps for unusually high pressures, very high vapor pressure hydrocarbons (typically above 200 psi), or offshore hydrocarbon condensate.

Boiler Feed Pumps

The most common application for double-case pumps is for boiler feed service in fossil-fueled power plants. These pumps must combine high efficiency with maximum reliability. Feedwater pump outages were estimated to have cost more than $408 million in replacement power alone in the United States in 1981 [3]. Several multi-million dollar efforts to reduce this cost have been implemented by users and manufacturers worldwide. These efforts have resulted in increased product knowledge that now can be applied to high-energy pumps, system design, and operation. Research in this area is continuing.

Charge Pumps

Oil refinery charge pumps handle liquids that are flammable and often toxic, at very high temperatures and pressures. Wide variations in viscosity of the feed stock or the presence of abrasives may add to pump design problems. In spite of inherent application problems, these pumps must combine maximum reliability with good efficiency.

Waterflood Pumps

Oil field water injection pumps operate at capacities to 5,000 gpm. Double-case pumps provide differential heads to 11,000 feet and discharge pressures to 8,000 psi from two pumps operating in series. This application is covered in more detail in Chapter 10.

Pipeline Pumps

The vast majority of pipeline pumps are of the horizontally split, multistage design, covered in Chapter 10. Double-case pumps are used only when unusually high pressures are required or when handling hydrocarbons near their supercritical condition.

Design Features

Removable Inner Case Subassembly

Modern double-case pumps have a fully separate inner case subassembly (including rotor). The inner case subassembly for a volute-type pump is shown in Figure 12-4. This subassembly can be removed, after disassembling the outboard cover, without disturbing the suction piping, discharge piping or the driver. It is common practice to have a spare subassembly available for replacement, thereby reducing maintenance turnaround time or the downtime caused by unscheduled outages.

If the pumped fluid is hot, time is needed to lower the temperature of the components before maintenance work can begin. Time to cool by ambient air is extended because the pump is normally well insulated. Forced liquid cooling can be helpful, but must be preplanned to avoid subjecting the pump to unacceptable thermal gradients.

In some designs the inner case subassembly includes the radial and thrust bearings. This feature further reduces downtime because the replacement rotor is aligned before the outage. A boiler feedwater pump of this construction, called "cartridge," "full cartridge," "pullout," or "cartridge pullout" design, is shown in Figure 12-5.

A saltwater injection pump with full cartridge pullout is shown in Figure 12-6. The configuration shown is said to save at least 40 manhours of labor, compared to conventional construction, each time the inner-case subassembly is replaced. This design features a springplate on the high pressure end to preload the internal gasket between the inner volute case and the outer barrel. This gasket seals the full differential pressure. The springplate design compensates for manufacturing tolerances to assure interchangeability among spare inner assemblies and also compensates

Figure 12-4. Inner case subassembly for a volute-type pump (courtesy BW/IP International, Inc. Pump Division, manufacturer of Byron Jackson/United™ Pumps).

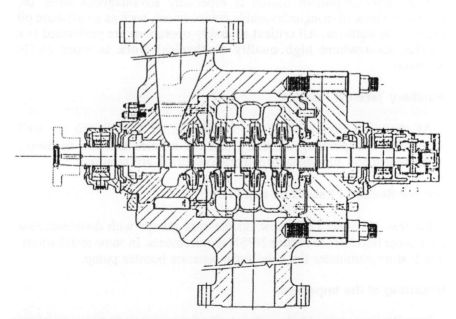

Figure 12-5. Boiler feedwater pump of cartridge pullout design to reduce maintenance turnaround time (courtesy Sulzer Bingham Pumps Inc.).

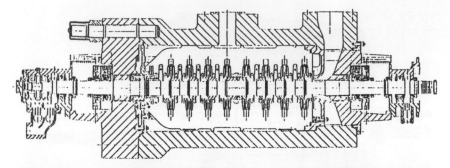

Figure 12-6. Water injection pump with full cartridge pullout, featuring a springplate on the high pressure end to preload the internal gasket that seals full differential pressure (courtesy BW/IP International, Inc. Pump Division, manufacturer of Byron Jackson/United™ Pumps).

for differential thermal expansion. Finite element analysis of the springplate assures that all design goals are achieved. This mechanical design, with the proper materials of construction, is also suitable for boiler-feed service.

The cartridge pullout design is especially advantageous when the pumps are located in an unfavorable environment, such as an offshore oil production platform. All critical assembly operations are performed in a service shop where high quality mechanical work is more easily achieved.

Auxiliary Take-off Nozzles

As shown in Figure 12-3, double-case boiler feed pumps are well suited for incorporating auxiliary take-off nozzles that are necessary when reheat and superheat attemperation sprays are required.

Double-Suction First-Stage Impellers

Figures 12-1 and 12-3 show double-case pumps with double-suction first-stage impellers to reduce NPSH requirements. In some installations, this feature eliminates the need for a separate booster pump.

Mounting of the Impellers

Impellers are assembled on the shaft with a shrink fit to prevent mechanical looseness under all operating conditions and are positioned axially by split rings on the suction side of each impeller hub. This is the preferred mounting.

An alternative mounting employs a stack with sliding fits along the shaft. A spacer sleeve is fit snugly between each pair of impellers and an outside locknut is used to secure the impellers and sleeves. Great care is required to assure that all spacer sleeve and impeller faces are parallel to each other, perpendicular to the shaft, and smooth. Small errors in parallelism or perpendicularity will misalign the stack.

Impeller Wear Rings

Impeller wear rings are generally specified for refinery pumps [1], but must be secured with great care on high-speed, high-pressure pumps. Because boiler feedwater is a relatively clean liquid and rapid wear or seizure is unusual, impeller wear rings are seldom used in large, high-speed double-case boiler feed pumps. These boiler feed pump impellers are designed with extra stock on the wearing surfaces. When worn, impellers are skim cut true, and the pump then fitted with case wear rings that are undersized to match the impeller wearing surfaces.

Shaft Seals

Shaft seal failure is the most common cause of unscheduled outage for double-case pumps. Selection of a shaft seal system designed for the application is therefore critical to reliable pump operation.

Face-Type Mechanical Seals. Oil refinery pumps almost universally use mechanical seals. Reliable seals for high temperature oil and light hydrocarbons now exist. New federal air quality laws limiting hydrocarbon emissions encourage the use of tandem-type seal systems. Figure 12-7 shows a tandem seal assembly with bellows-type seals for hot oil service.

In the United States, only small to medium-size (up to 7,000 hp and 5,000 rpm) boiler feed pumps use mechanical seals. In Europe, they are also used in medium to large boiler feed pumps. A typical boiler feed pump seal is shown in Figure 12-8.

Most waterflood pumps have mechanical seals, operating at ambient temperatures, but in corrosive liquid.

Mechanical seals are described in detail in Chapter 17.

Throttle Bushings. The most reliable shaft sealing system for large boiler-feed pumps consists of throttle bushings with a custom designed, cold (90°–120°F) condensate injection system. Pumps have operated for 40 years or more with their original throttle bushings.

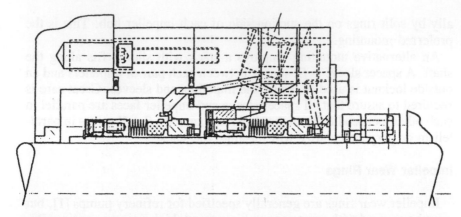

Figure 12-7. Tandem seal assembly for hot oil service (courtesy BW/IP International, Inc. Seal Division, manufacturer of BW Seals).

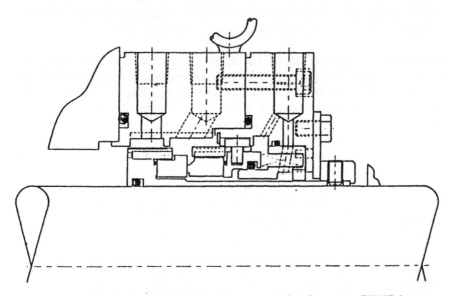

Figure 12-8. Shaft seal for boiler feed pump service (courtesy BW/IP International, Inc. Seal Division, manufacturer of BW Seals).

The throttle bushing bore, the shaft under the bushing, or both should be grooved. To obtain the desired effect, the following design parameters are varied: the groove cross section, the number of groove starts, the "hand" of the grooves, and the length of the grooved section. The grooves reduce leakage for a given running clearance and increase tolerance to solid particles in the feedwater. They also reduce the possibility of seizure if the pump is subjected to severe operating transients, such as flashing. Shaft sleeves under the throttle bushings are undesirable. They reduce the ability to resist seizure during severe temperature transients.

There are at least five types of condensate injection control systems [2]. The type of control is normally recommended by the pump manufacturer based on the purchaser's feedwater system design. A simple pressure-controlled system is shown in Figure 12-9. Temperature-controlled systems are more common. A drain-temperature control system is shown in Figure 12-10.

Two waterflood pumps such as the one shown in Figure 12-11 operate in series. The downstream pump has 4,000 psi suction pressure and 8,000 psi discharge pressure. Mechanical seals would not be practical for these pressures. Therefore the pumps are fitted with long throttle bushings that discharge into collection chambers with suitable drain connections. The cold leakage is expendable, and no re-injection system is needed.

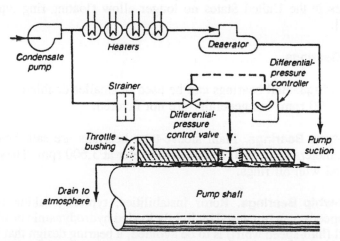

Figure 12-9. Pressure-controlled throttle bushing injection system (from Ashton [2]).

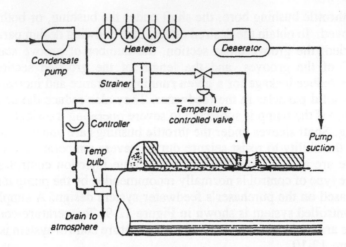

Figure 12-10. Drain-temperature controlled throttle bushing injection system (from Ashton [2]).

Floating-Ring Seals. Boiler feed pumps with floating-ring seals were common in the 1970s, but proved to fall short in reliability. A survey of boiler feed pumps in 1977 found 748 seal failures in 730 pumps with floating-ring or mechanical seals vs. 32 seal failures in more than 300 pumps with throttle bushings [5]. Specifications of most major architect/ engineers in the United States no longer allow floating-ring type shaft seals [3].

Radial Bearings

Ball Bearings. Ball bearings can be used in smaller double-case pumps below 4,000 rpm, but generally are not favored.

Sleeve-type Bearings. Plain sleeve-type bearings are satisfactory for shaft diameters up to 3.50 inches in diameter at 3,600 rpm. They can be lubricated with oil rings.

Anti-oil-whip Bearings. Rotor instabilities, typically taking place at higher speeds, can be caused by lightly loaded hydrodynamic bearings. If oil whirl (half-speed whirl) is to be avoided, a bearing design that is more stable at lower loads is required (see Chapter 19). Pressure dam bearings have been used successfully for shaft diameters up to 6.75 inches at

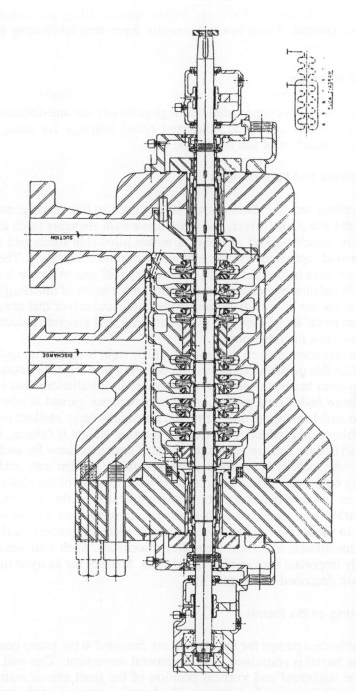

Figure 12-11. Double-case waterflood pump for 4,000 psi suction pressure (courtesy BW/IP International, Inc. Pump Division, manufacturer of Byron Jackson/United™ Pumps).

5,400 rpm. For larger shafts and higher speeds, tilting pad radial bearings are favored. These bearings require force-feed lubricating oil systems.

Thrust Bearings

Ring-oiled sleeve-type radial bearings generally use anti-friction thrust bearings. When force-feed lubricated radial bearings are used, thrust bearings should be of the tilting pad type.

Baseplates and Foundations

Baseplates and foundations must be designed so that misalignment between the pump, the driver, and other drive-train elements (such as gearboxes or variable-speed devices) is within allowable limits and so that they provide optimum dynamic support to minimize vibration. The traditional solution is to make the baseplate very stiff and to secure it rigidly to the foundation by bolting and grouting. A variation of this design technique is the use of sole plates under the pump and driver that are rigidly attached to the foundation and the use of very stiff concrete pedestals for the centerline mounted pump.

An alternative solution is to mount a fully rigid baseplate on springs or other flexible members to isolate the pumping unit from the foundation. This system has been used for many years in installations that include very large boiler feed pumps, and the experience gained is now being used to isolate pumping units from the potentially large motions of flexible offshore oil production platforms. Because weight is critical, honeycomb structures that are very light and stiff have been used for such baseplates. A sophisticated three-point mounting system can make the rotating equipment almost insensitive to large motions of the platform because no in-plane bending or torsion is generated by the deck motions.

In either case, large or high-speed pumping systems should be subjected to modal analysis and detailed finite element dynamic analysis to avoid mechanical and fluid (acoustic) resonances. Such analyses are especially important for variable-speed units. Some of the analysis methods used are described in Chapter 19.

Mounting of the Barrel

Double-case pumps for hot service are mounted at the pump centerline and the barrel is restrained from horizontal movement. This will assure that the horizontal and vertical position of the shaft axis is maintained during unit heatup and cooldown. The barrel is secured on the baseplate so that the thermal expansion is away from the coupling. This maintains the axial gap at the pump-to-driver coupling.

Design Features for Pumping Hot Oil with Abrasives

One of the most difficult double-case pump applications is the pumping of hot oil (500°F and above) with substantial quantities (2% or more) of entrained abrasive solids.

Surface Coating. The key to prolonged periods of operation without maintenance is the application of a hard surface coating, which may extend service life by a factor of 4 or more. The coating should have a minimum hardness of 60 Rockwell C. It should be applied to all wear surfaces, to all accessible hydraulic passages in impellers and inner cases, and to the outside of the impeller shrouds. The coating is typically applied with a high-velocity spray process that produces a strong mechanical bond. High coating density and proper coating thickness are critical.

Impeller and Case Wear Rings. Impellers designed with extra stock on the integral wear ring surfaces are generally preferable to replaceable impeller wear rings. Worn impeller wear ring surfaces can be re-coated and ground to size. They run against case wear rings that are coated in the bores and on the ends, and are sized to match the impeller running surfaces.

Keyways. Unless special design precautions are observed, rapid erosion occurs in keyways that are subjected to more than one stage of differential pressure. The two center stage impellers of opposed-impeller type pumps should be welded together at the hubs, and the key (or keys) terminated blind in a relief. A shrink-fit land is provided between the impeller bore and the shaft at the high-pressure end to seal against leakage and prevent erosion. Similar construction should be used for the sleeve under the throttle bushing of an opposed-impeller pump, or under the balancing drum of a pump with inline impellers.

Double-Case Pump Rotordynamic Analysis

The rotordynamic analysis requirements for a double-case pump depend on the size, rotational speed, and horsepower of the specific pump. Dry and wet critical speed analyses are adequate for small and medium-size pumps running at 5,500 rpm and below. This type of analysis is described in Chapter 19. Most current specifications only require critical speed analyses, and a full scale test with specified vibration limits.

Requirements of new supercritical power plants, new oil refinery processes, and high pressure oil field water injection facilities have increased the demand for predictably reliable, high speed, high horsepower double-case pumps. In addition to critical speed analyses, these pumps should be subjected to a rotor stability analysis as part of the design process. They also may be subjected to a rotor response analysis. At the present time, computer programs for rotor response analysis are available [7]. Response analysis includes consideration of excitation forces from both mechanical and hydraulic origins. Mechanical excitation forces from sources such as dynamic unbalance, misalignment, and shaft bow are well known. Hydraulic excitation forces are generated at the wear rings, long annular seals (such as balance drums or throttle bushings), and impellers. The magnitudes of hydraulic excitation forces (especially those generated by impellers) are less well known, but are believed to be much greater than mechanical excitation forces in large double-case pumps (which are precision manufactured to minimize mechanical forces). Some cutting edge research on hydraulic excitation forces has been conducted and is ongoing [7].

The Effect of Stage Arrangement on Rotordynamics

The opposed-impeller stage arrangement of volute-type pumps offers greater rotordynamic stability than the inline arrangement with all impellers facing in the same direction, which is common to diffuser-type pumps. This has been shown for a number of years [4] by critical speed analyses. A recent comparison, based on the stability analysis of an 8,000 rpm pump [6], is given in Table 12-1. Here an analysis of a pump with opposed-impellers is compared with the analysis of an "equivalent" inline impeller arrangement. Identical impeller forces and annular seal coefficients were used. With design clearances and smooth (not grooved) annular seals, the analyses showed stable operation and no subsynchronous whirling up to 14,000 rpm for opposed impellers, but only up to 8,000 rpm for inline impellers. The difference is attributed to the extra center bushing in the opposed-impeller design, where a strongly stabilizing Lomakin effect is generated. This advantage is reduced as the internal annular seals wear and clearances increase.

The Effect of Impeller Growth from Centrifugal Forces

Radial growth of high-speed pump impellers caused by centrifugal forces is significant [6]. The unsymmetrical impeller deformation caused by centrifugal forces, pressure loading, and shrink fit to the shaft for a four-stage, 8,000-rpm boiler feed pump is shown in Figure 12-12.

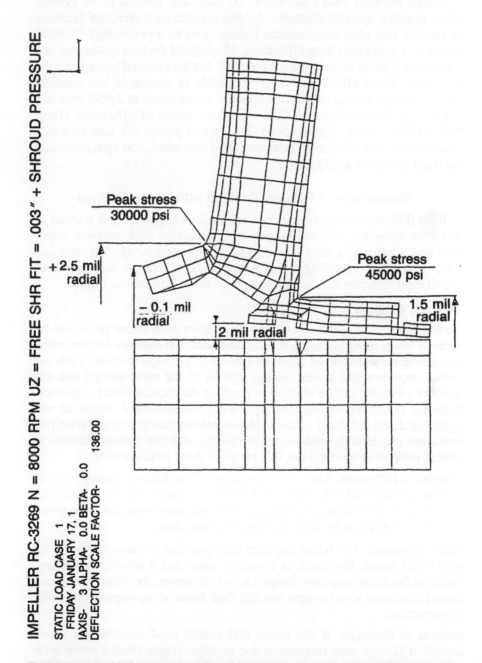

Figure 12-12. High-speed impeller deformation caused by centrifugal forces, pressure loading, and shrink fit (from Verhoeven [6]).

Reduced Annular Seal Clearance. Of particular interest is the growth of the impeller eye seal diameter. As pump speed increases, the decrease in annular seal clearance reduces leakage loss as a percentage of input power and improves pump efficiency. Mechanical friction losses and impeller disk friction losses also are reduced, but the reduced leakage loss is dominant. These effects are most noticeable in pumps of low specific speed. Factory testing of a 1,010 specific speed pump at 2,950 rpm and 7,000 rpm demonstrated an increase of six points of efficiency (from 49% to 55%) at the higher speed. Low speed pumps are less affected. The difference in efficiency between 1,800 rpm and 3,600 rpm operation for most pumps is negligible.

Comparison of Diffuser Casings with Volute Casings

Both diffuser-casing construction with inline impellers and stacked inner case assembly, and volute-casing construction with opposed impellers and horizontally split inner case assembly can be applied with success to critical centrifugal pump applications. There are, however, significant differences. A comparison of these differences follows.

Diffuser Casings

Lower Cost. The diffuser-casing construction with inline impellers results in lower manufacturing cost. Additionally, if damage occurs, a single casing segment can be replaced instead of a complete volute case. All series impellers and series casings can be of the same design and any number of stages can be stacked to produce the required head. Opposed-impeller construction requires right-hand and left-hand impellers and right-hand and left-hand volutes. Major pattern changes or separate patterns are required to produce inner casings with the desired number of stages needed to produce the full range of head requirements.

Precision Diffusers. The critical portions of the diffuser passages (those in which high fluid velocity is converted to pressure) can be investment-cast or machined, thus manufactured with less roughness and more precision than the typically sand-cast inner volute cases.

More Compact. The inline impeller configuration is more compact. All else being equal, the result is a shorter rotor and a shorter pump, and because the inline impeller design has no crossover, the inner casing and barrel diameters also become smaller than those of the opposed impeller construction.

Simple to Destage. If the pump differential head requirement is reduced, it is very easy to remove one or more stages from a pump with inline impellers. Pumps with opposed impellers can be destaged, but the process is more complex.

Volute Casings

Rotordynamic Stability. As illustrated in Table 12-1, the opposed-impeller configuration has better rotordynamic stability. Swirl brakes are not required except to solve very unusual application problems.

Dynamic Balance. Because the volute casings are horizontally split, the fully assembled rotating element is dynamically balanced in its final

Table 12-1
Subsynchronous Whirl and Stability Threshold Speeds
of Inline and Opposed Impeller Arrangements
for Different Conditions

Condition	Opposed Impeller Arrangement		Equivalent Inline Impeller Arrangement	
	Grooved Annular Seals	Smooth Annular Seals	Grooved Annular Seals	Smooth Annular Seals
100% design clearance	SSW 8,000 rpm, Unstable 9,000 rpm	SSW 14,000 rpm, Unstable 15,000 rpm	SSW 5,500 rpm, Unstable 6,500 rpm	SSW 8,000 rpm, Unstable 10,000 rpm
200% design clearance		SSW 9,250 rpm, Unstable 10,000 rpm		SSW 6,600 rpm, Unstable 7,400 rpm
300% design clearance		SSW 7,700 rpm, Unstable 8,500 rpm		SSW 5,500 rpm, Unstable 6,300 rpm
400% design clearance		SSW 5,150 rpm, Unstable 5,900 rpm		SSW 5,000 rpm, Unstable 5,600 rpm
100% design clearance with swirl brakes	SSW 12,000 rpm Unstable 13,500 rpm		SSW 9,000 rpm Unstable 10,500 rpm	

SSW = Subsynchronous Whirling

form. This is not possible with the diffuser-casing construction. The latter requires alternate assembly of impellers and diffuser casings and therefore dismantling and reassembly of the impellers on the shaft after dynamic balance. Exact restoration of dynamic balance after the rotor has been dismantled cannot be assured.

Sag Bore. The double-volute design lends itself to machining of the equivalent natural deflection of the rotating element into the bottom volute case half. The result is that all running clearances remain concentric because the shaft will operate in its deflected position. This is especially important when operating on turning gear in hot-standby condition. This refinement, known as sag bore, cannot readily be effected with the diffuser design with its multiplicity of concentric fits.

Running Clearance Check with Feeler Gauge. By placing the rotating element, including all rotating and stationary wear parts, into the bottom volute case half, all running clearances can be checked with a feeler gauge to verify that a reconditioned rotor has proper clearances, or to determine if wear has taken place in a used rotor.

Easier Rotor Replacement. A rotating element can be quickly removed from its volute case and a spare one installed. Stocking of a spare inner volute is optional because it is not considered a wearing part. Many users stock a spare rotating element only. In order to expedite disassembly and assembly time with a diffuser design, the user must purchase a complete inner case assembly because field assembly of the rotor and diffuser cases is a time-consuming procedure.

References

1. API Standard 610 *Centrifugal Pumps for General Refinery Services,* 7th ed., American Petroleum Institute, Washington, D.C., 1990.
2. Ashton, R. D. "Optimize b-f Pump Throttle Bushings by Close Match to Feedwater Circuit," *Power,* Sept. 1980.
3. *Feed Pump Hydraulic Performance and Design Improvement, Phase I: Research Program Design,* CS-2323 Research Project 1886-4, Electric Power Research Institute, Palo Alto, CA, 1981.
4. Gopalakrishnan, S. and Husmann, J. "Some Observations on Feed Pump Vibrations," *Proceedings of EPRI Symposium on Power Plant Feed Pumps,* Cherry Hill, N.J., June 1982.

5. *Survey of Feed Pump Outages,* FP 754 Research Project 641, Electric Power Research Institute, Palo Alto, CA, 1978.
6. Verhoeven, J. J. "Rotordynamic Considerations in the Design of High Speed, Multistage Centrifugal Pumps," *Proceedings of the Fifth International Pump Users Symposium,* Texas A&M University, May 1988.
7. Verhoeven, J. J. "Rotor Dynamics of Centrifugal Pumps, a Matter of Fluid Forces," *The Shock and Vibration Digest,* Vibration Institute, Volume 23, Number 7, July 1991.

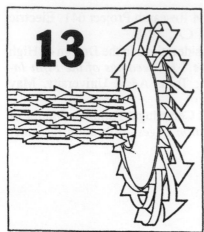

13

Slurry Pumps

by **George Wilson**
Goulds Pumps, Inc.

There are various types of centrifugal slurry pumps, which are identified by their capability to handle solids ranging in size, hardness, concentration, and velocity. An understanding of these important factors will lead to an optimum choice of pump design where the materials of construction and rotational speed are ideally matched to the process system.

Slurry Abrasivity

The abrasiveness of slurries is difficult to define due to the number of variables involved. It is dependent on the nature of the slurry being pumped and the materials of construction of the pump liquid end components.

Wear increases with increasing particle size. For example, the rate of erosion wear when pumping a silica sand slurry is approximately proportional to the average particle size raised to the power of 1.4.

Wear increases with concentration; the relationship is linear up to about 10% by volume. At higher concentrations, the rate of wear will level out due to the cushioning effect of the particles as they collide.

Wear increases rapidly when the particle hardness exceeds that of the metal surface being abraded. The effective wear resistance of a metal will depend on the relative hardness of the metal to that of the particle.

An approximate comparison of hardness values of common ores and minerals is given in Figure 13-1.

Wherever possible, the hardness of the pump liquid end metal components should exceed the particle hardness. It should be noted that the measurement of hardness is not the only criteria and that the structure of the metal material itself has to be considered. An example of this is the

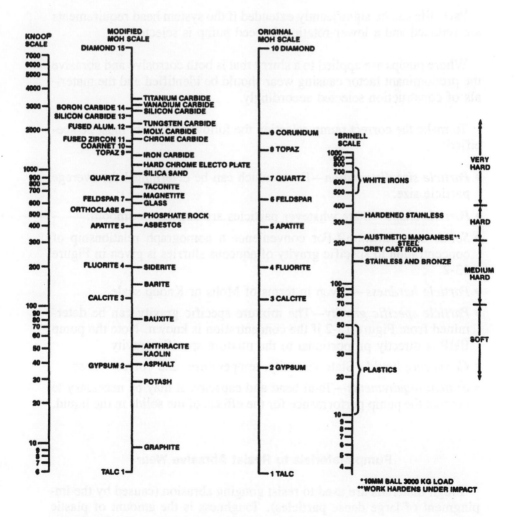

Figure 13-1. Approximate comparison of hardness values of common ores and minerals.

large degree of very hard chromium carbide (1800 Knoop hardness) precipitation that can be achieved in high chrome iron. Wear increases:

- When the particles are angular.
- With particle density.
- With increasing particle velocity such that the rate of wear is directly proportional to V^m where m can vary from 2.5 to 4.

Parts life can be significantly extended if the system head requirements are reduced and a lower-rotational-speed pump is selected.

Where pumps are applied to a slurry that is both corrosive and abrasive the predominant factor causing wear should be identified and the materials of construction selected accordingly.

To make the correct pump selection the following factors must be specified:

- *Particle size distribution*—From which can be determined the average particle size.
- *Particle shape*—State whatever particles are angular or smooth.
- *Solids concentration*—For convenience a nomograph relationship of concentration to specific gravity of aqueous slurries is given in Figure 13-2.
- *Particle hardness*—Given in terms of Mohs or Knoop scale.
- *Particle specific gravity*—The mixture specific gravity can be determined from Figure 13-2 if the concentration is known. Note the pump BHP is directly proportional to the mixture specific gravity.
- *Conveying liquid*—State viscosity, temperature, and corrosiveness.
- *System requirements*—Total head and capacity. It may be necessary to correct the pump performance for the effects of the solids in the liquid.

Pump Materials to Resist Abrasive Wear

Tough materials are used to resist gouging abrasion (caused by the impingment of large dense particles). Toughness is the amount of plastic deformation a material can withstand without fracture. Generally the larger the difference is between the yield and tensile strength, the tougher the material will be.

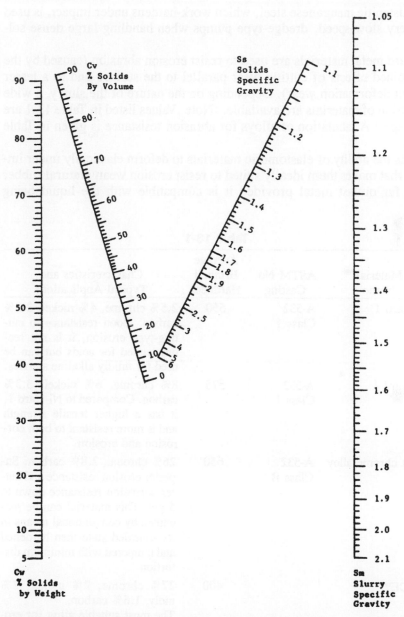

Figure 13-2. Nomograph of the relationship of concentration to the specific gravity in aqueous slurries.

Austenitic manganese steel, which work-hardens under impact, is used in very slow-speed, dredge-type pumps when handling large dense solids.

Hard metal materials are used to resist erosion abrasion (caused by the combined effects of cutting wear parallel to the surface and to a lesser extent deformation wear). Depending on the nature of the slurry, a wide selection of materials are available. (Note: Values listed in Table 13-1 are average.) A tabulation of alloys for abrasion resistance is given in Table 13-2.

It is the ability of elastomeric materials to deform elastically under impact that makes them ideally suited to resist erosion wear. Natural rubber will far outlast metal provided it is compatible with the liquid being

Table 13-1

Material Name	ASTM No. Casting	Average Brinnell Hardness	Characteristics and Typical Applications
Ni-Hard 1	A-532 Class 1	550	2.5% chrome, 4% nickel, 3.3% carbon. Good resistance to cutting-type erosion, it is not recommended for acids but can be used for mildly alkaline slurries.
Ni-Hard 4	A-532 Class 1	575	8% chrome, 6% nickel, 3.3% carbon. Compared to Ni-Hard 1, it has a higher tensile strength and is more resistant to both corrosion and erosion.
High chrome alloy	A-532 Class B	650	26% chrome, 2.8% carbon. Superior erosion resistance and better corrosion resistance down to 5 pH. This material can be machined by conventional means in its annealed state then hardened and tempered with minimum distortion.
*PACE™		400	27% chrome, 2% nickel, 2% moly, 1.6% carbon. The most suitable alloy for erosive, corrosive slurries in the range 1 to 11 pH. However its erosion resistance is inferior to high chrome iron.

Registered trademark of Abex Corporation

Table 13-2
Alloys for Abrasion Resistance
(Properties Sensitive to Carbon Content Structure)

Alloy	Properties
Tungsten carbide composites	Maximum abrasion resistance. Worn surfaces become rough.
High-chromium irons	Excellent erosion resistance. Oxidation resistance.
Martensitic iron	Excellent abrasion resistance. High compressive strength.
Cobalt base alloys	Oxidation resistance. Corrosion resistance. Hot strength and creep resistance.
Nickel base alloys	Corrosion resistance. May have oxidation and creep resistance.
Martensitic steels	Good combination of abrasion and impact resistance.
Pearlistic steels	Inexpensive. Fair abrasion and impact resistance.
Austenitic steels	Work hardening.
Stainless steels	Corrosion resistance.
Manganese steel	Maximum toughness with fair abrasion resistance. Good metal-to-metal wear resistance under impact.

(Left margin: vertical arrows labeled "Increasing Abrasion Resistance" and "Increasing Toughness")

pumped and the particle size is limited to fines below 7 mesh in size. At velocities above 35 ft/sec the rubber may not have sufficient time to flex and absorbs all the impact, and as result, wear will increase. Natural rubber is limited in temperature to 150°F or less.

Where oils are present, a synthetic rubber such as neoprene should be used; however the addition of fillers will have a detrimental effect on wear resistance.

Elastomer materials generally have good corrosion resistance, but care must be exercised to prevent the slurries from penetrating behind the casing and causing corrosive damage.

Natural rubber-lined pumps with a durometer hardness of 40 shore A are usually limited to about 120 feet total head. Higher heads can be generated if fillers are added to increase hardness.

Castable urethanes in the 90 shore A hardness range exhibit good tear strength and elongation properties and in certain applications have out performed both rubber and metal.

Ceramic materials in castable form have excellent resistance to cutting erosion but because of their brittle nature are unsuitable for direct impact. Silicone carbide refrax liners and impellers in the 9.5 original Mohs hardness range are commonly used for pumping fines where the impeller tip velocity is limited to less than 100 ft/sec.

Slurry Pump Types

There is no specific demarcation point where one pump design ceases to be effective and another takes over. Figure 13-3 shows a classification of pumps and materials according to particle size. It is important to note that the selection of the pump type and its materials of construction depend also on the abrasivity of the slurry and the total head to be generated.

The abrasivity of slurries can be divided into five distinct classifications to which limits on pump selection can be applied. Table 13-3 shows a pump selection guide for wear resistance.

Specific Speed and Wear

The majority of centrifugal pumps are conventionally designed to achieve the desired hydraulic performance at the highest efficiency and lowest cost when handling clear fluids in reasonably clean environments. Manufacturing limitations are not imposed on the configuration of the pump, since conventional materials such as cast iron, bronze, and stainless steel are used. Since wear is not a major consideration, the highest possible specific speed is chosen.

When a centrifugal pump is designed for a slurry service, the factors that predominantly influence the pump design are wear and materials of construction; efficiency is of lesser importance. To achieve these objectives the pump has to operate at a lower rotational speed and the impeller is typically a radial-flow type. This suggests that the pump must be of a low specific speed design in the range 600 to 1,800. Specific speed is defined in Chapter 2.

Since wear is a function of velocity it can be shown that for a given head and capacity, wear will increase with increased N_s.

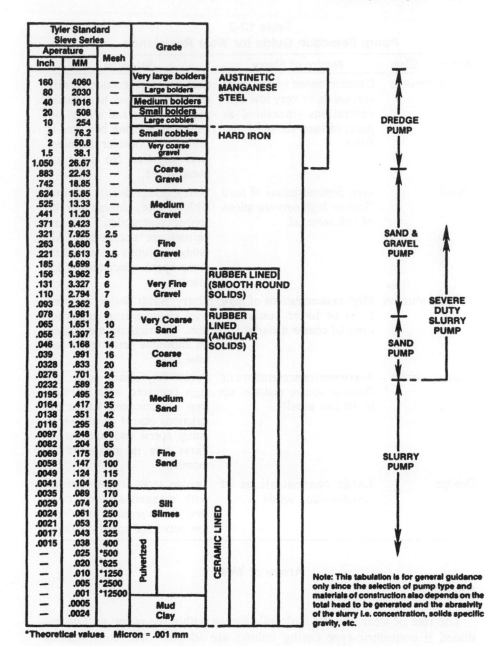

Figure 13-3. Classification of pumps according to solid size.

Table 13-3
Pump Selection Guide for Wear Resistance

Abrasion Class	Nature of Slurry	Selection
Mildly abrasive	Concentrations of relatively soft solids or very low concentrations (measured as ppm) of hard silt-sized particles.	Cast iron construction usually satisfactory, but hard-faced impeller rings and special attention to stuffing box area is justified. Consider stainless steel impeller. No limits on pump speed.
Abrasive	Low concentrations of hard fines or high concentrations of soft material.	Slurry pump design required with Ni-hard, chrome iron, or rubber construction. Open impellers are acceptable. Although no limits are placed on pump speed, discretion is advised.
Severely abrasive	High concentrations of hard fines or lower concentrations of coarse material.	Slurry pump design required with chrome iron construction. Restrictions are placed on allowable pump speed and total head.
Primary circuit	Maximum concentrations of fines or coarse material up to 10 mm usually.	Severe-duty slurry pump design required with chrome iron construction. Large restrictions placed on allowable pump speed and total head. Parts life is measured in months.
Dredge	Large concentrations of boulder-sized solids.	Dredge type design required with manganese steel construction to resist impact. Very low rotational speed required.

Areas of Wear

Casing

The rate of wear and the hydraulic forces within the pump will be reduced if concentric-type casing volutes are adopted over conventional spiral volutes. At "off" design point operation, the static pressure around the impeller's outside diameter will be relatively uniform, and turbulence in the vicinity of the cutwater will be effectively reduced as will the slurry velocity entering the casing throat. Recirculation flows from the

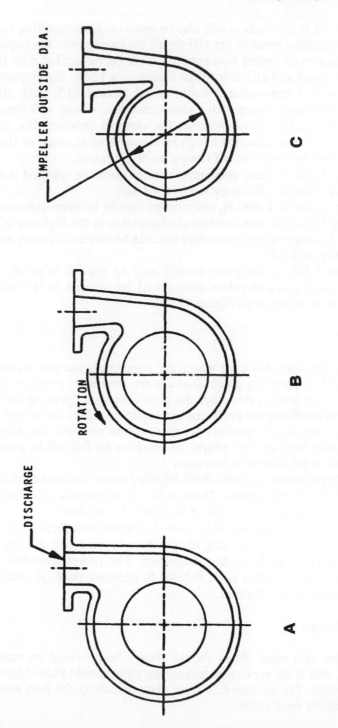

Figure 13-4. Casing configurations: (A) concentric volute; (B) semi-concentric volute; (C) spiral volute.

volute back to the suction will also be more uniform resulting in a reduction in localized wear in the vicinity of the casing near the impeller eye.

The degree of casing concentricity must be reconciled with the pump specific speed and efficiency. For example, a 12-in. slurry pump (N_s = 1,350) with a conventional volute could have a 82% peak efficiency, whereas the same pump with a semi-concentric casing will have an efficiency of 80.5%. Concentric casing designs produce flat efficiency curves that are sustained at a high level over a wide range of flows making it more amenable to off-design point operation.

Up to 1,200 N_s fully concentric casings could be adopted without too much sacrifice in efficiency (Figure 13-4a).

From 1,200 to 1,800 N_s the casings should be semi-concentric, progressing towards a spiral-volute configuration at the high end of specific speed. A compromise is therefore reached between efficiency and rate of wear (Figure 13-4b).

Above 1,800 N_s the pump should only be applied to mildly abrasive services and a spiral-volute casing will be utilized in the interests of higher peak efficiency (Figure 13-4c).

Impeller

Open impellers are used where the abrasion is not too severe as they have good air handling capabilities and are cheaper to produce. However, performance deteriorates when the front clearance opens up due to wear.

Closed impellers are preferred over open impellers for severe abrasive slurries since those impellers are more robust and will last longer. Also closed impellers are not nearly so sensitive to fall off in performance when the front clearance increases.

The requirement for extra thick impeller vanes can cause restrictions at the impeller eye and inlets. Three to five vanes are normal, depending on the pump specific speed and solids handling capability.

Pump out vanes are normally provided on the rear shroud. These vanes have the effect of minimizing the pressure at the pump stuffing box and reducing the axial hydraulic unbalance. The power absorbed by these vanes is not all wasted since it helps to generate head. A small drop in efficiency can be expected.

Wear Plates

Suction-side wear plates should always be provided on metal slurry pumps, and if the service is severe, the plate should extend into the suction nozzle. The suction-side wear plate is usually the part which needs replacement most often.

There is little to be gained by fitting a rear wear plate, provided the rear of the hard metal casing extends to the stuffing box. Experience has shown that the rate of wear in this area is not any greater than in the casing itself.

Bearing Frames

Usually the bearings are oil lubricated with a calculated life of over 50,000 hours. Slurry pumps are installed in dirty dust-laden atmospheres, and extra precautions have to be taken to seal the bearing covers and prevent the ingress of liquid and dust. In severe services, taconite seals are provided (i.e. double-lip type seals with grease cavities).

Sealing

Slurry pumps are often subjected to severe shock loading and shaft whip due to the presence of solids and system upsets. For these reasons soft compression packing is still favored as a means of sealing at the stuffing box.

The preferred method for packing a slurry pump is the "flush" seal shown in Figure 13-5a. Here the lantern ring is positioned in front of the packing rings and a copious supply of clean liquid is injected at a pressure higher than the prevailing slurry pressure in the stuffing box. The clean liquid acts as a barrier and prevents the ingress of abrasive particles that cause packing and sleeve wear. The disadvantage of this system is that large amounts of flushing water are required and the pumped product will be diluted. This system is recommended for severe abrasive services.

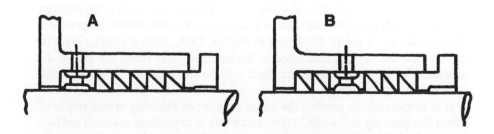

Figure 13-5. (A) Typical "flush-type" slurry pump stuffing box. Barrier flush prevents abrasive wear. (B) Typical "weep-type" stuffing box. It uses considerably less gland water but is much more susceptible to abrasive wear.

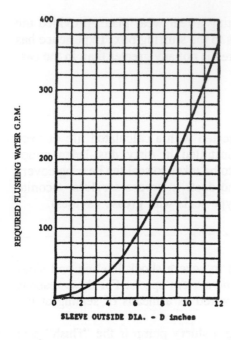

Figure 13-6. Approximation of flushing water requirements for "flush-type" slurry pump stuffing boxes where flush pressure is 15 psi above slurry pressure at the box. "Weep-type" stuffing boxes use about 5% of the water that "flush-types" use.

An alternative method for sealing is shown on Figure 13-5b. Here the lantern ring is positioned between packing rings. This configuration is called a "weep" seal. Again, clean liquid should be injected at a pressure higher than the prevailing slurry pressure near the stuffing box. Product dilution is significantly reduced compared to the "flush" seal design. However, the barrier so created is not very effective, causing abrasive particles to penetrate and cause wear. If the service is only mildly abrasive, then grease can be used in lieu of liquid.

An approximation of flushing requirements for a "flush" type packing arrangement for conventional throat restriction devices where no attempt has been made to curtail the use of flushing water and where the pressure differential is 15 psi is displayed in Figure 13-6. Such a restriction will have an annular radial clearance in the order of .007 times the sleeve diameter. The length of the throat bush will be about the same as the width of one turn of packing.

It is impossible to predict the exact amount of flushing water required when the packing is "weep" type, since this is dependent on shaft deflection and gland maintenance. However, under normal operating conditions, weepage would be in the order of 5% of the values stated in Figure 13-6 for "flush" packing arrangement.

In most cases, seals and flush requirements are provided in ignorance

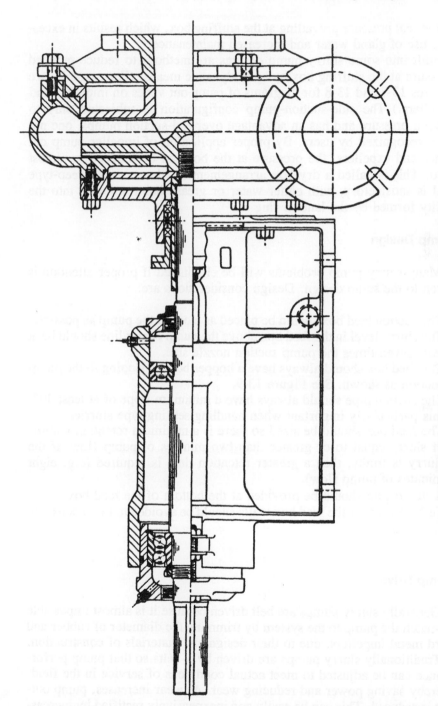

Figure 13-7. End-suction pump fitted with expeller (courtesy Goulds Pumps, Inc.).

of the real pressure prevailing at the stuffing box, which results in excessive use of gland water and increased maintenance.

Built into some slurry pump designs are methods to reduce pumped pressure at the stuffing box by hydrodynamic means. (For example, see Figures 13-7 and 13-8 for diagrams of pump out vanes on impellers and expellers.) The side-suction-pump configuration is subjected only to suction pressure and has an advantage over end-suction pumps, one not fully recognized by users. By proper application of impeller pump out vanes and expellers, the pressure at the box can be reduced to almost zero. This is called a dry box arrangement. In these cases, weep-type seal is satisfactory, with either water or grease being injected into the cavity formed by the lantern ring.

Sump Design

Many slurry pump problems will be eliminated if proper attention is given to the sump design. Design considerations are:

- The suction feed box should be placed as close to the pump as possible.
- The slurry level in the feed box above the pump center line should be at least seven times the pump suction nozzle size.
- The feed box should always have a hopper bottom sloping to the pump suction as shown. See Figure 13-9.
- The suction pipe should always have a minimum slope of at least 30°; this particularly important when handling settling-type slurries.
- The feed box should be sized so there is a minimum retention volume of slurry equal to or greater than two minutes of pump flow. If the slurry is frothy, then a greater retention time is required (e.g. eight minutes of pump flow).
- A dump gate should be provided at the bottom of the feed box.
- Turbulence near the feed box walls should be avoided to prevent excessive wear.

Pump Drive

Generally slurry pumps are belt driven because it is almost impossible to match the pump to the system by trimming the diameter of rubber and hard metal impellers, due to their design and materials of construction.

Traditionally slurry pumps are driven by V-belts so that pump performance can be adjusted to meet actual conditions of service in the field, thereby saving power and reducing wear. As wear increases, pump output is reduced. This can be easily and inexpensively rectified by increas-

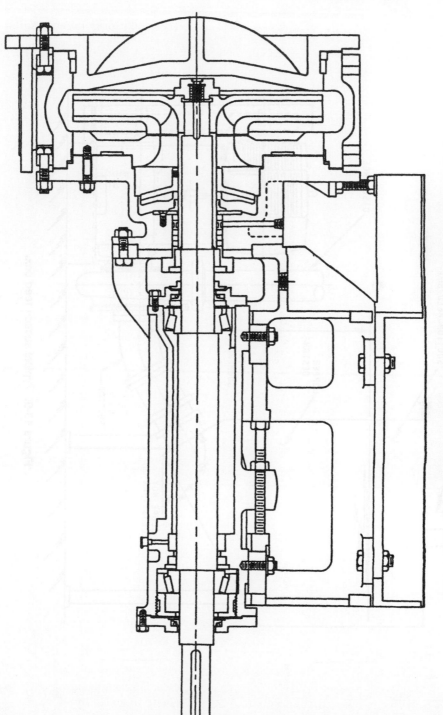

Figure 13-8. Side-suction pump fitted with expeller (courtesy Goulds Pumps, Inc.).

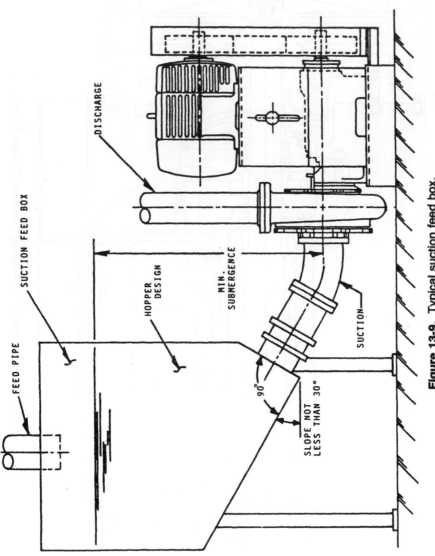

Figure 13-9. Typical suction feed box.

ing the pump speed by changing the sheave ratios. Three to five percent should be added to the motor BHP to compensate for belt losses. It is always good practice to add one more belt than is normally calculated to cover upset conditions and belt breakage.

Motors must be rated with an adequate margin to cover upset conditions such as high flow due to lower-than-expected system losses, higher concentrations, and start-up. At start-up, the concentration is often higher, and if the pump was not flushed out during the previous shutdown, the pump could be plugged with solids requiring high breakaway torques to get the impeller rotating. This undesirable condition happens all too frequently and can cause pump damage, excessive wear, and motor overload.

Under well-controlled systems, free from upset, the motor could be rated at 20% above the motor shaft BHP; however, this percentage could

Figure 13-10. Typical belt-driven, overhead-motor-mounted slurry pump.

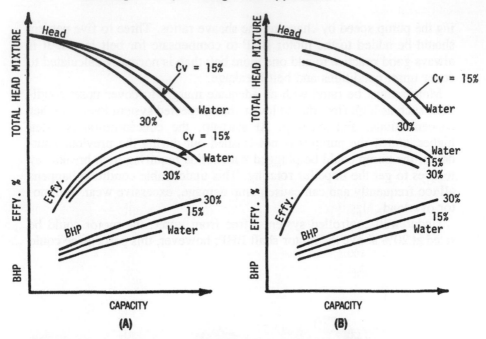

Figure 13-11. (A) Typical performance characteristic of nonsettling slurries; (B) typical performance characteristic of settling slurries.

go over 100% in badly controlled systems. For this reason large margins are built into the design of slurry pumps and motors. Usually the motor is mounted above the pump on an adjustable frame (belt adjustment) to save space and to safeguard against flooding. (Refer to Figure 13-10.)

The Effect of Slurries on Pump Performance

When centrifugal pumps are required to handle slurries, it is standard practice to publish pump performance curves based on clear water performance. Therefore, to predict the performance of pumps handling slurries of different characteristics, correction factors are applied.

When handling slurries, the pump performance is mainly affected by the solid particle diameter, specific gravity, and concentration. Very fine particles in a slurry can be "nonsettling" and cause it to behave as a homogeneous Newtonian liquid with an "apparent" viscosity. Slurries with very fine solids in suspension (usually less than 100 microns) will retain liquid-like characteristics at volumetric concentrations very near to the limiting voidage, and limits are related only to the effects of high viscosity. Figure 13-11a shows typical nonsettling slurry pump characteristics.

Where there exists a density difference between the conveying liquid and the solid particles, the particles will tend to settle. Usually slurries with a distribution of larger particles will be "settling," and the particles and the liquid will exhibit their own characteristics. As the liquid passes over the particles, energy is dissipated due to the liquid "drag" that reduces pump head and efficiency.

Actual tests indicate that for practical purposes, the amount of head derate will be the same as the efficiency derate. A typical pump performance characteristic for slurry mixture with coarse particles is shown in Figure 13-11b.

Performance correction factors for slurries are usually based on previous test data. In the absence of such data, reference should be made to the pump manufacturer.

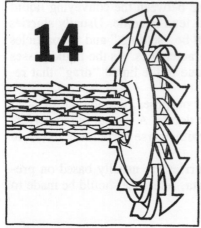

14

Hydraulic Power Recovery Turbines

by **Rolf Lueneburg** and
Richard M. Nelson
Bingham-Willamette, Ltd.

The potential for power recovery from high-pressure liquid-streams exists any time a liquid flows from a higher pressure to a lower pressure in such a manner that throttling occurs. As in pumping, this throttling can exhibit a hydraulic horsepower (HHP) and a brake horsepower (BHP), except that in throttling, these horsepowers are available rather than consumed. Hydraulic power recovery turbines (HPRT's) are used instead of throttling valves to recover liquid power.

The two main types of HPRT's are:

1. *Reaction*—Reverse running pumps and turbomachines in single- and multi-stage configurations with radial flow, (Francis), mixed flow, and axial flow (Kaplan) type runners. They come with fixed and variable guide vanes. The axial flow Kaplan propeller has adjustable runner blades.

2. *Impulse*—Most prominent is the Pelton wheel, usually specified for relatively high differential pressures and low to medium liquid flows.

This chapter will discuss reaction-type HPRT's; namely, the reverse-running pump (Figure 14-1) and machines specifically designed to run as HPRT's.

Basically all centrifugal pumps, from low to high specific speed and whether single- or multi-stage, radially or axially split, and in horizontal or vertical installations, can be operated in reverse and used as HPRT's. The discharge nozzle of the pump becomes the inlet of the turbine; the suction nozzle or bell of the pump becomes the outlet of the turbine, and

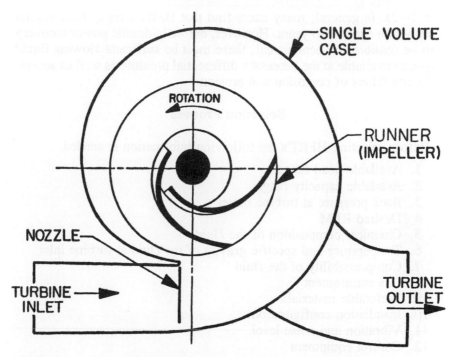

Figure 14-1. Typical single-volute-type pump used as a hydraulic turbine.

the impeller of the pump, rotating in reverse direction, becomes the runner of the turbine. Pumps are readily available, and many sizes are stock items. Reverse running pumps are an excellent alternative to conventional turbomachinery.

Centrifugal pumps operating as HPRT's have negligible operating costs. The installation costs are essentially the same as for an equivalent pump, and in terms of reliability and maintainability (R & M), they do have less maintenance costs because of their smoother and quieter operation. Also, since the efficiency of a pump operating as an HPRT is equal to or slightly better than the pump efficiency, the use of reverse-running pumps, or specially designed turbomachines, as primary or secondary drivers becomes very attractive.

The purchase price of an HPRT is generally approximately 10% greater than the price of a pump of equivalent design dimensions and metallurgy. This reflects the costs of the modifications that must be made to the impellers and volutes or diffusers, plus the complex testing that is required to verify the hydraulic performance of the finished machine. A single-stage HPRT may be profitable when as little as 30 BHP is recovered, while a multi-stage HPRT may be justifiable above 100 BHP (Figure 14-2). In general, many users find that HPRT's repay their capital cost within one to two years. However, before hydraulic power recovery can be feasible and economical, there must be sufficient flowing liquid capacity available at the necessary differential pressure as well as acceptable conditions of corrosion and erosion.

Selection Process

Before selecting HPRT's the following information is needed:
1. Available head range
2. Available capacity range
3. Back pressure at turbine outlet
4. Desired RPM
5. Chemical composition of the fluid
6. Temperature and specific gravity of the fluid at turbine inlet
7. Compressibility of the fluid
8. Gas entrainment
9. Preferable materials
10. Installation configuration
11. Vibration and noise level
12. Control equipment

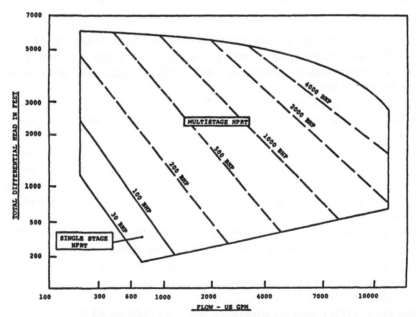

Figure 14-2. HPRT application range chart (from McClaskey and Lundquist).

The following criteria should be considered since they will help in specifying and classifying the HPRT.

Specific Speed

HPRT's are classified by their specific speed (N_s) which is a dimensionless quantity that governs the selection of the type of runner best suited for a given operating condition.

$$N_s = \frac{N \times BHP^{.5}}{H^{1.25}}$$

where   N   = Revolutions per minute
        BHP = Developed power in horsepower
        H   = Total dynamic head in feet across turbine at best efficiency point (BEP)

The physical meaning of specific speed is: Revolutions per minute at which a unit will run if the runner diameter is such that running at 1-ft head it will develop 1 BHP.

The customary specific speed form used for pumps for classification of impeller-type characteristics is also applicable for HPRT (basically for reverse running pumps). The values will be similar to those for pumps.

The impulse Pelton wheels have very low specific speeds as compared to propellers (Kaplan) having high specific speeds. Francis-type runners cover the N_s range between the impulse and propeller types (Figure 14-3).

Net Positive Discharge Head

Net positive discharge head required (NPDHR) applies to an HPRT as does NPSHR to a pump to preclude cavitation and its attendant physical damage effects. Some literature refers to the term "total required exhaust head" (TREH) rather than NPDHR.

Test data have indicated that the NPDHR or TREH of a machine for the turbine mode is less than the NPSHR of the same machine for the pump mode at the same flow rate. The available net positive discharge head (NPDHA) or total available exhaust head (TAEH) at the installation side of the HPRT has to be higher than or at least equal to the NPDHR or TREH. This applies only to the reaction-type HPRT, since the impulse-type is a free jet action and is therefore not subject to low-pressure areas.

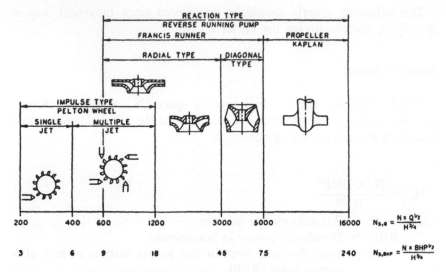

Figure 14-3. Turbine-type vs. specific speed. The ratio between $N_{S,Q}$ and $N_{S,BHP}$ is approximate only, since $N_{S,BHP}$ is a function of turbine efficiency.

Power Output and Affinity Laws

Power output is the rotational energy developed by the HPRT. Its value in BHP is calculated in a similar manner as for pumps except for the efficiency term.

$$BHP = \frac{Q \times H \times sp\ gr \times E_t}{3,960}$$

where Q = Capacity GPM
H = Total head in feet
$sp\ gr$ = Specific gravity
E_t = Overall Efficiency at the turbine mode

Variations in capacity, head, and BHP due to RPM (N) changes can be determined within reasonable limits by using the affinity laws, which normally are used for pumps but also apply to HPRT's (described in Chapter 2).

Configuration

The configuration of an HPRT is a function of the N_s, NPDHA or TAEH, BHP, RPM, installation requirements, and preferable vibration and noise levels. Similar to pumps, the specific speed, N_s, controls the number of stages for a given head capacity and RPM. The specific speed also specifies either single- or multi-stage HPRT's, which can be either the horizontal or vertical configurations.

The available net positive discharge head or total available exhaust head of the system for a given capacity will limit the RPM of the HPRT and determine whether the runner will be the single- or double-eye construction. Space requirements will specify the length of a horizontal unit or will call for the vertical installation. Energy-level requirements will limit the BHP per stage and may result in use of multi-stage units.

Turbine Performance Prediction

Prediction by Approximation

The performance characteristics of centrifugal pumps operating as hydraulic turbines may be approximated from pump performance characteristics. Typically, the capacity and head at the best efficiency point (BEP) will be greater for the turbine operation than for operation as a pump. The amount of shift from pump performance generally varies according to the specific speed. From tests, curves are developed that give ratios versus specific speed that are used to give the percent shift from pump to turbine performance.

Another procedure that is used to estimate the turbine performance from known pump performance characteristics is to simply divide the pump capacity and head values at the BEP by the pump efficiency at that point. This will give a rough approximation of the turbine head and capacity at the turbine BEP. Since there can be considerable error using these estimating procedures, they should only be used for preliminary selection of candidates for a particular application.

Prediction by Analysis

The performance characteristics of centrifugal pumps operating as hydraulic turbines and other turbomachines used exclusively as hydraulic turbines can be readily predicted with reasonable accuracy by use of a relatively simple analysis procedure. Only minor adjustments are to be expected to obtain the required performance on actual tests compared to the predicted performance by analysis. A modern computer program may be used to perform the calculations, print out the results, and also plot the performance curve.

The prediction procedure generally consists of accounting for the various components that compose the total head characteristics. These are friction losses, absorbed head, shock loss, and outlet loss. Power losses due to internal leakage, disc friction, and mechanical losses are also calculated or estimated as appropriate. The calculations are made using the required turbine speed, flow capacities, viscosity, specific gravity of the fluid, and various combinations of mechanical data required for certain multi-stage turbines. There are many publications that cover basic theory and design of pumps that show how to calculate the head and loss components for pumps. These also apply to hydraulic turbines.

Friction losses. Certain components of the total dynamic head are attributed to friction losses. These are due to flow through the cases, volute nozzles, diffusers, guide vanes, and runners (impellers) as appropriate. These losses may be simply calculated as the resistance to the incompressible flow of fluid in a pipe, using appropriate friction factors, length to diameter (or hydraulic radius) ratios, and the velocity head.

Absorbed head. The absorbed head is derived from the well-known Euler's equations and velocity triangles, which have general validity for all conditions of flow through turbomachines. Refer to Figure 14-4 for illustration. In practice, the true velocities of flow and direction are never known. The idealized velocity triangles of the Euler head equation assume perfect guidance of the flow by the vanes. It is known that there is a deviation of the fluid from the vane direction, which is the phenomenon called "slip." This is a consequence of the nonuniform velocity distribution across the runner channels, boundary-layer accumulation, and any separation.

Actual prediction of "slip" cannot be predetermined in a practical manner. However, it has been found that "slip" factors used for pump design applied to the turbine outlet vectors produce good results. The absolute velocity at the runner inlet (C_1) is the average velocity of the liquid at the nozzles with a free vortex correction applied to account for the distance from the nozzles to the runner. The nozzles are the highest velocity throat areas of the volute cases, diffusers, or guide vanes as appropriate for the turbine construction.

Shock loss. The shock loss component of the total dynamic head is calculated as the velocity head ($V_s^2/2g$) due to the mismatch of the absolute

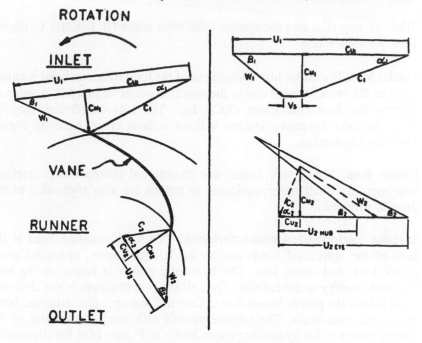

ROTATION

INLET

VANE

RUNNER

OUTLET

C = Absolute Velocity of Flow (Ft/Sec)

U = Peripheral Velocity of Runner (Ft/Sec)

W = Relative Velocity of Flow (Ft/Sec)

β = Vane Angle

V_s = Shock Velocity (Ft/Sec)

g = Acceleration Gravity

SG = Specific Gravity of Liquid

E_t = Overall Efficiency

Q = Total Flow (GPM)

H = Total Dynamic Head (Ft)

HP_L = Horsepower Losses (Leakage, Disc, Mechanical)

H_A = Absorbed Head = $\dfrac{U_1 C_{u_1} - U_2 C_{u_2}}{g}$ = Ft

Output BHP = $\dfrac{H_A \times Q \times SG - HP_L}{3960}$ = $\dfrac{Q \times H \times E_t \times SG}{3960}$

Figure 14-4. Turbine theory.

inlet velocity (C_1) and the runner inlet vane angle ($\beta 1$). Refer to Figure 14-4 for illustration.

Outlet loss. The outlet loss component of the total dynamic head is calculated as the velocity head due to the absolute outlet velocity (C_2) times an appropriate loss coefficient ($KC_2^2/2g$). The loss coefficient may be taken as unity for many designs without serious error. Refer to Figure 14-4 for illustration.

Power loss. The power losses due to internal leakage, disc friction, bearings, and shaft seals applicable to pumps are also applicable to hydraulic turbines.

Turbine performance characteristics. The total dynamic head is the sum of the individual heads due to the friction losses, absorbed head, shock loss, and outlet loss. The hydraulic power is based on the total available energy to the turbine. The turbine output power is the absorbed head minus the power losses due to internal leakage, disc friction, bearings, and shaft seals. The turbine overall efficiency is the ratio of the output power to the hydraulic power. Refer to Figure 14-4 for illustration of terms.

Predicted performance vs. test results. Figure 14-5 shows a typical comparison of the turbine performance characteristics determined by the preceding calculation procedure and the results obtained by actual test. Identical nozzle sizes were used.

Turbine Performance Prediction by Factoring

The performance characteristics of a hydraulic turbine may be quite accurately predicted by size factoring from a known performance at a specified specific speed. The rules that apply to pumps (described in Chapter 2) also apply to turbines.

Optimizing and Adjusting Performance Characteristics

The inlet and outlet velocity triangles as illustrated by Figure 14-4 are used to predict, adjust, and optimize the turbine performance. These give an instant picture of whether the turbine performance characteristics are expected to be optimum, satisfactory, marginal, or unsatisfactory. The optimum overall efficiency will generally be achieved when the shock loss at the inlet to the runner is near zero and the absolute velocity (C_2) at the outlet from the runner is near a minimum value.

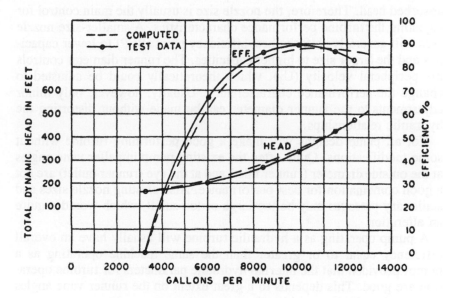

```
INPUT DATA FILE NAME:
BOOKP
14x16x26  HT HSA
IMPELLER NO.   1    NO. OF STAGES   1
VOLUTE DRG.  6685 IMPELLER DRG.   7131
NOZZLE AREA 23.00 VORTEX DIST. 30.00
TURBINE RPM= 1170

*****************************************
     GPM     HEAD      BHP      EFF.
  TOTAL TURBINE PERFORMANCE
    3000    163.1     -8.8     -7.1
    4000    173.7     38.6     22.0
    6000    211.2    198.9     62.1
    8000    270.7    446.0     81.5
   10000    352.0    779.5     87.7
   11000    400.9    978.5     87.9
   11500    427.4   1086.0     87.5
   12000    455.3   1198.9     86.9
   13000    515.1   1440.6     85.2
```

Figure 14-5. A computed-vs.-test HPRT performance (courtesy of Bingham-Willamette Company).

The inlet velocity (C_1), which depends on the nozzle size, is critical to the turbine performance characteristics. It can produce a significant change to the turbine performance by its effect on the shock loss and the absorbed head. Therefore, the nozzle size is usually the main control for adjusting the turbine performance characteristics. A smaller-size nozzle area will generally shift the best efficiency point (BEP) to lower capacities and the larger size to higher capacities. The runner diameter controls the peripheral velocity (U_1), which theoretically could be adjusted to change the performance characteristics. Usually, however, only minor adjustments to the runner diameter can be made without distorting the hydraulic relationships.

Not all pump designs will make a good performing turbine without some modifications. Quite often the existing pump impeller vane angles at the outside diameter (runner inlet) and at the eye (runner outlet) are not a good combination for best performance. Also existing nozzle sizes and stationary passages may be too large or too small, which would require an alteration.

A pump operating as a hydraulic turbine will usually have an overall efficiency equal to or greater than the same machine operating as a pump, provided that the internal hydraulic parameters for turbine operation are good. This depends to a great extent on the runner vane angles and nozzle velocity considerations.

The overall efficiency of a turbine at capacities near the best efficiency point usually is improved by shaping the inlet ends of the runner vanes to a bullet-nose-type configuration and slightly rounding the inlet edges of the runner shrouds. The improvement is 1% to 2% at BEP capacity (100% capacity) and still achievable at $\pm 20\%$ of BEP capacity (80% to 120% capacity). The reason is the reduced turbulence of the runner inlet. The effect of surface finish on friction losses and the effect of leakage losses may be readily evaluated by the turbine performance prediction procedure.

Design Features (Hydraulic and Mechanical)

Reverse-Running Pump

Most centrifugal pumps in the low- to medium-specific speed range $(N_s,Q = 600$ to $5,000$ or $N_s,BHP = 9$ to 75 (see Figure 14-3)) are suitable and capable of operating as HPRT's. Because of the reverse rotation, one has to check that the bearing lubrication system and threaded shaft components, such as impeller locking devices, cannot loosen. However, most pumps nowadays are designed to withstand reverse rotation.

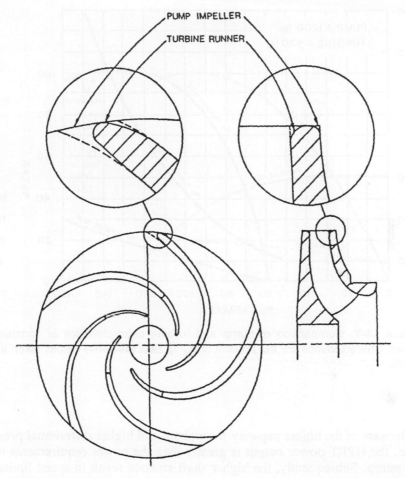

Figure 14-6. Rework of pump impeller for operation as an HPRT runner.

Trimming the runner (impeller) diameter, as is done for pumps, to shift the performance characteristics is not normally done for hydraulic turbines. The turbine runner diameter is selected for optimum running clearance.

The inlet ends of the turbine runner vanes are ground to a bullet-nose shape and the inlet edges of the runner shrouds are rounded slightly to preclude excessive turbulence for efficiency considerations (Figure 14-6). But before operating a pump as an HPRT at the same speed, one has to remember the change in performance characteristics. A comparison is shown in Figure 14-7.

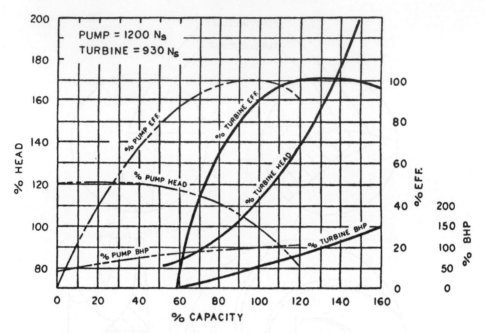

Figure 14-7. Comparison of pump and turbine characteristics at constant speed. Characteristics are the percent of pump best efficiency values taken as 100%.

Because of the higher capacity throughput and higher differential pressure, the HPRT power output is greater than the power requirements of the pump. Subsequently, the higher shaft stresses result in speed limitation unless the allowable stress or diameter of the shaft has been increased.

If a multi-stage pump is selected to operate as an HPRT, a heavier shaft can be installed, since most pump manufacturers build their multi-stage pump line with standard and heavy-duty shafts because of a larger number of stages, higher specific gravities, or high-speed applications. Single-case pumps are limited to the maximum pressure they can withstand. For high-pressure HPRT applications, either the allowable maximum working pressure of the pump case has to be increased or a double, or "barrel-type," case has to be selected. Barrel-type cases are also used for high-energy and low-specific-gravity-type HPRT's.

Another important check is evaluating the adequacy of the bearing design. Depending on specific speed, some pumps when operating as HPRT's at the same speed will have twice the differential pressure and

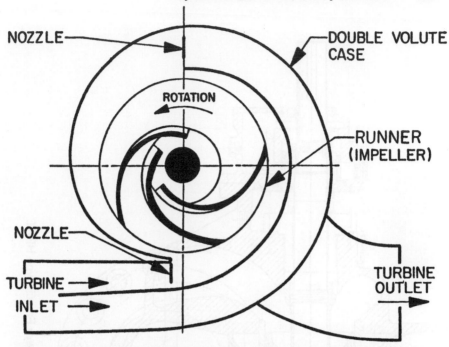

NOZZLE

DOUBLE VOLUTE
CASE

ROTATION

RUNNER
(IMPELLER)

NOZZLE

TURBINE
INLET

TURBINE
OUTLET

Figure 14-8. A typical double-volute-type pump used as a hydraulic turbine.

radial force across the impeller or runner. This results in increased radial loading of the bearing and could present a problem with end-suction pumps; especially, if the case is of the single-volute design. Stronger bearings will increase the bearing life; however, the resulting greater shaft deflection at the impeller wear rings and seal faces could decrease wear ring and seal life and increase the vibration level. A pump case in double-volute design (Figure 14-8), which results in radial balance, will solve these problems. Besides an increase in radial load there will be a change in axial thrust. If required, a change in wear-ring or balancing-device diameter will reduce the axial thrust to an acceptable level.

In general, pumps are built to customer specifications such as API 610. Therefore it is natural that the same specifications will apply to pumps operating as HPRT's. These customer requirements cover: running clearances, limitations for horizontal split-case pumps in relationship to operating temperature and specific gravity, nondestructive examination (NDE), shaft sealing and seal flushing, bearing life, lubrication and cooling system, baseplate design, material selection for corrosion and erosion protection as well as nozzle loading, and noise and vibration level to name a few. With all these considerations in mind, a well-designed pump will operate smoothly, quietly, and reliably as an HPRT.

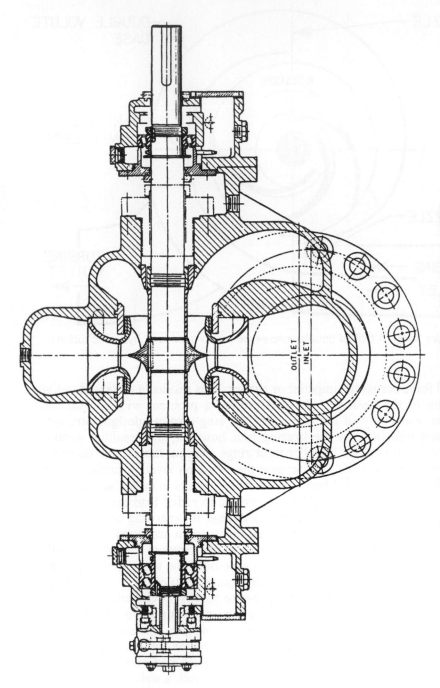

Figure 14-9. A single-stage double-eye Francis-type HPRT.

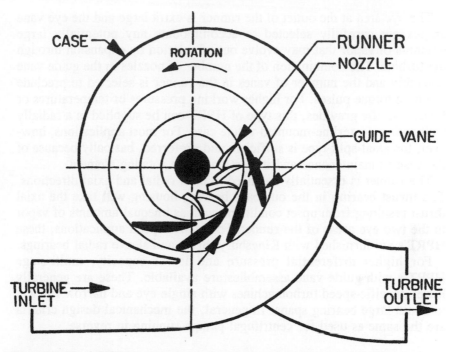

Figure 14-10. HPRT construction using single-volute-type case, turbine runner, and guide vanes.

Turbine Design with Fixed Guide Vanes

Most pump manufacturers offer, in addition, different lines of turbo-machines designed specifically to be applied as HPRT's. They are basically developed to cover a range of performance generally not available with a reverse-running pump. One of these HPRT's is shown in Figure 14-9.

This HPRT features an axial-split single-volute-type scroll case, a single-stage double-eye Francis-type runner, and a removable guide vane component (Figure 14-10), that controls the inlet flow to the runner.

This type of HPRT has a fixed performance characteristic. However, the design can accommodate seasonal or "plant turn-down" flow-capacity conditions by changing to a different guide vane assembly or by using an adjustable guide vane assembly for optimum performance. The runners used in this type of HPRT are quite different from the conventional pump impeller in that it has a greater number of vanes and generally larger vane angles.

The eye area at the outlet of the runner is extra large and the eye vane angles are carefully selected to accommodate any potentially large amounts of vapor that may evolve out of solution by expansion through the turbine. The combination of the number of nozzles in the guide vane assembly and the number of vanes in the runner is selected to preclude in-phase torque pulses. For higher working pressures or temperatures or lower specific gravities, this type of HPRT can be supplied in a radially split and/or centerline-mounted volute case. For most applications, however, the axial-split-type is sufficient and preferred, basically because of the ease of maintenance and inspection of the rotating element.

The runner is essentially balanced in both radial and axial directions. The thrust bearing in the outboard bearing housing will take the axial thrust resulting from upset conditions such as unequal amounts of vapor in the two eye areas of the runner. For higher-speed applications, these HPRT's are furnished with Kingsbury thrust and sleeve radial bearings.

For higher differential pressure and lower capacity, multi-stage HPRT's with guide vane assemblies are available. These are generally lower-specific-speed turbomachines with single eye and narrow runners to avoid large bearing spans. In general, the mechanical design criteria are the same as used for centrifugal pumps running in reverse.

Turbine Design with Internally and Externally Adjustable Guide Vanes

Specially designed HPRT's include the feature of an adjustable guide vane assembly, which can be furnished for a single-stage or multi-stage HPRT. The method of adjusting the guide vane assembly is made possible by an internal or external design feature. The advantage of this variable vane assembly is the capability of operating more efficiently over an extended flow range compared to an HPRT or a reverse running pump with fixed inlet guide vanes.

The performance characteristics of an HPRT can be varied over a considerable range by changes to the velocity of the liquid passing through the guide vane assembly. For optimum performance, this is best accomplished by changes to the flow-cross-section area formed by the vanes and the side walls of the assembly when aligned at a proper angle to the runner. A decrease in the flow-cross-section area will generally shift the optimum efficiency to a lower flow range.

The typical performance characteristic curve for an HPRT with fixed guide vanes is illustrated in Figure 14-11. The hydraulic turbine is essentially like an orifice in a fixed-pressure-differential system. The operating point will be where the particular head, capacity, speed, and power relationship is satisfied.

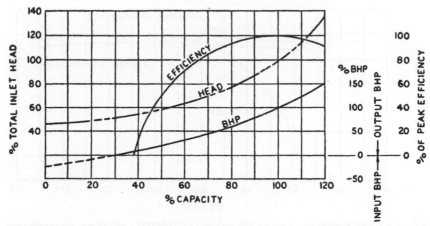

Figure 14-11. Typical performance characteristics for an HPRT with fixed guide vanes.

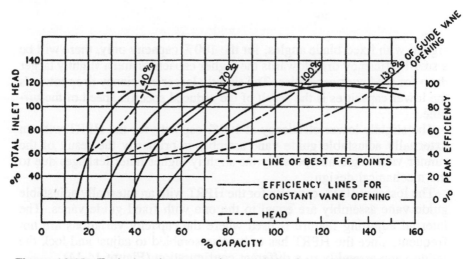

Figure 14-12. Typical performance characteristics for an HPRT with an adjustable guide vane assembly.

The typical performance characteristic curves for an HPRT with an adjustable guide vane assembly are shown in Figure 14-12. The adjustable guide vane assembly performs, as one can see, as a variable orifice in a large number of fixed-pressure-differential systems. The vane setting will control the flow at a certain differential pressure.

In Figure 14-13, the best efficiency points of each vane setting or opening are connected to a single line and compared with the efficiency curve of a hydraulic turbine with fixed inlet vanes. Because the runner is de-

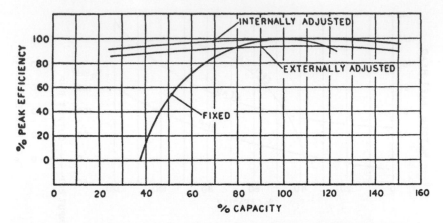

Figure 14-13. Efficiency comparison between HPRT's with fixed, internally adjustable guide vanes and those with externally adjustable guide vanes.

signed with fixed blade angles, for the 100% capacity only, there will be a small efficiency drop towards the higher capacities and a slightly larger drop for the lower capacities. The slightly larger efficiency drop towards the lower capacities is the result of the reduced specific speed of the turbine.

The slightly lower efficiency at 100% capacity for the HPRT with an externally adjustable guide vane asembly is primarily the result of the runner vane angle and possibly the slightly higher inner leakages due to the mechanical design.

The losses at 100% capacity for the HPRT with an internally adjustable guide vane assembly are equal to the one with fixed guide vanes. The internal adjusting feature is used where the capacity variations are not frequent, since the HPRT has to be disassembled to adjust and lock the guide vane assembly to a different configuration (Figure 14-14).

The external adjusting feature makes it possible to vary the guide vane setting during operation of the HPRT. According to the available capacity, a level controller or capacity indicator sends an air or electric signal to the turbine-actuator, which in turn changes the setting of the guide vane assembly until the flow through the openings equals the available capacity and the signal stops. Continuous capacity changes will result in continuous resetting of the guide vane assembly, thus making it possible to operate the HPRT always at its BEP (best efficiency point), if the differential pressure across the turbine remains about constant.

The externally adjustable guide vane assembly features are illustrated in Figure 14-15. The design incorporates the conventional principle of

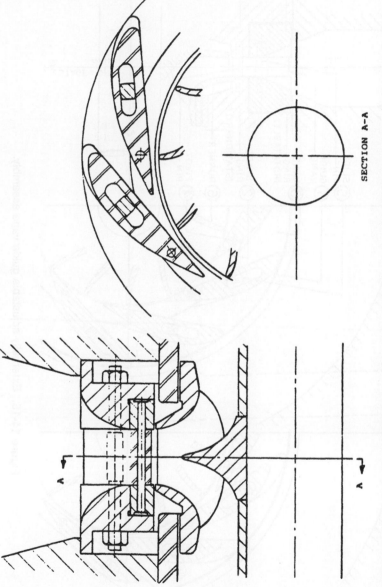

SECTION A–A

Figure 14-14. Internally adjustable guide vane assembly (courtesy of Bingham- Willamette Company).

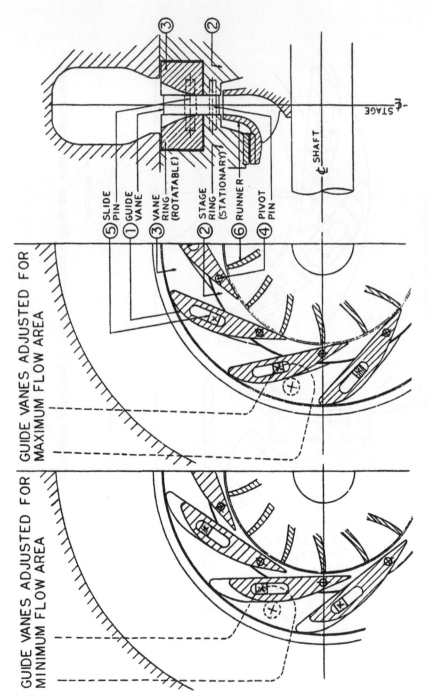

Figure 14-15. Externally adjustable guide vane assembly.

tilting the guide vanes (Part #1) about a pivot pin (Part #4) parallel to the runner (Part #6) shaft axis to vary the velocity of the liquid flowing through the assembly at a proper flow orientation angle relative to the runner. Each guide vane is held in position by the pivot pin and by a slide pin (Part #5), which moves the guide vane by its position in the slot through the vane.

The pivot pins are located in the two stationary vane rings or stage pieces (Part #2) and the slide pins are assembled to the two rotatable vane rings (Part #3). The stationary and rotatable rings establish the width of the inlet opening. They are the side walls of the vane assembly. The operating position of the vanes and the resultant through-flow cross-section area is dependent on the angular position of the rotatable vane ring in relation to the stationary rings.

Between the guide vanes, the rotatable vane rings are shaped in a manner to achieve the correct velocity increase for each through-flow cross-section area.

The rotatable vane rings perform the additional function of avoiding undesirable vane flutter, by a clamping action due to developed differential pressure. A reduction in pressure occurs in the flow passages due to the increase in velocity of the fluid, while the pressure acting on the outward side areas of the rotatable rings is essentially the same as at the entrance to the vane passages.

Because of the relatively large outward side areas of the rotatable rings, the clamping force is higher than the different hydraulic forces that act on the guide vanes and could cause vane flutter. However, the force is not restricting the adjustment of the guide vane position during operation.

The cross-section of this HPRT is shown in Figure 14-16. The turbine is built basically like a multi-stage pump with standard bearing housings. The runner eyes of each stage face all in the same direction and a drum takes care of balancing the axial thrust.

Figure 14-17 shows the crossunder in the bottom half. There are no crossovers. The top half contains the yoke assembly, which moves up and down and creates the rotational position of the rotatable rings and subsequently the resultant through-flow cross-section area of the guide vane openings.

A crossbeam as shown in Figure 14-18 connects the yokes for synchronous travel. Individual setting of through-flow areas for each stage is possible by adjusting the nuts on the crossbeam. If required, through-flow areas for each stage can be adjusted differently to allow for an increase in the specific volume for compressible liquids, when the pressure reduces from stage to stage. This is another feature to achieve optimum performance. The up-and-down movement of the beam can be achieved by an electric or pneumatic actuator that is mounted on top of the beam cover.

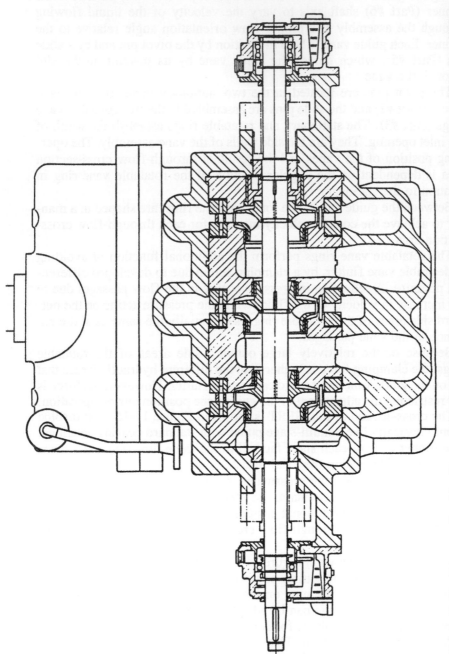

Figure 14-16. Cross section of a three-stage HPRT with externally adjustable guide vanes (courtesy of Bingham-Willamette Company).

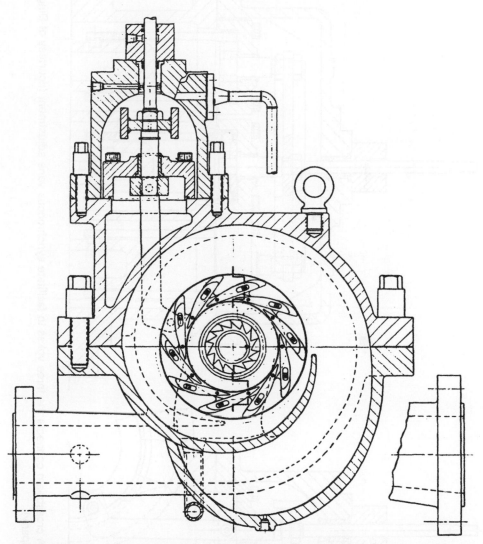

Figure 14-17. Cross section of a three-stage HPRT illustrating the guide vane adjusting mechanism (courtesy of Bingham-Willamette Company).

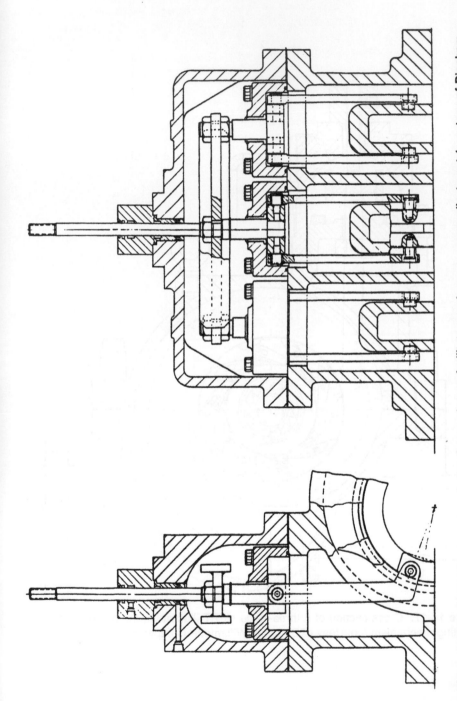

Figure 14-18. A crossbeam connects the three yokes to facilitate synchronous vane adjustment (courtesy of Bingham-Willamette Company).

Operating Considerations

The product handled by a hydraulic turbine may be a single-phase liquid, a multiphase liquid-gas mixture, or a slurry composition.

Hydraulic turbines have been extensively used for two-phase, liquid-gas flow streams where there is a potential for a substantial amount of gas released as the product passes through the turbine. There may also be small amounts of "free" gas at the turbine inlet. With a decrease in pressure, gas is subject to be released from the liquid with a resultant increase in volumetric flow. The effects of the potential vaporization at the various turbine stage pressures is evaluated to assure proper turbine performance. Generally, this may be accomplished by limiting the two-phase flow velocities at the runner (impeller) outlet eye to a reasonable value. It is also appropriate to give consideration to the runner (impeller) design to assure proper vane angles and eye sizes to accommodate any potential vapor release from the fluid stream. Actual field experience known to the author has shown that calculated two-phase flow velocities at the turbine outlet runner eye up to 150 ft/sec can be accommodated with no adverse effects. This velocity is suggested as a guideline for HPRT's whether they be single- or multi-stage types. Using this limit, the two-phase flow rate by volume can be at least three or four times the single-phase flow rate for many applications.

Theoretically higher output horsepower should be achieved by gas expansion through the turbine since the increase in volume means more work done. However, many reports have indicated that the expected additional power has not been realized. One explanation may be that the product passes through the turbine too fast for vapor-equilibrium to be obtained. For example, consider the time it takes for the carbon dioxide to escape from a bottle of carbonated beverage when the cap is removed. It does not all escape instantly. Another reason may be the fact that as the gas expands, the product velocity increases, and causes additional losses to occur.

For multi-stage hydraulic turbines, the nozzles may be sized differently from stage to stage to accommodate any theoretical increase of the volumetric flow as the pressure is reduced.

Performance Testing

Performance tests for hydraulic turbines may be accomplished by use of a centrifugal pump to furnish the head and flow capacity necessary to drive the turbine and to verify the turbine performance throughout its operating range. An induction motor excited by AC power from the utility system is used as an induction generator to absorb the output from the hydraulic turbine when gain over synchronous speed is achieved and to drive the hydraulic turbine at low flow capacities where input HP is required.

The output and input HP is determined by use of a wattmeter with efficiency curves for the induction machine. A torque meter may also be used to measure the power. A venturi meter or an orifice is used to measure the flow capacity. Dead weight testers and Bourdon-type gauges are used to measure the head. The RPM change from synchronous speed is counted by use of a strobotac and stop watch.

The turbine test may be performed at a reasonable reduced speed to facilitate testing. The performance at normal speed is then determined by applying the "affinity laws". The availability of a drive pump with sufficient head and flow capacity is a determining factor for the test speed. Cavitation tests are needed for hydraulic turbine performance characteristics. This is best determined by reducing the turbine outlet pressure and observing any resulting changes in the total dynamic head of the turbine, the power, the capacity, or efficiency. Measurements of noise, pulsation, and vibration accompanying the operation of the turbine during the cavitation test should be recorded.

Applications

Any continuous process where high pressure liquid or partially gas-saturated media is let down to a lower pressure across a reducing device is a potential application for an HPRT. Such potentials are:

- In pipeline service on the downside of high mountain ranges to keep the pipeline full and avoid excessive pressures.
- In bleeding products from a high-pressure point in the pipeline to storage.
- In geopressured-geothermal zones where high-temperature water is at a very high pressure. The formation pressure may exceed 10,000 psig, while pressure at the surface may approximate 2,000 to 6,00 psig depending on flow rates.

The early HPRT applications were basically in the noncorrosive and nonerosive service. Modern plants use HPTR's nowadays in mildly severe services such as:

- In hydrocracking operations, where boosting pressure of charge stocks to the 1,500–2,000 psi operating pressures used in modern hydrocracking processes requires large quantities of energy. Effluent from the reactor is still at high pressures, however, so that much of this energy can be recovered if an HPRT is incorporated in the drive train.

- In the gas processing industry where crude gas is scrubbed by a high pressure fluid medium such as potassium carbonate or amine in order to remove unwanted components. For the purpose of regeneration and recycling, the pressure has to be reduced; in other words, possible energy recovery has been made available.

The pressure can be reduced by using pressure breakdown valves; however, the differential pressure will be converted into thermal energy, which is either wasted or very uneconomical to recover. A relatively efficient method for pressure reduction and energy recovery is by the use of HPRT's.

HPRT's will convert the differential pressure into rotational energy, which can be utilized in helping to drive the centrifugal pump that returns the regenerated medium to the absorber. Both major types, namely, the reaction and impulse types, are used in the gas processing industry. Figure 14-19 shows the operational system using a reverse-running pump with fixed guide vanes. Since in a recycle system the recovered energy is smaller than the required energy to drive the pump, an electric motor or steam turbine on the other side of the pump is used to cover the energy difference and to maintain as a second function a constant RPM of the entire train.

The desired flow can be obtained either by changing speed of the assembly (steam turbine drive) or by throttling the pump output (motor drive), which means loss of energy. Unfortunately, the operating behavior of the standard reverse-running pump with fixed guide vanes requires a controllable throttling inlet valve for reduced capacity and a bypass line for increased capacity. Both represent additional energy losses (see Curve "A" and "B" in Figure 14-20).

Figure 14-21 illustrates the system using an HPRT with variable guide vanes.

The losses of the system in Figure 14-19 are avoided. The function of the inlet throttling valve (reduced capacity) and the bypass (increased capacity) are served by the variable inlet guide vanes installed in the HPRT, which satisfy the following purposes:

- Regulation of the capacity by varying the cross-section area of the guide vanes depending on the level in the absorber.
- Feeding the medium to the runner in a definite direction.
- Complete or partial conversion of the differential pressure into kinetic energy.

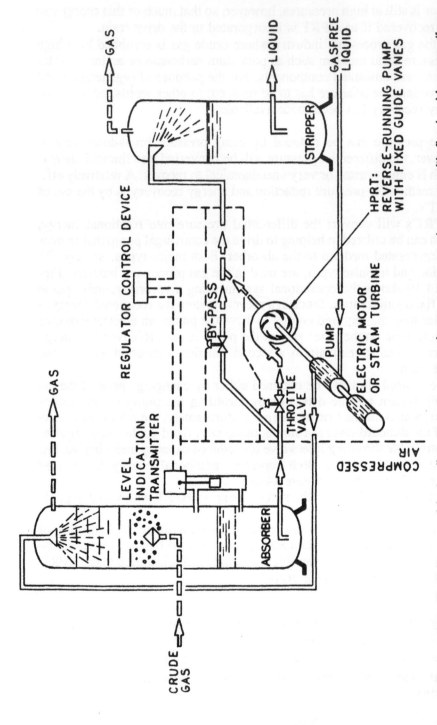

Figure 14-19. Flow diagram of hydraulic energy recovery using a reverse-running pump with fixed guide vanes (from Franzke).

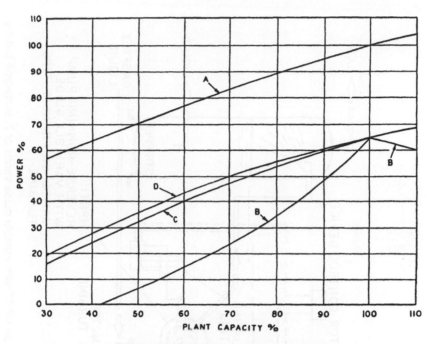

Figure 14-20. Power vs. plant capacity.

● Tripping the HPRT in the event of failure of compressed air, oil pressure, or power.

As a result of the improved supply of medium, the HPRT will still generate energy when the plant capacity drops below 40%, whereas the reverse running pump with fixed guide vanes will start to consume energy. (See Curve "A" and "C" in Figure 14-20).

Installation of a Pelton-type HPRT is shown in Figure 14-22. The pump-driver train arrangement is identical to the ones with the reaction-type HPRT's. However, a horizontal-level controlled tank has to be placed immediately below the turbine outlet to avoid paddling of the wheel in fluid flowing from the wheel. Paddling of the wheel in the medium results in high energy losses and strain being placed on the turbine. The efficiency of a properly installed Pelton-type HPRT is slightly higher at lower flows than the one of the HPRT with adjustable guide vanes, since the effect of the lower specific speed is not as significant (see Curve "A" and "D" in Figure 14-20).

HPRT's find many more applications such as services in hydropower stations and cooling towers and as reversible machines for pumped stor-

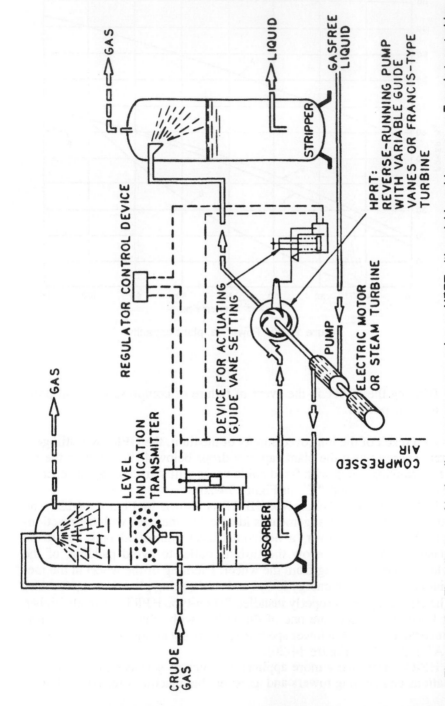

Figure 14-21. Flow diagram of hydraulic energy recovery using an HPRT with variable guide vanes or Francis-type turbine (from Franzke).

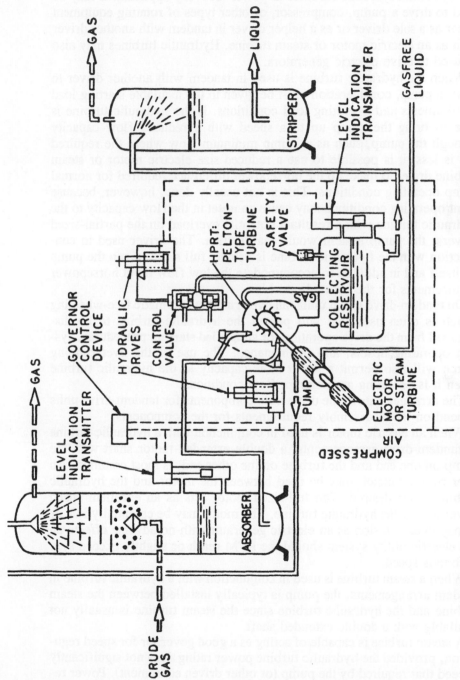

Figure 14-22. Flow diagram of hydraulic energy recovery using a Pelton-type HPRT (from Franzke).

age systems. Hydraulic turbines in power recovery applications may be used to drive a pump, compressor, or other types of rotating equipment either as a sole driver or as a helper driver in tandem with another driver such as an electric motor or steam turbine. Hydraulic turbines may also be used to drive electric generators.

When the hydraulic turbine is used in tandem with another driver to drive a pump, consideration must be given to the available starting load requirements and operating load conditions. If the hydraulic turbine is able to bring the pump up to a speed with a reduced flow capacity through the pump, such as at pump minimum flow where the required HP is less, it is possible to use a reduced size electric motor or steam turbine driver to make up the horsepower difference required for normal pump operating conditions. This is not usually done, however, because plant operating conditions may cause an upset in the flow capacity to the hydraulic turbine with a resultant potential overload on the partial-sized drivers; the pump system would malfunction. The driver used in conjunction with the hydraulic turbine is usually full sized to run the pump by itself and in addition to accommodate the low flow input horsepower requirements for the hydraulic turbine.

On tandem-drive pump units, an over-running automatic free-wheeling clutch is often used that will permit the hydraulic turbine to be disengaged from the drive operation for simplified start-up procedures, system operating upsets, and maintenance. The use of the over-running clutch will also permit a lower flow capacity to the hydraulic turbine when it is operating at minimum flow conditions.

The arrangements of the drive train components for tandem-drive units depend on the disassembly requirements for the components.

When an electric motor is used in conjunction with a hydraulic turbine in tandem-drive arrangements, a double-extended motor shaft with the pump on one end and the turbine on the other end, is most common. An over-running clutch may be used between the motor and the hydraulic turbine when desired. The full-sized motor acts as an excellent speed governor for the hydraulic turbine. The motor may be essentially idle or it may even function as an electric generator with no adverse effects on the electric utility system should the RPM reach or slightly exceed synchronous speed.

When a steam turbine is used in conjunction with a hydraulic turbine in tandem arrangements, the pump is typically installed between the steam turbine and the hydraulic turbine since the steam turbine is usually not available with a double extended shaft.

A steam turbine is capable of acting as a good governor for speed regulation, provided the hydraulic turbine power rating does not significantly exceed that required by the pump (or other driven equipment). Power recovery may be realized by using a hydraulic turbine to drive an electric generator of either the synchronous or induction type. For the smaller systems, the induction-type generator is attractive for economic considerations.

A squirrel-cage induction machine becomes an excellent power generator when it is excited by AC power while the shaft is rotated above synchronous speed. Frequency of the generated power is that of the excitation; shaft speed determines only the amount of power consumed or delivered. If the shaft is rotated much faster than synchronous speed, the machine can burn out. But the system tends to be self-regulating because the shaft becomes increasingly harder to rotate as speed increases above synchronous.

When the induction machine is excited by AC power from a utility system, power is fed back into the power grid as the speed reaches and surpasses rated synchronous speed. The power grid provides the excitation voltage needed by the induction machine for both motor and generating action.

When an induction generator must work without a source of AC power, excitation can be supplied by residual magnetism and capacitors connected phase to phase. A storage battery can be used to provide a current pulse through one of the windings and thus leave sufficient flux to start generation.

Generation occurs when the capacitor current exceeds the excitation current of the windings. Generation stops when the shaft speed is lowered to the point where capacitive reactance exceeds that of the winding or when the load absorbs too large a portion of the capacitor current.

Operation and Control Equipment

As the flow through the HPRT increases from the no-flow condition, the fluid velocity through the runner gradually imparts to the runner not only enough energy to overcome internal friction but also to permit some net power output. This point usually occurs at about 40% of design flow or capacity. As in any turbine driver, the machine will speed up until the load imposed on the shaft coupling equals the power developed by the turbine. The hydraulic turbine must operate to satisfy its own head-capacity-speed-horsepower relationship within the available head and imposed speed limits.

Consider a power recovery turbine operating as the only driver. If more liquid is allowed to flow to the power recovery turbine than is needed to produce the horsepower required, the turbine will speed up and try to handle the liquid; at the same time the driven pump or compressor will speed up. In speeding up, the turbine will produce more shaft horse-power, which the driven pump or compressor must absorb at the new speed. Finally, the horsepower will be balanced, but the speed of the driven unit may be off design.

If speed control is necessary, throttling some of the turbine's driving fluid across a valve bypassing the turbine allows it to satisfy the horse-power-capacity-speed requirements of the driven unit. If the amount of fluid available to the turbine is less than that needed for the design conditions of the driven unit, the turbine will slow down and try to shed some of the load. Here speed control can be achieved by throttling the available pressure so that the turbine sees only that portion of the available head needed to satisfy its head-capacity-speed relationship at the desired speed.

When a power recovery turbine is combined with a makeup driver, except at a single point, the recovery turbine always requires either flow bypassing or inlet pressure throttling. The balance point is always determined by the power-speed characteristics of the driven unit. If the driven unit can use all the generated horsepower, such as a floating electric generator would, capacity control and pressure throttling may not be needed. When a speed-controlling, variable-horsepower helper driver such as an electric motor or steam turbine is used it will hold the speed constant and make up just enough horsepower to permit the power recovery pump turbine to satisfy its head-capacity curve at virtually any flow rate.

Split-range liquid-level controllers are typically used to regulate the available flow to HPRT's. The split-range liquid-level controllers and pressure-control valves are usually furnished and installed by the purchaser. The pressure-control valve is usually located at the inlet side of the hydraulic turbine to prevent an excess pressure condition from occurring at the turbine shaft seals by a closed valve. Also, the low shaft sealing pressure usually results in a lower initial cost and reduced maintenance. The signal from the controller is used to adjust the pressure-control valve when too much head is available for the capacity and speed and to bypass excess capacity from the system when more liquid is available to the HPRT than needed to satisfy the relationship. When an HPRT is provided with adjustable guide vane nozzles for performance variation, a proportional range controller will provide the operator signal to appropriately adjust the guide vane setting for optimum conditions.

An over-speed trip device is often furnished with the hydraulic turbine. This device is typically used to provide a signal to operate an over-speed alarm or to close the pressure-control valve for minimum flow turbine operation. The sensing device may be a pneumatic or electronic transmitter or a mechanical trip mechanism installed to sense the turbine shaft speed.

If the hydraulic turbine should operate at runaway conditions (zero torque) due to no load, the turbine shaft speed will generally increase to within the range of 120% to 155% of the normal design speed with 100% normal design head. The overspeed amount depends on the specific speed characteristics of the machine. Should an upset condition occur where there is a large amount of vapor present with a loss of liquid level and with full differential pressure across the turbine, a very high runaway speed could occur. This is due to the low-density vapor producing a high differential head and a high-volume flow.

HPRT's should be brought up to full operating speed as rapidly as possible, because they not only fail to generate power but actually consume power until they attain about 40% of the design capacity.

The installation of the previously mentioned over-running automatic free-wheeling clutch between turbine and the driven pump or compressor is a good solution. The to-be-driven machine does not have to turn until fluid is available to the HPRT, which is not connected to the to-be-driven unit until it tries to run faster and puts out power. Using this arrangement, the start-up sequence can be selected so that the HPRT goes from zero speed to full operating speed along the zero torque curve.

Conclusion

In view of the significant power savings possible by use of power recovery turbines, energy users should take advantage of every opportunity to investigate the economics involved. Justification is based on the value of the energy saved during a projected life of the turbine versus the projected cost of purchasing, installing, and maintaining the machine for the same period of time.

The effects of changes to the operating conditions, such as available flow capacities and differential pressures for the HPRT's and driven machines need to be considered. Since the most commonly used turbine types have fixed performances, changes to the operating conditions may cause a significant change to the power output from the turbine unless modifications to the turbine internal nozzle sizes are made. HPRT's with internally or externally adjustable guide vane assemblies are desirable when changes to performance characteristics are expected.

Another consideration for selecting a hydraulic turbine as a driver in place of an electric motor or steam turbine is the fact that the hydraulic turbine does not have the incremental costs in energy. Experience with HPRT's in actual operating installations shows that these machines are very reliable, they perform the design requirements, and the operating costs are minimal. The hydraulic and mechanical performances are readily predictable.

Current inquiries show that there is a significant potential for hydraulic turbines during the eighties and surely beyond. Indications are that sizes much larger than those currently in use will be needed. Also, the types in demand will include the adjustable guide vane nozzles, diagonal flow types, vertical types, and possibly the combination turbomachine with the turbine and pump unit in the same case.

References

Evans, Jr. F. L., *Equipment Design Handbook for Refineries and Chemical Plants,* 2nd edition, Gulf Publishing Company, 1979.

Jennet, E., "Hydraulic Power Recovery Systems" *Chemical Engineering,* April 8, 1968.

Lueneburg, R., Bingham-Willamette Ltd., HPRT Session, Pacific Energy Association Annual Fall Meeting, October 1983.

McClaskey, B. M., and Lundquist, J. A., "Hydraulic Power Recovery Turbines" ASME Publication 76-Pet-65.

Nelson, R. M., Bingham-Willamette Co., "Introduction to Hydraulic Turbines" *BWC* (April 1981).

Purcell, J. M., and Beard, M. W., "Applying Hydraulic Turbines to Hydrocracking Operations," *The Oil and Gas Journal* (November 20, 1967).

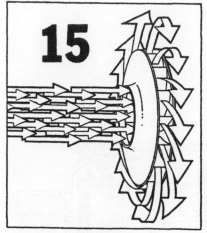

15

Chemical Pumps Metallic and Nonmetallic

by **Frederic W. Buse**
Ingersoll-Rand Company

Chemical pumps are designed for many processes and products that are not normally handled by pumps designed for a single product or process such as general water pumps, boiler feed pumps, cooling water pumps, or petroleum industry pumps. The chemical processes vary from acids, alkalies, toxics, reducing agents, oxides, slurries, organics, or inorganics causing corrosion, erosion, galvanic action, or leaching to occur on the pumps and piping system and any other product in the process.

To handle this variety of conditions, the pumps employ various materials such as 316 stainless steel, ductile iron, alloy 20, titanium, Hastelloy B and C. The continuous development of nonmetallics also make these pumps available in vinyl esters, epoxies, PVC, or with linings of teflon. Some pumps employ carbon, ceramic, and glass bodies or linings.

ANSI Pumps

Specifications

Most chemical pumps in the United States in the past 25 years have been developed according to the ANSI B73.1M and .2M specifications for horizontal and vertical pumps respectively (Figures 15-1 and 15-2). These specifications were initially developed in 1955 and were published in 1962. The current specifications were published in 1991. Besides safety criteria, the main objective for these specifications was to establish dimensional standards and interchangeability of various size pumps

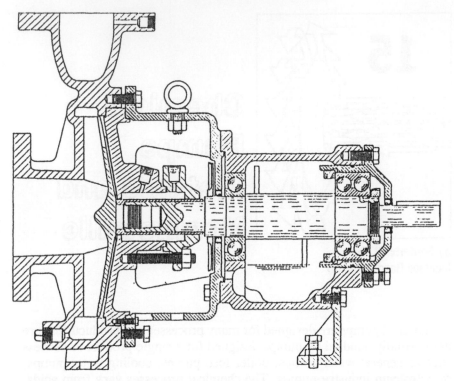

Figure 15-1. ANSI overhung single-stage pump (courtesy of Ingersoll-Rand Company).

within a given envelope (Figure 15-3A). The ANSI B73.1M specification has a dimensional designation of AA to A120 that covers 19 various size pumps. Its dimensional standards not only cover the pump itself, but also cover the pumps on bedplates. ANSI B73.2M has a dimensional designation of 2015 to 6040 that covers 15 various size pumps (Figure 15-3B). This dimensional standard became an important criterion for chemical plant designers because they could rely on the pump envelopes for dimensional accuracy when laying out the piping and foundations for the pumps. This eliminated the need for certified drawings of the pump assembly or pump bedplate combination from the pump suppliers. It also eliminated the need for extra inventory for spare parts because spare pumps could be purchased from various pump manufacturers with the assurance that they would fit into an existing piping system.

The hydraulic range of these pumps at a synchronous speed of 3600 RPM is 2000 gallons per minute and over 800 feet. At 1800 synchronous speed, the range is from 3500 gallons per minute to 250 feet (Figures 15-4 and 15-5).

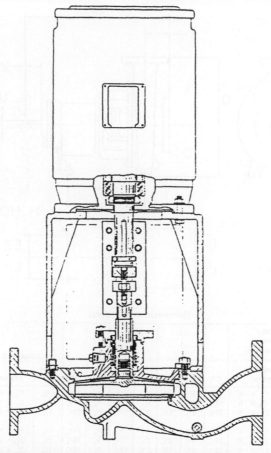

Figure 15-2. ANSI vertical in-line overhung single-stage pump with rigid coupling (courtesy of Ingersoll-Rand Company).

The specifications also stipulate that the pumps have centerline discharge casings and should be pulled from the rear rotor design that allows disassembly without disconnecting the suction or discharge nozzles. To maximize mechanical seal life, the specifications require a .005-inch shaft deflection limit at the impeller centerline due to dynamic deflection and a maximum full indicator run out at the stuffing box face of .002 inches.

The specifications also require that there be a minimum bearing life of 17,500 hours due to the defined maximum imposed hydraulic loads and that the suction and discharge flange pressure-temperature limits comply to a minimum of ANSI B16.5 Class 150 (Figures 15-6 and 15-7).

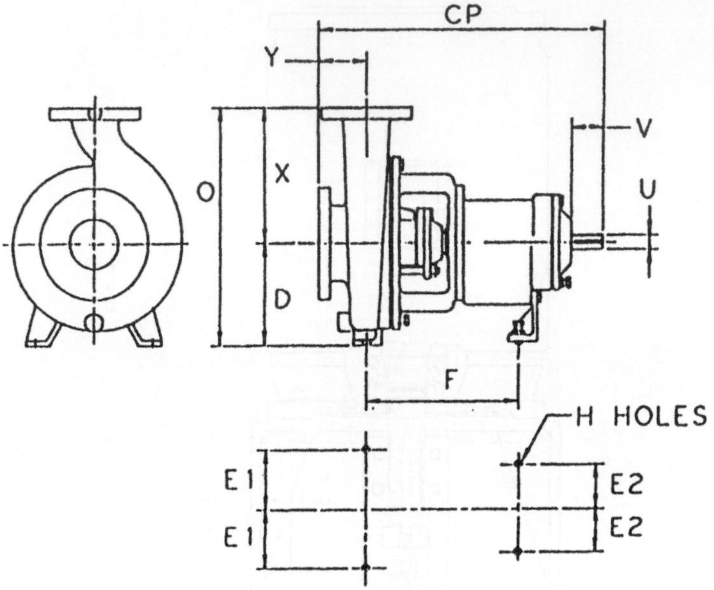

(Dimensions in Inches)

Dimension Designation	Size, Suction × Discharge × Nominal Impeller Diameter	CP	D	2E1	2E2	F	H	O	U [Note (1)] Diameter	U [Note (1)] Keyway	V Minimum	X	Y
AA	1 1/2 × 1 × 6	17 1/2	5 1/4	6	0	7 1/4	5/8	11 3/4	7/8	3/16 × 3/32	2	6 1/2	4
AB	3 × 1 1/2 × 6	17 1/2	5 1/4	6	0	7 1/4	5/8	11 3/4	7/8	3/16 × 3/32	2	6 1/2	4
A10	3 × 2 × 6	23 1/2	8 1/4	9 3/4	7 1/4	12 1/2	5/8	16 1/2	1 1/8	1/4 × 1/8	2 5/8	8 1/4	4
AA	1 1/2 × 1 × 8	17 1/2	5 1/4	6	0	7 1/4	5/8	11 3/4	7/8	3/16 × 3/32	2	6 1/2	4
A50	3 × 1 1/2 × 8	23 1/2	8 1/4	9 3/4	7 1/4	12 1/2	5/8	16 3/4	1 1/8	1/4 × 1/8	2 5/8	8 1/2	4
A60	3 × 2 × 8	23 1/2	8 1/4	9 3/4	7 1/4	12 1/2	5/8	17 3/4	1 1/8	1/4 × 1/8	2 5/8	9 1/2	4
A70	4 × 3 × 8	23 1/2	8 1/4	9 3/4	7 1/4	12 1/2	5/8	19 1/4	1 1/8	1/4 × 1/8	2 5/8	11	4
A05	2 × 1 × 10	23 1/2	8 1/4	9 3/4	7 1/4	12 1/2	5/8	16 3/4	1 1/8	1/4 × 1/8	2 5/8	8 1/2	4
A50	3 × 1 1/2 × 10	23 1/2	8 1/4	9 3/4	7 1/4	12 1/2	5/8	16 3/4	1 1/8	1/4 × 1/8	2 5/8	8 1/2	4
A60	3 × 2 × 10	23 1/2	8 1/4	9 3/4	7 1/4	12 1/2	5/8	17 3/4	1 1/8	1/4 × 1/8	2 5/8	9 1/2	4
A70	4 × 3 × 10	23 1/2	8 1/4	9 3/4	7 1/4	12 1/2	5/8	19 1/4	1 1/8	1/4 × 1/8	2 5/8	11	4
A80	6 × 4 × 10	23 1/2	10	9 3/4	7 1/4	12 1/2	5/8	23 1/2	1 1/8	1/4 × 1/8	2 5/8	13 1/2	4
A20	3 × 1 1/2 × 13	23 1/2	10	9 3/4	7 1/4	12 1/2	5/8	20 1/2	1 1/8	1/4 × 1/8	2 5/8	10 1/2	4
A30	3 × 2 × 13	23 1/2	10	9 3/4	7 1/4	12 1/2	5/8	21 1/2	1 1/8	1/4 × 1/8	2 5/8	11 1/2	4
A40	4 × 3 × 13	23 1/2	10	9 3/4	7 1/4	12 1/2	5/8	22 1/2	1 1/8	1/4 × 1/8	2 5/8	12 1/2	4
A80 (2)	6 × 4 × 13	23 1/2	10	9 3/4	7 1/4	12 1/2	5/8	23 1/2	1 1/8	1/4 × 1/8	2 5/8	13 1/2	4
A90 (2)	8 × 6 × 13	33 7/8	14 1/2	16	9	18 3/4	7/8	30 1/2	2 3/8	5/8 × 5/16	4	16	6
A100 (2)	10 × 8 × 13	33 7/8	14 1/2	16	9	18 3/4	7/8	32 1/2	2 3/8	5/8 × 5/16	4	18	6
A110 (2)	8 × 6 × 15	33 7/8	14 1/2	16	9	18 3/4	7/8	32 1/2	2 3/8	5/8 × 5/16	4	18	6
A120 (2)	10 × 8 × 15	33 7/8	14 1/2	16	9	18 3/4	7/8	33 1/2	2 3/8	5/8 × 5/16	4	19	6

NOTES:
(1) *U* may be 1 3/8 in. diameter in A05 through A80 sizes to accommodate high torque values.
(2) Suction connection may have tapped bolt holes.

Figure 15-3A. ANSI pump dimensions (from ASME B73.1M-1991 by permission of the American Society of Mechanical Engineers).

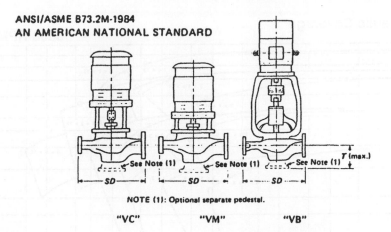

ANSI/ASME B73.2M-1984
AN AMERICAN NATIONAL STANDARD

"VC"   "VM"   "VB"

NOTE (1): Optional separate pedestal.

T (max.)

VERTICAL IN-LINE
CENTRIFUGAL PUMPS FOR CHEMICAL PROCESS

DIMENSIONS, in.

Standard Pump Designation[1]	ANSI 125, 150, 250, or 300 Flange Sizes		SD +0.10 −0.08	T (maxi-mum)
	Suction	Discharge		
VC, VB, VM				
2015/15	2	1½	14.96	
2015/17	2	1½	16.93	6.89
2015/19	2	1½	18.90	
3015/15	3	1½	14.96	
3015/19	3	1½	18.90	7.87
3015/24	3	1½	24.02	
3020/17	3	2	16.93	
3020/20	3	2.	20.08	7.87
3020/24	3	2	24.02	
4030/22	4	3	22.05	
4030/25	4	3	25.00	8.86
4030/28	4	3	27.95	
6040/24	6	4	24.02	
6040/28	6	4	27.95	9.84
6040/30	6	4	29.92	

NOTE:
(1) Pump Designation: defines design, flange sizes, and SD dimension [e.g., VC, VB 50-40-380).

Figure 15-3B. ANSI pump dimensions (from ASME B73.2M-1991 by permission of the American Society of Mechanical Engineers).

Hydraulic Coverage

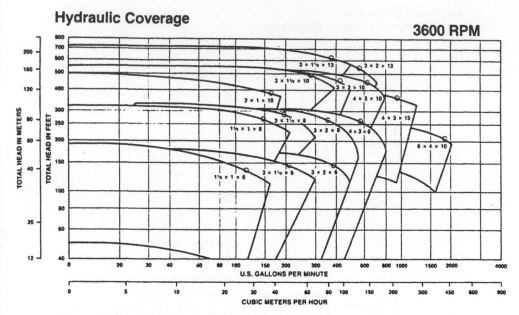

Figure 15-4. Typical 60 cycle-3600 rpm performance chart for ANSI B73.1 pumps (courtesy of Ingersoll-Rand Company).

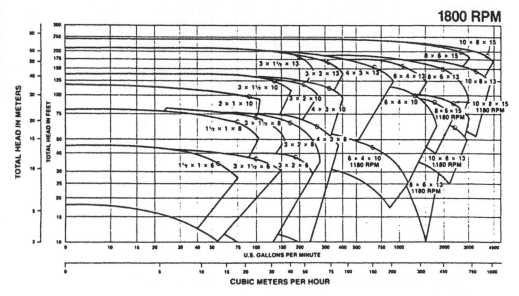

Figure 15-5. Typical 60 cycle-1800 rpm performance chart for ANSI B73.1 pumps (courtesy of Ingersoll-Rand Company).

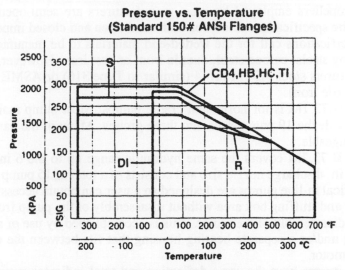

Figure 15-6. Pressure versus temperature for 150 pound ANSI flange.

Material Specifications	
DI	Ductile Iron
S	316 Stainless Steel
R	Alloy 20
CD4	CD4MCU
HB	Hastelloy B
HC	Hastelloy C
TI	Titanium

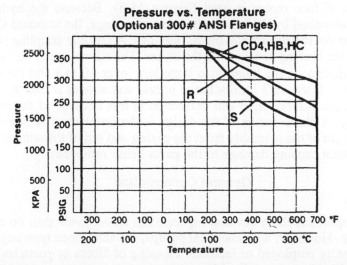

Figure 15-7. Pressure versus temperature for 300 pound ANSI flange.

The impellers employed by most manufacturers are semi-open even though the specifications allow for both semi-open and closed impellers. The specifications call for the wetted-end materials to be manufactured from alloy steels, carbon steel, ductile iron, or cast iron. However, most manufacturers stock ASTM A744 (similar to Type 316) or ASME A395 (cast ductile iron).

ANSI B.73.1M is for horizontal cradle pumps. Most pump manufacturers divide the 19 pump sizes into three groups. Many of the parts are interchangeable.

ANSI B.73.2M covers the same hydraulic range up to the 6 in. suction × 4 in. discharge nozzle size and consists of a total of 15 pump sizes. The vertical in-line pumps are designed so a user can obtain access to the impeller and stuffing box area without disassembly of the pump from the line nor disassembly of the motor. This is accomplished by use of a rigid coupling and/or a separate bearing housing that fits between the casing and the motor.

These pumps have the same deflection and total indicator runouts as the horizontal pumps. When using a rigid coupling, these pumps employ a P-face motor. This motor is called an in-line motor (NEMA MGI-18.620). It was developed in conjunction with NEMA and employs double-row or back-to-back deep groove thrust bearings to absorb the radial and axial thrust developed by the pump. The construction of this motor is such that for a radial load of 25 lbs. at the end of the motor shaft the radial deflection shall be no more than .001 in. with an axial load of 50 lbs., the shaft movement is limited to .0015 in.

When a separate bearing housing design is employed for the vertical pumps, a C-face motor is used (Figure 15-8). Because the hydraulic thrust is absorbed by the bearing housing's bearings, the standard C-face motors do not require special thrust bearings. A flexible coupling is used between the bearing housing and the motor. The advantage of the bearing housing design on the larger size impellers (over 10 in.) and the larger size motors (used on 3 in. discharge nozzle and above) is that the pump shaft system is more rigid and deflection is less because of the smaller overhang. A disadvantage of the design is the problem of removing the extra weight of the assembly from the casing and out of the support head area without causing damage to the parts being removed.

General Construction

Impeller

Semi-open impellers develop higher axial thrust loads than do closed impellers. However, with chemical pumps, the semi-open type impellers are normally employed to facilitate cleaning of fibers or particles often contained in the process liquid.

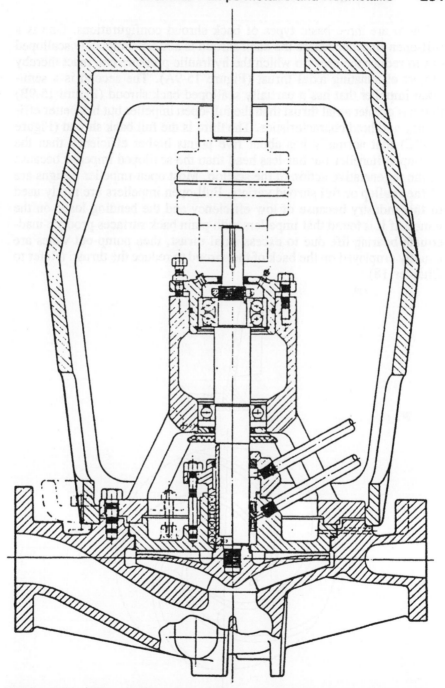

Figure 15-8. ANSI vertical in-line overhung single-stage pump with bearing housing (courtesy of Goulds Pumps, Inc.).

There are three basic types of back shroud configurations. One is a full-open impeller where the back shroud is almost completely scalloped out to reduce the area on which the hydraulic pressure can react thereby almost eliminating axial thrust (Figure 15-9A). The second is a semi-open impeller that has a partially scalloped back shroud (Figure 15-9B) that has greater axial thrust than the full-open impeller but has better efficiency and head characteristics. The third is the full back shroud (Figure 15-9C) that normally has about five points higher efficiency than the scalloped impeller but has less head than the scalloped impeller because of the regenerative action of the scallop. Most open-impeller designs are of the scallop or full shroud variety. Full-open impellers are rarely used in this industry because of low efficiency and the bending loads on the vanes. If it is found that impellers with plain back surfaces produce inadequate bearing life due to excess axial thrust, then pump-out vanes are usually employed on the back of the shroud to reduce the thrust. (Refer to Chapter 18).

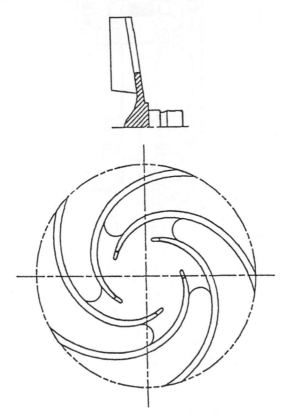

Figure 15-9A. Fully scalloped open impeller.

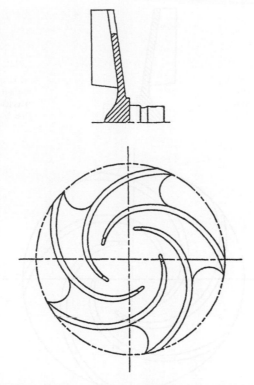

Figure 15-9B. Partially scalloped open impeller.

Casings

Casings for the ANSI chemical pumps have centerline discharge and suction both in the horizontal and vertical pumps. This makes it easier for laying out the piping in a system as well as reducing the nozzle loading. This is because the centerline nozzle eliminates the moment arm from the centerline of the casing to the centerline of the nozzle that exists with tangential discharge (Figures 15-10A and 15-10B).

On the horizontal casing, the centerline discharge results in a cutwater being approximately 30° off the centerline allowing the casing to be self venting. The casings are designed so that the rotor can be removed from the back without disturbing the suction and discharge piping. The gaskets of the casing are atmospheric confined so that the internal pressure cannot push the gasket out, as could occur with a full, flatface gasket. The flanges are 150 ASME flatface with an option of a raised face for steel and alloy material casings. There is the option of 300 lb. flanges for both the suction and the discharge on the steel and alloy casings. In the chemi-

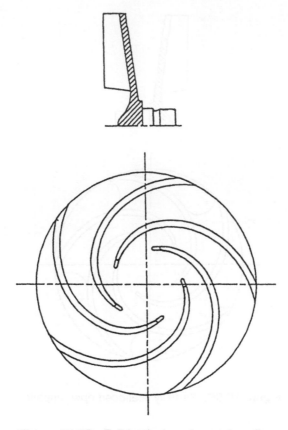

Figure 15-9C. Full back shroud open impeller.

cal industry, to prevent localized erosion-corrosion many customers do not want any holes in the casings; therefore, vents and drains are offered as options.

The running surface of the casing that is adjacent to the front face of a semi-open impeller is designed so that the clearance between the impeller and case ranges from 0.010 to 0.020 inches depending on the manufacturer, pump size, and material. When this surface is machined on an angle relative to the centerline of the pump, the surface has to be concentric with the centerline within .0010 inches. If quality control is not adhered to, the wear surface will have wider clearance on one side relative to the other resulting in inconsistent performance. The surface has to be machined on an angle of plus or minus 3 minutes of a degree to maintain performance. Instead of putting this surface on an angle, some manufacturers machine it so it is perpendicular to the centerline of the casing.

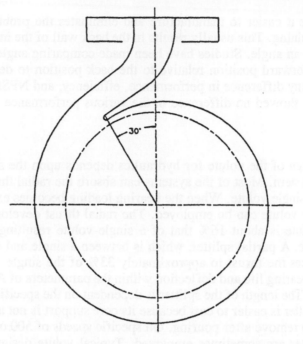

Figure 15-10A. Casing volute with centerline discharge nozzle.

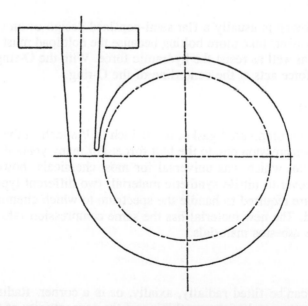

Figure 15-10B. Casing volute with tangential discharge nozzle.

This makes it easier to manufacture and eliminates the problem of the angle machining. This usually results in the back wall of the impeller being cast on an angle. Studies have been made comparing angles of 0° to 8° in the forward position relative to the back position to determine if there was any difference in performance, efficiency, and NPSH. The investigation showed no difference in the various performance criteria.

Volute

The design of the volute for hydraulics depends upon the stiffness of the shaft system. Most of the systems can absorb the radial thrust developed by a single volute. When the bearing loading becomes excessive, a full double volute can be employed. The radial thrust developed by the double volute is about 16% that of a single volute resulting in longer bearing life. A partial splitter, which is between a single and double volute, reduces the thrust to approximately 33% of the single volute and results in bearing life and deflection within the parameters of ANSI specifications. The length of the splitter is dependent on the specific speed. A partial splitter is easier to cast because its core support is not as long and is easier to remove after pouring. On specific speeds of 500 or less, circular volutes are sometimes employed. Typical volute designs and the method used to calculate radial load are described in Chapter 5.

Gasketing

The gasketing is usually a flat semi-confined design or an O-ring design. Flat gaskets take more bolting because the bolt load must compress the gasket as well as resist the hydraulic force. With the O-ring only the hydraulic force acts at the centerline of the O-ring.

Flat Gasket

The flat uncompressed gasket is $1/32$ inch to $1/16$ inch in thickness and has 27% compression due to the bolt force. For many years it was made from asbestos which was universal for most chemicals; however, with the changeover to nitrile synthetic material, two different types of base materials are required to handle the spectrum to which chemical pumps are applied. The new material has the same compression rate and hardness as the asbestos materials.

O-Rings

O-rings can be fitted radially, axially, or in a corner. Radial O-rings require more control of the machining of the concentricity of the casing and the casing cover relative to the axial O-ring. Because the axial O-ring has a greater sealing diameter, it requires additional bolting. Corner O-rings take more tolerancing but are a good compromise between radial and axial machining. The O-rings are made out of EP (ethylene propylene) for hot water, buna for hydrocarbons, viton for general chemicals, and Kalrez for highly corrosive chemicals. They are usually color coded to designate materials.

Casing Covers

Stuffing Box

The casing cover, sometimes called a stuffing box extension, encloses the back end of the casing. The casing cover also includes the stuffing box or seal chamber. Originally, the ANSI standards required that the minimum stuffing box packing size be $5/16$ in., $3/8$ in., or $7/16$ in. depending on pump size. The stuffing box was designed to handle both mechanical seals and packing; however, through years of experience, it was found that the mechanical seal's outside diameter had too small clearance between it and the bore of the box. This limited the amount of cooling that a seal could obtain, especially in double seals. So even though there was cooling injection into the gland and out of the box for double mechanical seals, there were frequent failures at the outboard seal. As a result, the specification includes optional large bores that only accommodate mechanical seals. This should give adequate cooling of the box to increase the life of the seal. If a customer requires a box for both seals and packing, he will require a box to the original specification. The taps into the stuffing box may be $1/4$ in. minimum, but $3/8$ in. is the preferred NPT size.

Depending on fluid temperature, the option of a cooled or heated stuffing box is usually offered. The type of mechanical seals offered on these pumps are single seals, double seals, and tandem seals. Mechanical seals used in ANSI pump applications are discussed in detail in Chapter 17.

Frame

The frame for the horizontal pump is composed of the support head and bearing housing. Depending on pump size, this can be one integral component or two separate pieces. Some manufacturers refer to the bearing frame housing as the bearing housing. The bearing frame or housing consists of the housing, the shaft, bearings, bearing end cover, flinger, and feet. On pump sizes AA and AB, the feet are usually cast integral with the bearing housing support head combination.

Support Head

The support head is the member that aligns and fixes the casing to the bearing housing. ANSI requires the support head to be made of ductile iron or carbon steel. This requirement stems from concern that a system upset could subject the pump to excessive pressure and result in catastrophic failure of the cast iron support head. Support heads are also offered in stainless steel as an optional feature. This is done for three reasons:

- To reduce corrosion due to leakage from packing or mechanical seals.
- To reduce thermal conductivity in very high or very low temperature applications.
- To have a material that has high impact properties for temperatures below 40°F.

On vertical in-line pumps, the support heads are larger than horizontal pumps because they must allow the rotor to be passed out of the support head during disassembly and also must adequately support the weight of the vertical motor. When a bearing housing vertical in-line is employed, the support head is at least 50% higher in height than with the rigid coupling design. Depending on size and motor horsepower, the support heads are made out of cast iron, ductile iron, or fabricated steel. When the vertical support heads are made out of ductile iron or carbon steel, extra care has to be taken in machining the toleranced dimensions because of the release of residual stresses in these materials.

Bearing Housing

This is usually made out of cast iron. After machining, it is protected internally with a rust preventative such as a paint or clear material to prevent rust particles from forming internally in the housing during storage or shut down. The housing is designed to hold a reservoir of oil that is approximately a half pint on the small pumps, and 3 to 4½ pints on the large pumps. The housings have vents, drains, and a tap for an oiler. The vent and drain should be a minimum of ½ in. so the oil can readily flow during filling or draining. The vent should be designed in a way that water cannot enter into the housing. Sometimes the vent is made up of a pipe that comes up through the bottom of the housing; other times it is composed of a nipple and cap with a sixteen-hole or commercial vent. The oiler is located so that movement of the oil from the rotation of the shaft does not prevent the oil from entering into the housing. Oilers are supplied with a glass or plastic bubble; glass is specified in a refinery type of service.

Shaft

The shaft has to be designed to take the radial, axial, cyclic, and torsional loads. The axial load, depending on suction pressure, can be in either direction. Shaft diameters are rated by horsepower per 100 RPM. The shaft material can be 1020, heat treated 4140 or 316. In each case the material should be reviewed for imposed stresses. Refer to Chapter 16 for methods of calculations.

Impeller Attachment

On chemical service pumps, impellers are usually attached with a male or female threaded connection. Threaded connections are used between the impeller and shaft because they can be more readily sealed than when a key design is used, which inherently has an additional joint. Specifications do allow both types of connection.

Bearings

Specifications require that when the maximum hydraulic load from the largest impeller at a given speed is applied to the bearings, they will have a minimum L10 life of 17,500 hours. The designer and user should carefully select original or replacement bearings because the interchangeable dimensional envelope of an AFBMA bearing does not ensure that the size or number of balls are the same from one manufacturer to another, thereby resulting in a possible change in the rating of the bearing. It should also be remembered that the life of a particular design will change with suction pressure.

Most chemical pumps use semi-open impellers, and operate with a .010 to .020 inch gap between the impeller and casing wall. To maintain this clearance, the bearings should have .0015 to .002 of an inch axial input end play. Double-row bearings have an assembly end play of .002 to .003 of an inch. Back-to-back bearings have an assembly end play of .005. However, the back-to-back bearings are more forgiving to misalignment when radial loads are applied to the bearings. The life of a back-to-back bearing is about twice that of the comparable double-row bearing. Back-to-back bearings are usually offered as an optional feature on chemical pumps. This applies to both horizontal and vertical pumps.

Lubrication of the Bearings

The bearings are normally lubricated with either oil bath or grease. The oil level is usually at mid-ball. If it is higher than mid-ball, churning usually occurs resulting in foam in the bearing housing and an increase in temperature. Disk flingers are used to splash oil within the housing. Roll pins are also used on the shafts to splash oil. In this case, the initial oil level is below the balls.

With oil lube, the temperature on the outside skin of the housing is about 20° cooler than the temperature of the outer race. Skin temperature of the small pumps ranges between 110°F and 130°F, and on the large pumps between 140°F and 165°F. At skin temperatures above 185°F, the unit should be shut down and inspected to determine the cause of high temperature. It takes approximately 45 minutes to one hour for the temperature of a cradle to stabilize. With roll pin splash, the temperature is 20°F less than the oil bath.

Grease bearings are either replacement grease or seal-for-life bearings. The problem with the replaceable grease is that excess grease is usually put into the bearing's cavity causing sharp increase in temperature and drop in life.

The seal-for-life bearings come with either shields or seals. Shield bearings are adequate for the majority of applications. In general, the temperature of grease bearings is approximately 20°F less than the oil lubrication. The subject of bearing lubrication is discussed in detail in Chapter 20.

Mounting the Bearing

Finish and dimensions of the shaft should meet the bearing manufacturer's recommendations. Typically, the bearing is .000 to .0005 inches tight on the inner race and .0005 to .001 inch loose on the outer race. Refer to the bearing manufacturer's catalog for recommended fit. The bearings can be pressed on to the shaft, but it is usually better to heat them to prevent excess stress. They can be heated in an oven or put in an oil bath up to 240°F. When put in an oven, the bearings should be laid flat and should not touch each other. For field installation, the bearing can be set over a light bulb to expand the inner race. If the bearing is too loose on the outer race, the race will spin within its housing. The bearings are secured to the shaft with a snap ring or a bearing lock nut. The bearing lock nut is secured with a tab washer. This is preferred to a lock nut with a nylon type pellet. The outer race axial movement is restricted in two ways: either within a separate end cover or by having a snap ring in the outer race clamped between the bearing housing and the end cover.

Clamping Between the Housing and End Cover

With this type of a mount (Figure 15-11), there are fewer tolerances and fits to cause misalignment than when used in a separate end cover. However, to obtain axial adjustment, the end cover has to be removed to add additional shims to either side of the snap ring to obtain the impeller clearance. This will require a complete shut down of the pump and disassembly of the end cover.

End Cover Mount

The end cover is a separate piece that slides within the bearing housing. The end cover has an O-ring to prevent oil leakage from the housing. The fit between the end cover and the bearing housing bore is .001 to .002 inches. Expertise is required in machining of the end cover with automatic machines that can exert excessive jaw pressure. This will cause distortion resulting in incorrect bearing fits. The end cover has an oil return passage line that can be either cast or machined so oil can flow back into the reservoir.

The bearing is held in the end cover by a snap ring, a lock ring, or a solid ring (Figure 15-12A). When using snap or spiral flex rings, care has to be taken that the radius on the outer race of the bearing is maintained within the tolerance. If this radius is too large, the axial load will concentrate on the inner diameter of the snap or circle ring causing it to deflect in a cantilever action reducing the clearance between the impeller and the casing.

A solid ring is used to prevent this problem. The solid ring is either threaded into the bearing end cover or screwed on with a series of small bolts (Figure 15-12B). Axial adjustment of the impeller to casing is obtained by two sets of three bolts. One set of bolts is threaded into the end cover so that when they are turned they go against the bearing housing causing the end cover to move back toward the coupling. The other set is screwed into the bearing housing so when they are tightened and the others are loosened the end cover goes toward the impeller.

End Sealing

Ends of the bearing housings have to be sealed to prevent external liquids from entering into the housing as well as oil from leaking out. This

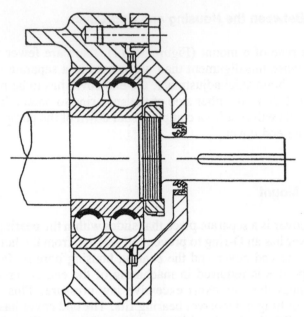

Figure 15-11. Thrust bearing positioned by a snap ring in the outer race.

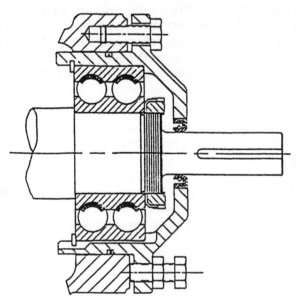

Figure 15-12A. Thrust bearing positioned with snap ring in the bore of an end cover.

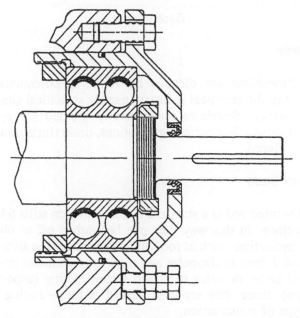

Figure 15-12B. Thrust bearing positioned with a solid ring in bore of an end cover.

is usually done with closures. Closure lips usually point out to prevent liquid from entering the housing especially during wash down. Closures can be supplied with garter-type springs or leaf-type springs. The garter-type spring is usually easier to assemble with this type of design. It is important that the finish of the shaft be within the manufacturer's recommendation, usually 16 RMS or less. A light film of castor oil should be applied to the closure for ease of assembly. There should be a 30° chamfer on the shaft for assembly of the closures over the shaft.

On the outer diameter of the closure, even though boring is correct and stamped casement is correct, the pieces are not always completely round. Therefore, it is recommended that some type of adhesive be put between the outer diameter of the closure and the housing to prevent leakage. When it is known that there is excess environmental water or vapor or contamination, magnetic-type closures or labyrinth-type closures are used. They can fit in place of the clipper-type of closure.

Bedplates

Standard Beds

Bedplate dimensions are dictated by ANSI specifications. Various types are used in the chemical industry, namely structural channel, bent plate, and castings. Bedplates are generally grouted for applications above 25 horsepower. For certain applications, drain rim channels or cast channels are offered.

Stilt-Mounted Beds

The stilt-mounted bed is a standard bed mounted on stilts 6 in. to 8 in. above the surface. In this way they can be washed off to obtain clean drainage in applications such as food and paper mills. The stilts should be a minimum of 1 inch in diameter and made of a stainless material. On some channel beds, in order to obtain secure footing (especially on a rough surface), three stilts are used instead of four—ending up with a milk stool type of construction.

Spring-Mounted Beds

This is a stilt-mounted bed with springs attached to the stilts that enable the entire assembly of pump, motor, and bed to move when external loads are applied to the nozzles. This is done in lieu of using piping expansion joints and loops.

Noncorrosive Beds

Noncorrosive beds are used in a corrosive atmosphere where it is known that steel beds will corrode in a short period of time. These beds are offered in the same sizes of the ANSI beds. They are made by epoxy coating steel beds or from nonmetallic. The nonmetallics are made from a form using resin transfer molding or are made as a solid mass, the thickness being that of the height of the bed. When grouting is required, the grout hole is usually cut with a saber saw or hole saw.

Flinger

The flinger is installed on the shaft with a press fit in front of the support head or cradle wall. The flinger is usually made out of elastomer or polymer. Its function is to prevent excess packing or seal leakage from entering into the bearing housing.

Other Types of Chemical Pumps

With continuous development of structural composite materials, pump manufacturers are offering various types of nonmetallic pumps in the ANSI envelope. ANSI specifications do not encompass nonmetallic pumps and there are no national standards for pressure, temperature, or limitations. These pumps can be horizontal cradle pumps, self-priming pumps, or submersible sump pumps. The design of these pumps will be discussed later in this chapter.

Sealless Pumps

Another type of chemical pump is referred to as a sealless pump, but more properly should be called a vapor tight or leakproof pump. These pumps have no mechanical seals; therefore, there are minimum risks due to seal failure or pump liquid being exposed to the atmosphere. Some manufacturers offer pumps up to 500 horsepower; however, the majority of vapor tight pumps are offered below 10 horsepower. Some companies manufacture the casing dimensional envelopes to be the same as the B73.1.

The two popular drives are magnetic drive and canned motor pumps (Figures 15-13 and 15-14). In both cases the normal limiting factors are the size of particles that can be pumped through their mechanism. The length of life of these types of designs depends on the type of bearings that are used. Excessive wear or seizure of the bearing results in downtime of the whole unit. Surveys of failures show that most are caused by flashing of the bearing lubricant rather than to the presence of particles. Bearings are sleeve journal and flat plate thrust or conical combination for journal and thrust. Monitors indicating bearing wear can be supplied to prevent catastrophic failures. The magnetic drive usually uses permanent magnets and can operate at higher temperature limits before cooling is required. The magnetic drive allows the use of a standard type motor to drive the magnets.

The canned motor pump has the advantage of being one complete unit for the pump and motor; thus, it is shorter than the magnetic drive. The outside liner of a canned motor pump is reinforced by the stator of the motor resulting in high allowable pressures. Maximum viscosity for this type of pump is 150 centipoise. This value can be much less depending on the size, torque, and speed of the equipment.

Initial capital expenditure of sealless pumps is higher than standard horizontal pumps with mechanical seals; however, manufacturers of this type of equipment suggest that this can be amortized in a short period because mechanical seal maintenance cost will be eliminated.

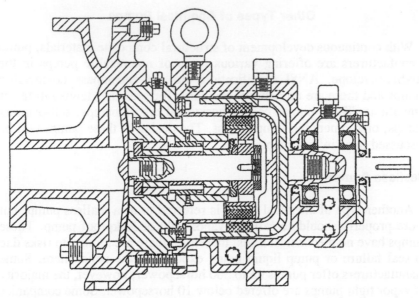

Figure 15-13. Overhung single-stage magnetic drive pump (courtesy of Ingersoll-Rand Company).

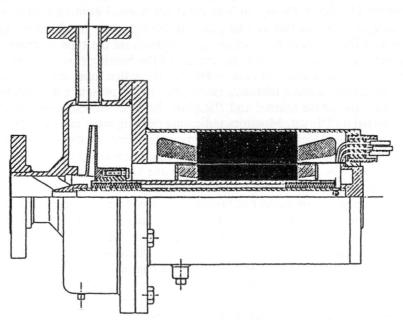

Figure 15-14. Overhung single-stage close-coupled canned motor pump (courtesy of Pacific Pumps Division of Dresser Industries).

Sump Pumps

Submersible or immersible sump pumps are a line of chemical pumps that have been derived from the combination of parts of the horizontal and vertical pumps (Figure 15-15). These pumps fit into wet sumps that may be 3 ft to 20 ft deep.

These pumps employ the casing and impeller of a horizontal pump and the support head, motors, and sometimes a casing cover of the vertical pump. To prevent critical speed frequencies, the shaft is supported by line bearings usually having a centerline to centerline distance of 5 ft for 1750 RPM and 3 ft for 3550 RPM. The column supporting the bearings and the discharge pipe is made of compatible material for the liquid in the sump. The bearings are either product lube, grease lube, or external lube which has been centrifuged or filtered. Typical applications are shown in Table 15-1.

The depth of the pump relative to the sump is called setting. Setting goes from the bottom of the strainer to the bottom of the mounting plate.

The sump pump is usually required at the mounting plate; therefore, when the impeller is selected, the frictional loss through the discharge pipe as well as the static head above the minimum liquid level has to be added to what is required at the mounting flange. These additional hydraulic losses may require a motor larger than would normally be used for the same selection as a horizontal pump. These pumps also require a minimum submergence to prevent vortexing or entrained air from entering into the suction (Figure 15-16).

There are no standards for the location in the sump of pumps of this type. Many users employ the suggested applications shown in the Hydraulic Institute Standards for sump design. The mounting plates for these pumps are usually plain carbon steel or carbon steel with an epoxy coating on one side. Sometimes stainless steel is used.

Sometimes it is desirable to pump the liquid to a level below the suction of the pump. A tailpipe is used to achieve this; however, the liquid level has to be above the impeller centerline when the pump is started (Figure 15-17). The tailpipe allows the liquid level to be pumped down to as much as 10 ft below the end of the flange of the suction pipe. The use of a tailpipe reduces the cost of the initial pump. The disadvantage, however, is that air can be pulled into the back of the casing thus reducing the overall performance of the pump.

Self Priming

The self-priming chemical pumps are also an offshoot of the horizontal pumps (Figure 15-18). These are usually available in 316 or ductile iron.

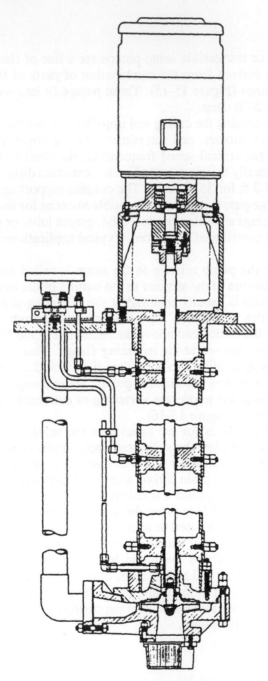

Figure 15-15. Vertical nonmetallic sump pump (courtesy of Ingersoll-Rand Company).

Table 15-1
Vertical Metallic Sump Pump Bearings

Bearing Material	Max Temp °F	Min Temp °F	Liquid	Shaft Mat'l
Bronze	180	– 20	Water & compatible liquids	Carbon steel
Iron	180	– 39	Water & compatible liquids	Carbon steel
Rubber	160	39	Abrasive with liquids compatible to rubber	316
Carbon	350	– 65	Acids, chemicals hydrocarbons	316
Teflon grease lube	180	0	Chemicals	316
Teflon product lube	350	– 100	Not compatible with teflon	316

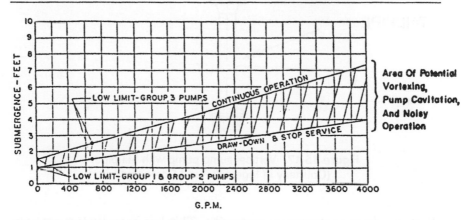

Figure 15-16. Minimum submergence versus g.p.m. for vertical sump pumps.

Most of these pumps are designed like a horizontal pump except they have a self-priming casing. These pumps are used in mine dewatering (which is usually acidic), or in refinery service where a vertical immersion pump may not be used because of the space limitations.

Nonmetallic Pumps

In the past 30 years, the demand for pumps with greater corrosion resistance than is offered by the high alloy stainless steels and nickel-based

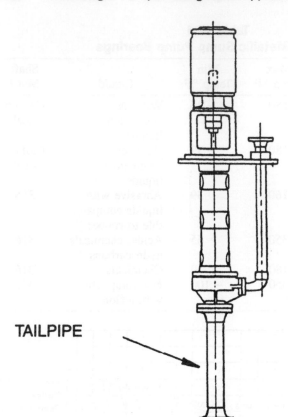

TAILPIPE

Figure 15-17. Vertical sump pump with tail pipe.
(Courtesy of Ingersoll-Rand Company)

alloys has been continuously increasing. This has been evident not only in the chemical industry, but in other industries that use chemicals in their processes. This demand has been met with various expensive alloys such as titanium, alloy 20, zirconium, and many types of hastelloys. Although these materials improved corrosion resistance, they caused other problems.

- Foundries experienced difficulties in castings.
- Existing patterns did not compensate for different shrinkage.
- Machinability was reduced.
- Delivery was longer due to foundry problems.
- The alloys were more expensive.

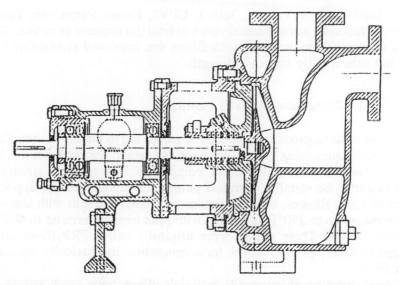

Figure 15-18. Self-primer overhung single-stage (courtesy of Goulds Pumps, Inc.).

- Quality was more difficult to obtain, and it was difficult and costly to comply with NDE (nondestructive examination) requirements.
- Alloy 20 and Hastelloy-C pump case defects exposed during hydrotest required excessive weld repair.

The extensive efforts made to overcome these problems included:

- Creation of new patterns to satisfy metal shrinkage.
- Development of Teflon-lined pumps.
- Development of other linings, such as polypropylene, epoxy, glass, and Kynar.

Although lining the inside of pumps solved many of the earlier problems, new ones were created because linings:

- Were difficult to produce and apply to the complicated pump casing areas.
- Would not properly adhere to metals.
- Would buckle, cold flow, or fail for other strength reasons.

Armored Pump

To overcome the difficulties experienced with pump linings, the armored pump was developed with complete pump casing and other wetted parts produced from carbon, Teflon, CPVS, Kynar, Ryton, etc. These were protected by outside metal plates to hold the required pressure. The various resins were produced with fillers that improved moldability but did not substantially improve strength.

Reinforced Composite Material Pumps

The next development in nonmetallic pumps led to improved manufacturing techniques using thermo resin without armor. Successful resins include glass-reinforced thermoset composites. These have strengths equivalent to the metallic chemical pumps and are suitable for applications of acids, alkalies, oxidizing agents, solvents, and salts with temperature ranges up to 250°F as normal with peak temperatures up to 400°F (Figure 15-19). These pumps were originally called FRP (fiber reinforced polymer) pumps but the term *composites* has basically replaced that label.

Proper selection of composite materials offers many combinations to improve corrosion resistance, lightweight, flame retardation, low-cost magnetic transparency, and complexity of art design. The terms *reinforced plastics* or *composites* generally include two large groups of organic compounds that differ in their make-up. These are thermosetting polymers and thermoplastics.

How does a designer choose between thermoplastics and thermosets? With the present state of the art, the chemical compatibility, maximum applicable temperatures, and consistent quality are about the same for both processes. The differences are listed in Table 15-2.

Thermosetting Polymers

Thermosetting polymers for pump use are reinforced with fiberglass or carbon fibers. During the molding cycle, these materials undergo a chemical change that is irreversible. The resulting material will not soften or become pliable with heat. They have four basic chemistries: polyesters, phenolics, vinyl esters, and epoxies. Each has its own set of advantages, manufacturing processes, and mechanical and chemical properties. The fibers are either continuous or short fibers and are the key in developing the temperature range and corrosion resistance of the final part. There are many manufacturing processes for thermosets and they are often every bit as critical to the final part performance as the selection of the proper polymer and reinforcement combination. Compression molding, transfer molding, resin transfer molding, cold molding, and extrusions are among the most commonly used processes.

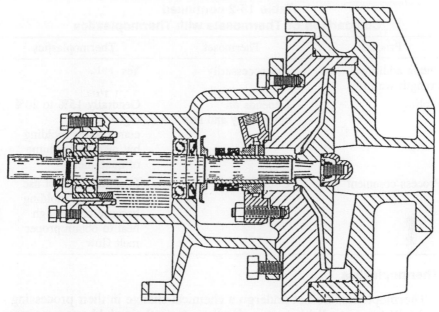

Figure 15-19. Nonmetallic overhung single-stage pump built to ANSI dimensions (courtesy of Ingersoll-Rand Company).

Table 15-2
Comparison of Thermosets with Thermoplastics

Process	Thermoset	Thermoplastics
Average range of molded thickness	.030 to 2.0	.06 to .38
Weight range of material per molded piece	1 to 500 pounds	.5 to 5 pounds
Glass content range by volume	50%–60%	30%–40% (glass degrades when put through the auger of the injection machine)
Length of glass fiber	.25 inch	.060 inch
Strength	Not uniform throughout (anisotropic)	Basically uniform
Minimum annual quantities for design criteria	1,000	10,000

Table 15-2 continued
Comparison of Thermosets with Thermoplastics

Process	Thermoset	Thermoplastics
Obtain additional strength with ribs	Not necessarily	Yes
Tooling	Depends on complexity and size	Generally 15% to 20% higher than compression molding but offset by volume of quantities
Process comment	Compression	Injection. Cannot use compression molding because not enough heat to obtain proper melt flow

Thermoplastics

Thermoplastics do not undergo a chemical change in their processing and will become pliable upon reheating above their yield temperature. Thermoplastic materials are available in a wide range of strengths and application envelopes. They can be divided into fluoropolymers (PFA-PTFE), engineering plastics (LCP-PPS), and general plastics (ABS acrylics, polyethylene, PVC, and polypropylene). Thermoplastic processes such as injection molding, vacuum forming, extrusion, and blow molding offer the design engineer many selections for optimum cost considerations. Selecting a suitable composite requires a complete understanding of the end use application as well as a familiarity with the polymer's physical, chemical, and processing properties. Although direct replacement without design changes is feasible, more often the use of a nonmetallic is optimized by a well-informed specialist.

Table 15-3 shows a general comparison of various resins applications.

Manufacturing Techniques

Two methods used in manufacturing the casing, casing cover, and impeller of nonmetallic pumps are compression molding and resin transfer.

Compression Molding

This process uses matched metal dies that have cored heat transfer passages to control the temperature of the process. The base resin is mixed with appropriate amounts of chopped glass, fillers, and chemical catalysts, inhibitors, and release agents to make a batch. This batch can be set aside in plastic containers for a shelf life of approximately 30 days.

Table 15-3
Resin Performance

Resin Type	Strong Acids	Alkalies (Caustic)	Oxidizing Agents (Bleaches)	Organics (Solvents)	Temp. Limit
General Purpose Polyester (Fiberglass Boats & Bathtubs)	Poor	Very Poor	Very Poor	Poor	160°F
Isophthalic Polyester (Structural Applications)	Fair	Poor	Poor	Fair	190°F
Anhydride Polyester	Excellent	Poor	Poor	Good	275°F
Bisphenol A Polyester	Good	Good	Fair	Poor	250°F
Epoxy	Poor	Excellent	Poor	Excellent	190–250°F
Conventional Vinyl Ester	Good	Good	Fair	Fair	210°F
High-Performance Vinyl Ester Dow Derakane 470 IR GRP Materials	Excellent	Good	Good	Excellent	300°F

When a piece is to be made, a portion of the batch for the piece is measured within an ounce of what is required. If there is too little, the die will not be completely filled; if there is too much, the piece will have an extra thick parting line and will not meet specifications. When the portion of the batch is put into the die, the die closes and compresses the batch at a temperature of approximately >300°F for 10 minutes.

The atmospheric condition to which the whole molding machine is subjected should be controlled for temperature and humidity to obtain the proper quality of the piece. The design engineer has to work closely with the tooling engineer to make sure there is proper flow of material and the path of glass or reinforcement is in the proper location. Experience has shown that reinforcing ribs on casings can be detrimental to the strength because the glass will form a continuous path within the rib producing a knit line. (Knit lines are a result of material coming from two directions and meeting.) This is usually a weak point. In many cases the piece will be stronger by eliminating ribs where it was thought they would be beneficial.

The pieces made from the compression molding process are consistent from one piece to another, both in dimensions and in quality. Poor quality from this process can be a result of (1) a bad mix of batch, (2) batch that is too old, (3) temperature within the die that was not controlled, (4) temperature of the atmospheric conditions that were not controlled, (5) excess humidity within the atmospheric conditions, (6) too little or too much batch, (7) time under compression that was not held as specified.

Resin Transfer

The tooling for resin transfer can be less costly than that of compression modeling, but the number of pieces that can be obtained from the die will be less. In this process, the two halves of the die are separated and reinforced cloth is cut to shape and put into the upper and lower envelope portions of the die. Core made of beeswax is then set within the die. The die is closed and vacuumed and brought up to temperature. A valve is then opened to allow the resin to flow into the die. When the die is filled, it is allowed to cool from 12 to 24 hours. The piece is then removed and set into an oven. The beeswax is then melted and is recovered. The resulting cavity gives the desired shape of the core.

The disadvantage of the resin transfer pieces is that there is a knit line where the two halves of the die meet; therefore, the way the reinforced cloth is put into the die is extremely important to obtain the proper strength of the piece. Pieces made from this process usually do not have the strength of a comparable compression molded piece. Pieces also do not have the consistency of the compression molded piece due to the hand lay up of the cloth. This process is usually used for larger types of pieces where the allowable tolerances are greater than with the compression molded piece. The internal finish for hydraulic passages with this process is not as smooth as obtained with the compression molded piece.

The problems of quality in this process are:

- The quality of tooling that is used to substantiate life of the part for consistency.
- The type of reinforcing glass.
- The method in which the glass is put in the die.
- The amount of glass cloth that is put in the die.
- The quality of resin.
- The control of vacuum to allow the resin to come into the die.
- The quality of the beeswax that is reused from one piece to another.
- The length of time that the piece is allowed to solidify.
- The temperature that is used to melt out the core.

Design Stresses

Designers should be made aware that the stresses advertised in the sample ASTM bars will not necessarily be equivalent to the stresses of the molded piece. This will be verified by the molder as well as the material supplier. When designing with metals, a designer can use the same stress throughout the piece; however, this is not the case with nonmetallic parts. When designing the casing, it is advisable to use different stress levels for the suction nozzle, discharge nozzle, volute, and the wall of the casing. This also applies to the tensile, compression, and hoop stress. Likewise, the modulus of elasticity that is used for the design will change from one process to another. Experience has shown that design values of the actual piece or structural part may be ⅕ to ¹⁄₁₀ that of the test bar. The modulus of elasticity is between 1 to 2 million.

Pressure vs. Temperature

Unlike metallic pumps, there are no standards for pressure-temperature ratings of the flanges. Presently, the ratings change from manufacturer to manufacturer and material to material. Good design practices have demonstrated that the pressure reinforced vinyl ester capability at ambient temperature of the flanges can be equivalent to that of the metal flanges using the same dimensions as the metal flanges. The pressure-temperature gradient is a linear factor and basically degrades above 100°F.

Because heavy wall vinyl ester material is a good insulator, the temperature gradient from the liquid side of the pump case to atmosphere is basically 100°F. This is based on tests of heat soaking the vessel for 24 hours. This allows the manufacturer to pump higher temperatures without excess bearing temperatures. This is also good for the user since there will be little loss of heat from the fluid while passing through the pump.

NPSHR

As with metallic pumps, NPSHR is established by using 3% head loss as a criterion. However, it should be recognized that pumps operating at 3% drop in head or relatively close to this mode of operation in incipient cavitation. Where damage might not be apparent on metallic pumps, it will be observed after a period of time on nonmetallic pumps. It is suggested that the NPSHR offered by the manufacturer be 3 to 5 feet higher than metallic pumps in order to give equivalent life to the nonmetallic counterparts.

General Construction of Nonmetallic Pumps

Most nonmetallic chemical pumps presently being offered for the same hydraulic range as the ANSI pumps are being built to the ANSI dimensional standards envelope. Consequently, most manufacturers are using the same support head and bearing housing construction as on the metallic pumps. This allows the user to be able to interchange the bearing housing parts from metallic pumps to nonmetallic pumps. To obtain additional strength, some manufacturers employ back-up rings that are either separate pieces bolted to the support head or they have support heads that include a back-up ring. The nonmetallic pumps were initially designed with integral nozzles, but there were many molding problems. Some manufacturers resorted to molding separate nozzles and then either molded them to the casing or adhered them to the casing. This was found to be a problem when applying external nozzle loads. With the advancement of materials and dies, many manufacturers now mold the nozzle integral with the casing without incurring nozzle loading problems. Because the materials have moduli that are between 1/15 and 1/30 that of standard metallic materials, it is advisable not to put excess nozzle loadings on composite casings.

Nozzle Loading

There are no standards for the nozzle loads on ANSI pumps, and the manufacturer's specifications are usually referred to for the maximum load. The criterion used by the manufacturer for maximum nozzle loads is usually the movement on the coupling end of the shaft. This may be .0050 to .0100 depending on the size of the pump. This deflection can be caused by:

- The movement of the entire assembly when load is put on.
- Movement of the feet of the casing relative to the bedplate due to the friction force between the two.
- The movement of the bearing housing relative to the bedplate.
- Internal movements causing rubbing of the impeller against the casing.
- Deflection of the bedplate surface relative to the driver shaft.

With nonmetallic pumps the allowable nozzle loads are much less than with the metallic pumps because the casing feet move or deflect under a much lighter load. When nonmetallic beds are employed with either a metallic or nonmetallic pump, the movement of the top surface of the bed is the weak member of the assembly resulting in low allowable nozzle loads. This will occur if the bed is grouted or ungrouted, especially if the force is along the X axis or a moment around the Z axis.

Bolting

As threaded studs will impose tension in the composite case during assembly, it is preferable to use through bolting. This leaves the casing in compression rather than in tension. When through bolting cannot be used, the bolts or studs are fastened into stainless steel or alloy inserts and are molded into the piece. The inserts are gnarled and grooved on the outside diameter to prevent twisting or pulling within the piece when torque is applied to the fastener. The inserts are usually a class 3 fit on the inside diameter for the fasteners. A blind end insert is used to give a positive stop for the studs. When inserts are used, it is best to mold them within the piece rather than post insert them. When they are molded in the piece they should be located at least 1/8 in. below the finished surface so that when machining is being done, the cutting tool does not have an interrupt cut against the insert resulting in weakening of the mounting of the insert. When inserts are used, care has to be taken by the designer that there is proper flow of the composite material to avoid a path for leakage during hydrostatic testing.

Gaskets

With nonmetallic pumps, most main gaskets are O-rings. These can be either round or square cross sectional O-rings. O-rings result in less bolt loading on the main bolts. If gasket surface requires final machining, then it is recommended that the surfaces be coated with the base resin to prevent wicking of the pump fluid through the exposed ends of the glass reinforcement resulting in leakage of the gasket.

Back-up Support for Bolting

To reduce the bolt head or nut loading, it is recommended that when washers are used their diameter should be at least three times the diameter of the bolt. Casing covers usually have inserts for the gland studs as well as inserts for jacking bolts to aid in the disassembly of casing covers. Inserts require optimum strength to absorb radial and axial forces and must be compatible with the atmospheric conditions and in many cases with the liquid being pumped.

Stuffing Box Area

If the glands are made of a composite material, they must be capable of withstanding the torque that is applied without creeping. Depending whether an inside or outside seal is used, the gland may need additional reinforcing with either a metallic back up or extra strength reinforcing cloth. When designing the stuffing box area, the heat transfer of the injection fluid around the seal should be considered. The larger this area, the better the life of the mechanical seal.

The shaft sleeve can be a separate piece that is usually made by injection molding or it can be made integral with the impeller. There are advantages and disadvantages to both. When integral with the impeller, the entire impeller sleeve mechanism needs to be replaced if something goes wrong with the sleeve. When a separate shaft sleeve is used, there is an additional sealing surface between the impeller and the sleeve to prevent fluid from coming in contact with the shaft. When using nonmetallic sleeves, mechanical seals with teflon wedges should not be employed because of the excess fretting. Also, the designer has to be concerned with the extrusion from holding force of set screws on soft nonmetallic sleeves. Split clamping rings using a radial type of set screw are sometimes used to prevent damage to the shaft sleeve.

Mechanical Seals

Because of the corrosive properties of the fluids being used within nonmetallic pumps, many pumps use outside mechanical seals. As a result, the only wetted pieces are the stationary seat and the compatible rotating surface. The remaining springs and secondary seals are external to the stuffing box. However, care should be taken that if an outside seal fails it could be catastrophic. It is recommended that a seal guard be employed when outside mechanical seals are used. This subject is discussed in detail in Chapter 17.

Impellers

Many of the materials that have to be used for the liquids being pumped cannot be readily adhered or mechanically attached to themselves. Therefore, it is difficult to obtain closed impellers and consequently, most impellers are open vane design. Another basic problem with the nonmetallic pumps is the attachment of the impeller to the shaft. Depending on the speed and horsepower, most nonmetallic pumps use more than one key for attachment due to the stress levels of the material. Many impellers are attached by using threaded inserts that are molded within the impeller. The problem here is that care has to be taken that excess stress doesn't occur around the surface of the molded insert that would result in a weak surface between the two materials. Another method of attachment is a multi-keyed or polygon shape that does not require an internal insert because the stresses of the material are distributed throughout its circumference relative to the shaft surface. The disadvantage of the polygon attachment is that there are more surfaces that have to be sealed to prevent external fluid from attacking the shaft.

The sealing of the impellers with either the insert or a polygon fit is similar to that used in the metallic pumps. Most manufacturers will employ the same sealing mechanisms for the two types.

Nonmetallic Immersion Sump Pumps

Typical applications include wet pit chemical waste handling, effluent handling, and liquid transfer operations where broad corrosion resistance is required. These pumps are made of the same basic materials as horizontal nonmetallic pumps, either vinyl ester or epoxy. The hydraulics cover the same basic range as the horizontal pumps and in many cases, the casing impeller, and casing cover are the same parts as used in the horizontal pumps.

The shaft material is 316, alloy 20, Hastelloy B or C, or titanium, depending on the liquid being pumped. Optional shafts of 316 coated with various materials such as kynar are also available. The use of pultruded nonmetallic shafts is being investigated to eliminate all metallic parts for this type of application.

The column supporting the wet end to the mounting plate is a one-piece construction with inserted bearings or a multi-construction of short columns with flanges and the bearing support sandwiched between the flanges of the column. The column material is usually the same base material as the pump and impeller. The bolting of the casing and the columns can be of a nonmetallic material compatible with the fluid.

Bearings are made out of teflon or carbon with spiral flutes. The lubrication is either external or clean product lube. Clean liquid for lubrication is one that has less than 5 micron particle size. The lubrication to each bearing should be at least one half GPM at 160°F temperature or less and at a pressure of approximately 25 psig. Carbon bearings are furnished when external lubrication or injection pressure is not adequate.

Figure 15-20 shows when to supply carbon or teflon bearings based on particle size in the fluid and the flush pressure available to these bearings. It also shows when cyclone separators are required and what flow for a given flush pressure is obtainable from the separators. The lower bearings are usually twice as long as the line bearings to absorb the radial thrust developed by the impeller. A gap or relief hole is placed between the throat bushing of the casing cover and the bearing itself so that dirty liquid under pressure will be relieved of pressure and not be forced into the bearing clearances resulting in short life.

BEARING APPLICATION CHART — PRODUCT LUBE

Particle Size (2)	Flush Pressure (1)		
	>10 PSIG	>15 PSIG	>25 PSIG
< 10 Micron (Clean)	Carbon	Teflon	Teflon
> 10 Micron < 400 Mesh (Fine)	Carbon	Carbon	Teflon
> 400 Mesh < 20 Mesh (Coarse)	Note 3	Carbon (4)	Teflon (4)

1. Discharge pressure at mounting plate. Min flow of ¼ GPM per bearing is required. ½ GPM is recommended.
2. Particle sizes are as follows:
    10 micron = .0004 in.
    400 mesh (fine) = .0015 in.
    20 mesh (coarse) = .0328 in.
3. External flush only at 25 PSIG (Teflon bearings)
4. Cyclone separators required. Refer to chart below.

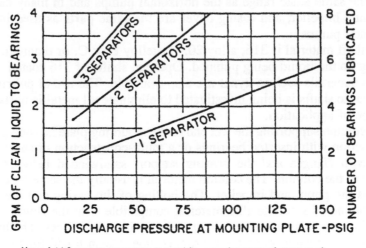

Note: Add 2 gpm per separator to total flow requirements of pump to allow for flow taken by separators.

Figure 15-20. Bearing and lubrication for nonmetallic vertical sump pumps.

A lip seal or closure is installed in the mounting plate where the shaft passes through preventing gases and vapors from escaping out of the sump. If the sump is under pressure or has toxic fluid, a mechanical seal is employed. Likewise, a gasket is placed between the mounting plate and pit cover. A strainer is placed at the bottom of the casing. It is made out of a polypropylene material, the net area of which should be three times the entrance area of the suction nozzle.

Driver

The pump shaft is connected directly to the motor by a rigid adjustable coupling, and in-line motors are used to absorb the axial thrust. If a normal thrust motor is used, then a separate thrust bearing is used within the support head to absorb the axial thrust.

Level Controls

With chemical sump pumps, the level control is usually encased to prevent the fluid from coming in direct contact with the switching mechanism.

Stilling Tubes

If it is anticipated that swirl or vortexing will exist within the sump, the level controls are mounted within a stilling tube that indicates the true level of the liquid within the sump.

Mounting Plates and Pit Covers

These plates are made out of the same base material as the pumps. Depending on the size, the pit covers may require reinforcement of steel angles or channels. These steel pieces are encapsulated so they are not exposed to the atmosphere.

Processes

Chemical pumps handle a variety of liquids that could be concentrated, diluted, or just a trace. The concentration could be from 5% to 50% or the liquid could be in parts per million. Temperature could be hot or cold, and a pump can be sold to handle maximum, minimum, or normal temperature. The range of temperature could cause thermal shock and can vary as much as 150°F from the cold to hot application. The rate of activity of the fluid changes approximately two to three times for each 18°F change in temperature. The liquids could have either an alkaline or acidic pH level. Liquids may contain solids that might cause erosion, corrosion, or settling problems that would result in clogging. The liquid may have entrained air that would make a reducing solution into oxidation or it could have inhibitors to reduce corrosion or accelerators to increase corrosion. The impurities could lead to something called discoloration or solution breakdown.

Final pump material selection is a collective decision based on input from the pump designer, plant operator, material supplier, and available technical literature.

Pump Corrosion

The types of corrosion encountered in a chemical environment fall into eight typical categories.

- General uniform corrosion at a uniform rate over entire surface, either very slow or very rapid.
- Crevice corrosion that is a localized form from small stagnant solutions in areas such as threads, gasket surfaces, or drain holes. Crevice corrosion is caused by a differential in concentration of metal ions and oxygen added to the main body. This causes an electrical current to flow, causing the damage.
- Pitting is localized. It is manifested as small or large holes usually produced by chlorines.
- Stress—corrosion occurs at cyclic stress on shafts.
- Intergranular corrosion—usually occurs in the presence of heat.
- Galvanic action.
- Erosion-corrosion—corrosion plus mechanical wear such as cavitation.
- Selective—that is, leaching corrosion or degraphitization usually not found in chemical pumps.

Pump Materials

The typical material of construction to combat corrosion is either a 304 or 316 stainless steel that is superior to austenitic or ferritic steels. Other materials would be composite plastic such as PTFE and FEP. Usually fiber reinforced plastic is used for strength and chemical resistance. This includes vinyl esters, epoxies, polypropylenes, and phenolics. Ceramic or glass is avoided because of low mechanical properties.

Some of the process liquids found in chemical plants are listed along with pump materials used in the various environments.

Chlorine

- 65% of the chlorines are used for organic chemicals such as vinyl chloride, pesticides, fluorcarbons.
- 15% of the chlorines are used for producing pump and paper.
- 10% is for inorganic chemicals.
- The remaining 5% is for sanitation, potable water, and waste water that are used in municipal water works and sewage plants.
- 50% of caustic soda is used for the chemical industry.

- 15% is pulp and paper.
- The remaining is in aluminum, rayon, cellophane, petroleum, soaps, and foods.

The elecrolytic plants produce chlorine and caustic soda using a range of 250 to 1000 GPM and about 20 to 30 pumps. They also have 15 to 20 peripheral transfer pumps for hydrochloric acid, diluted H_2SO_4, and sodium hypochlorite. These plants use pumps of titanium, nickel alloys, cast iron alloy 20, CD-4MCu, and nonmetallics such as vinyl ester.

Sodium-Hypochlorite

NaOCl is a byproduct of the chlorine and caustic process. It is found in the bleach plants of paper mills with a concentration of 12% to 20%, in commercial bleach such as household Chlorox, and OEMs for swimming pool chlorination. The electrolysis of sea water (brine) with a concentration of 1% to 3% is used for bleach plants, pulp mills, bleach plants for textile mills, and municipal waste treatment to kill bacteria. These plants use large amounts of sea water for cooling to prevent pipes from fouling with algae. It is also used in the pretreatment of desalinization intake and the pretreatment of secondary recovery brine in oil production. Pumps are alloy 20, titanium, and nonmetallic.

Hydrochloric Acid—HCl

Also known as muriatic acid, it usually requires Hastelloy B or titanium pumps. The primary consumption is pickling of steel for use with oil well acidizing where acid increases the permeability of wells by dissolving part of the limestone and dolomite formations. High purity aluminum chloride produced by aluminum hydroxide and HCl is used in pharmaceutical and cosmetic usage.

In food processing, HCl is used in the manufacture of sodium slutamate and gelatin for conversion of cornstarch to syrup or adjusting the pH value in breweries.

Sulfuric Acid—H_2SO_4

With a 97% concentration, cast iron pumps are used. Eighty percent or less concentration is either alloy 20 or vinyl ester nonmetallic pumps. It has a very high boiling point, therefore a minimum loss is incurred at elevated temperatures. It is an excellent drying agent in the manufacture of chlorine gas. It is used for making fertilizers and explosives.

Ferrous and Ferric Chloride

This is a byproduct from acid pickling of iron and steel. It is used for etching reagent in copper clad printed circuit boards, for electronics, or as a chemical coagulant for water treatment on waste water. The pumps are either titanium or vinyl ester.

Chlorinated Hydrocarbons

Chlorinated hydrocarbons are used to produce PVC, chloroform, carbon tetrachloride, solvents, flame retardants, insecticides, adhesives, pharmaceuticals, metal cleaning, and dry cleaning. The pumps are usually 316 or ductile iron.

Ethylene and Propylene Glycol

Ethylene glycol is permanent antifreeze. It is also used in alkaline resins for coating in brakes, shock absorbers, and latex paints. Propylene glycol is used in the manufacture of unsaturated polyester resins. It is used to make cellophane, tobacco moisture retention material, brake fluids, and food additives. These pumps are either 316 or ductile iron.

Synthetic Glycerine

It is used for making resins, cosmetics, cellophane, tobacco, food beverages, and explosives. The pumps can be nonmetallic material.

Corn Syrup

Used in corn oil process. Usually 316 or nonmetallic materials.

Dyes

Used for such things as the ink in ball point pens and dying silk and wools. Usually nonmetallic pumps.

Pesticides

Insecticides such as malathion for controlling insects; herbicides for controlling plant growth. These materials are solvents and nonmetallic pumps are usually not suitable for this application.

Sodium Chlorite—Na ClO₃

This is used for bleaches, herbacides, explosives, and rocket fuels. Pumps can be in 316 or nonmetallic.

Pulp and Paper

Sulfite process: usually acids and bleaches. Sulfate (which is Kraft) is used in making strong cardboard containers and wrappings. Usually use titanium or vinyl ester pumps.

Metal Finishing

Electronic plating, cleanings such as oil, rust, and scale. Usually use ductile iron pumps. Plating usually use nonmetallic pumps to prevent stray currents. Waste treating of the vent plating baths are usually 316 or nonmetallic pumps.

Carbon Steel Pickling

Sulfuric acid was replaced by hydrochloric acid to give a better finish. Use 316, alloy 20, or nonmetallic pumps.

Stainless Steel Pickling

Nitric hydrofluoric acid solutions process is not suitable for nonmetallic pump application.

Desalinization and Water Purification

Desalinization and water purification are done through distillation or reverse osmosis. Usually a simple plant used for marine or power plants is used for distillation. The size of the plant can be scaled up or down. It can tolerate a wide latitude of feeder water quality, but requires high temperatures and results in more corrosion and more maintenance. It is usually only efficient using low pressure steam. The temperature is from 70° to 250°F with a vacuum as low as 25 inches of mercury. The pumps are usually 316 or alloy 20.

The reverse osmosis process using a membrane is simple, compact, more efficient, and uses ambient temperature. The membranes are sensitive to metal pick ups. It can purify dirty well water, brackish water with 200 to 600 psi. If it is used for desalinization of salt water, the pressures are from 800 to 1000 psi. Miscellaneous pumps can either be 316 or nonmetallic.

Secondary Oil Recovery (Waterflood)

When no convenient surface water is available, deep well turbine pumps are used to pump the water containing calcium sulfite to the surface. This is filtered or chemically treated before reinjection to prevent clogging of the pores in oil sand. The oil and brine mixture then comes to the surface where it is separated. The water is a corrosive brine with hydrogen sulfide from the contact with oil. The brine is chemically treated and filtered for reinjection or disposed of as waste. Chemical pumps of 316 or nonmetallic can be used to transfer the brine.

Mining—Copper Leaching and Uranium Solvent Extraction

Copper ore is normally less than 1% copper. Flotation or leaching or solvent extraction is used to upgrade the ore. Copper emerges as a copper sulfate and then is pumped to an electrolytic cell where it is plated out. Pump material, either 316 or nonmetallic, is used at ambient temperature.

For uranium, a solvent extraction is settled and filtered. The clarified solution is mixed with kerosene and organic amines. The solution is stripped and uranium is precipitated as uranium oxide. Nonmetallic pumps or 316 for low head or ambient temperature are used.

Industrial Waste Treatment

These vary with great quantities due to the nature of the product and process that they drain. The range of fluids is from a discharge of great volumes of cooling to small but concentrated baths of inorganic or organic substances. The pump material is of ductile iron, 316, or nonmetallic. Some of these waste treatments are wastes containing mineral impurities, steel pickling, copper bearing wastes (where very small amounts of copper—less than 1 mg per liter—will interfere with life in a stream or biological sewage treatment works). Wastes containing chromates or cyanides are used for electroplating and electrolytic operations where the maximum is less than 1 mg per liter. They are also used for gas and coke plant wastes, oil-field brines (which are petroleum refinery wastes, mining wastes, or wastes containing organic impurities). They are also used for milk processing, meat packing, brewery and distillery, vegetable and fruit processing, textiles (such as wool, cotton, silk, linen, and dyes), laundries (which have soap, bleaches, dirt and grease), tanneries, and paper mills (which have black, green, and white liquors). The pumps could be ductile iron, stainless, alloy 20, titanium, or nonmetallic.

Material Selection

A selection of corrosion resistant materials is shown on Table 15-4.

Table 15-4
Thermoplastics Corrosion Resistance Guide

Key: A Acceptable   Q Questionable   NR Not Recommended

Media	200°F Poly-Phenylene Sulfide	Penton	200°F Kynar	200°F Teflon	200°F Poly-sulfone	Noryl	200°F Nylon	200°F Polycar-bonate	316 SS	Carbon Steel	Alu-minum 3003
Media	200°F Poly-Phenylene Sulfide	200°F Penton	200°F Kynar	200°F Teflon	200°F Poly-sulfone	Noryl	200°F Nylon	200°F Polycar-bonate	316 SS	Carbon Steel	Alu-minum 3003

* *Polyphenylene sulfide grades containing glass fiber and/or mineral fillers will be less chemical resistant than indicated.*

Information taken from Phillips Petroleum Corrosion Resistance Guide

References

ASME B73.1M—1991, American Society of Mechanical Engineers.
ASME B73.2M—1991, American Society of Mechanical Engineers.
Fegan, D., "Everything You Always Wanted to Know About Sealless Pumps," October 8, 1986, presented at American Institute of Chemical Engineers, Memphis Section.
Hydraulic Institute Standards, 14th Edition.

Part 3

Mechanical Design

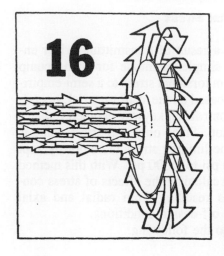

16

Shaft Design and Axial Thrust

Shaft Design

The pump rotor assembly consists of the shaft, impellers, sleeves, bearing or bearing surfaces, and other components such as balancing disks, shaft nuts, and seals that rotate as a unit. The primary component of the rotor assembly is the shaft. The pump shaft transmits driver energy to impellers and consequently to the pumped fluid. This section will be concerned primarily with the sizing of the pump shaft.

The pump shaft is subject to the combined effects of tension, compression, bending, and torsion. As a result of the cyclic nature of the load, when shaft failures occur they are almost exclusively fatigue-type failures. Therefore, the first consideration in sizing the shaft is to limit stresses to a level that will result in a satisfactory fatigue life for the pump. The degree of detail involved in the stress analysis will be dependent upon the intended application of the pump. The analysis can be a simple evaluation of torsional shear stress at the smallest diameter of the shaft or a comprehensive fatigue evaluation taking into consideration the combined loads, number of cycles and stress concentration factors.

Sizing the shaft based on stress is not the only consideration. Shaft deflection, key stresses, fits for mounted components, and rotor dynamics must be evaluated by the designer. The analytic tools available range from simple hand calculations to sophisticated finite element computer programs. The following sections are intended to present the fundamental considerations with which the designer can begin the design of the pump shaft. In some situations, satisfying these fundamental requirements can be considered adequate for a complete shaft design. In other, more critical services, further analysis is required before finalizing the design.

Shaft Sizing Based on Peak Torsional Stress

The stress produced in the shaft as a result of transmitting driver energy to the impellers is torsional. A simple technique for sizing pump shafts is based on limiting the maximum torsional stress to a semi-empirical value. The limiting-stress value is based on the shaft material, operating temperature, and certain design controls on keyway geometry, diameter transitions, and type of application. Since only one stress value is calculated due to one type of load, the limiting stress is obviously kept low. Typical values range from 4,000 psi to 8,500 psi. With this method of shaft sizing, no attempt is made to calculate the effects of stress concentration factors, combined stresses resulting from radial and axial loads, or stresses due to start-up and off-design conditions.

The peak torsional stress is equal to the following:

$$\tau = \frac{Tr}{J} \tag{16-1}$$

The torque is calculated from the maximum anticipated operating horsepower. Special attention should be given to pumps operating with products having a low specific gravity. Shop performance testing will generally be conducted with water as the fluid, and overloading the shaft may occur; hence, performance testing at reduced speed may be required. The shaft diameter used for calculating the stress should be the smallest diameter of the shaft that carries torsional load. For most centrifugal pumps the shaft diameters gradually increase toward the center of the shaft span. This is necessary to facilitate mounting the impellers. As a result the coupling diameter tends to be the smallest diameter carrying torsional load. It is a good design practice to ensure that all reliefs and grooves are not less than the coupling diameter. On some designs, such as single-stage overhung pumps, the smallest shaft diameter is under the impeller, in which case, this diameter shall be used for calculating shaft stress. Reliefs and groove should not be less than this diameter.

Example

Determine the minimum shaft diameter at the coupling for a 4-stage pump operating at 3,560 RPM where the maximum horsepower at the end of the curve is 850 bhp. Use 4140 shaft material with a limiting stress value of 6,500 psi.

Solution

$$\tau = \frac{Tr}{J}$$

$$J = \frac{\pi D^4}{32}$$

$$P = T\omega$$

Hence • $\dfrac{Tr}{J} = \dfrac{Pr}{\omega J} = \dfrac{Pr}{\dfrac{\omega \pi d^4}{32}} = \dfrac{P16}{\omega \pi D^3}$

Solving for D:

$$D = \left(\frac{P \times 16}{\omega \pi \tau}\right)^{.333}$$

$$D = \left(\frac{850 \text{ bhp} \times 550 \dfrac{\text{ft-lb}}{\text{hp-sec}} \times \dfrac{12 \text{ in.}}{1 \text{ ft}} \times 16}{3{,}560 \dfrac{\text{Rev}}{\text{min}} \times \dfrac{2}{\text{Rev}} \times \dfrac{1 \text{ min}}{60 \text{ sec}} \times \pi^2 \times 6{,}500 \dfrac{\text{lb}}{\text{in.}^2}}\right)^{.333}$$

$$D = \left(\frac{850 \text{ bhp} \times 321{,}000}{3{,}560 \text{ RPM} \times 6{,}500 \text{ psi}}\right)^{.333} = 2.276 \text{ in.}$$

For design round up to nearest $\frac{1}{8}$-in. increment: D = 2.375.

Example

A 2-stage pump has been designed with a 2⅝-in. shaft diameter at the coupling. The maximum horsepower at 3,560 RPM is 900. What is the maximum operating speed for a limiting stress of 7,000 psi?

Solution

Using the pump affinity laws described in Chapter 2:

$$\frac{bhp_2}{bhp_1} = \left(\frac{N_2}{N_1}\right)^3$$

or

$$bhp_2 = bhp_1 \left(\frac{N_2}{N_1}\right)^3$$

From the previous example:

$$\tau = \left(\frac{bhp \times 321{,}000}{N \times D^3}\right)$$

Substituting and solving for N:

$$N_2 = \left(\frac{\tau \times N_1^3 \times D^3}{bhp_1 \times 321{,}000}\right)^{.5}$$

$$N_2 = \left(\frac{7{,}000 \times 3{,}560^3 \times 2.625^3}{900 \times 321{,}000}\right)^{.5}$$

$$N_2 = 4{,}450 \text{ RPM}$$

Shaft Sizing Based on Fatigue Evaluation

Pump shafts are subjected to reversing or fluctuating stresses and can fail even though the actual maximum stresses are much less than the yield strength of the material. A pump shaft is subject to alternating or varying stresses as a result of the static weight and radial load of impellers, pressure pulses as impeller vanes pass diffuser vanes or cutwater lips, driver start-stop cycles, flow anomalies due to pump/driver/system interaction, driver torque variations, and other factors. In order to perform a fatigue analysis, it is first necessary to quantify the various alternating and steady-state loads and establish the number of cycles for the design life. In most cases, the design life is for an infinite number of cycles; however, in the case of start-stop cycles, the design life might be 500 or 1,000 cycles depending on the application.

Once the loads have been defined and the stresses have been calculated, it is necessary to establish what the acceptable stress values are. The use of the maximum-shear-stress theory of failure in conjunction with the Soderberg diagram provides one of the easier methods of determining the acceptable stress level for infinite life (Peterson; Shigley; Roark). Since the primary loads on pump shafts are generally torsion and bending loads, the equation for acceptable loading becomes:

$$\left(\frac{\tau a}{.5Se} + \frac{\tau m}{.5Sy}\right)^2 + \left(\frac{\sigma a}{Se} + \frac{\sigma m}{Sy}\right)^2 \leq 1 \qquad (16\text{-}2)$$

Where τa and σa = alternating stress components

τm and σm = mean stress components

Se = fatigue endurance limit for the shaft material corrected for the effects of temperature, size, surface roughness, and stress concentration factors

Sy = yield strength for the material at the operating temperature

A safety factor is generally applied to Se and Sy to account for unanticipated loads. Equation 16-2 is applied at the location(s) where stresses are the highest.

There are circumstances where it is not necessary to have infinite-life for certain loads. The designer must review all the operating modes and possible upset conditions before a load is classified as a finite-life load. Loads that might be placed in this category are start-stop cycles, and off-design flow, speed, or temperature transients. If the event has an anticipated occurrence of less than 1,000 cycles, it can be considered as a static load with no effect on fatigue life, providing the stresses are less than the material yield strength. For loading conditions of more than 1,000 cycles, but less than 10^7 cycles, the designer has the option to perform a cumulative fatigue damage analysis.

Example

It has been determined that the maximum stresses occur at an impeller-locating ring groove shown in Figure 16-1. The steady state torque is 28,000 in.-lb. The bending moment due to radial hydraulic load is 10,700 in.-lb. Due to axial thrust, there is a tensile force in the shaft of 20,000 lb. Are the stresses at the locating ring acceptable?

Solution

The steady-state loads are the torque and axial load. The alternating bending stress is:

$$\sigma = \frac{Mr}{I}$$

$$I = \frac{\pi D^4}{64} = 3.330 \text{ in.}^4$$

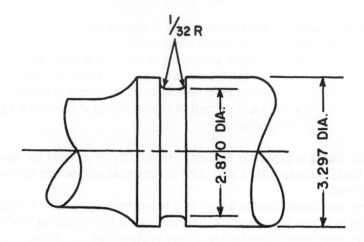

Figure 16-1. Shaft section.

$$\sigma = \frac{10,700}{3.330} \text{ in.-lb.} \times \frac{2.870}{2} = 4,610 \text{ psi}$$

Yield strength of 4140 steel is 80,000 psi.
The endurance limit for 4140 steel is 52,500 psi. This limit must be adjusted to account for the service condition (Shigley).

$$Se = K_a K_c K_e S'e$$

For 99% reliability use $K_c = .814$. For ground surface finish use $K_a = .9$. For stress concentration due to locating ring groove radius:

$$K_e \cong \frac{1}{K_t}$$

where K_t is the stress concentration factor from Peterson.

$$K_t = 3.75$$

$$K_e = \frac{1}{3.75} = .267$$

$$S'e = 52,500 \text{ psi}$$

$$Se = .9 \times .814 \times .267 \times 52,500 = 10,300 \text{ psi}$$

Torsional stress:

$$\tau = \frac{Tr}{J}$$

$$J = \frac{\pi D^4}{32} = 6.661 \text{ in.}^4$$

$$\tau = \frac{28,000 \text{ in.-lb}}{6.661 \text{ in.}^4} \times \frac{2.870}{2} \text{ in.} = 6,030 \text{ psi}$$

Axial stress:

$$\sigma = \frac{P}{A}$$

$$A = \frac{\pi D^2}{4} = 6.469 \text{ in.}^2$$

$$\sigma = \frac{20,000 \text{ lb}}{6.469 \text{ in.}^2} = 3,090 \text{ psi}$$

Hence from Equation 16-2:

$$\left(\frac{6,030}{.5 \times 80,000}\right)^2 + \left(\frac{4,610}{10,300} + \frac{3,090}{80,000}\right)^2$$

$$= .0227 + .237 = .260 < 1$$

Stresses are satisfactory.

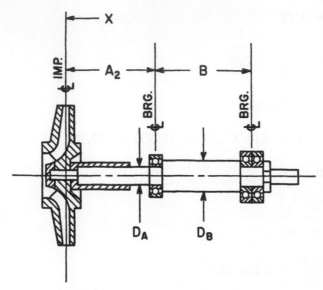

Figure 16-2. Schematic representation of horizontal overhung pump.

Shaft Deflection

Shaft deflection calculations are usually performed on single-stage overhung pumps to establish a relative measure of shaft stiffness. Deflection calculations are also performed on horizontal multi-stage pumps when the potential of galling at wear ring or sleeves exists during start-up or coastdown. The calculation for single-stage overhung pumps assumes that the loading on the shaft consists of weight (for a horizontal pump) and the dynamic radial thrust due to the pump hydraulics.

Figure 16-2 schematically represents a typical horizontal overhung pump (Figure 16-3). The deflection at any point along the shaft between the impeller and the radial bearing can be calculated from the following equation.

$$Y = \frac{W}{3E}\left(\frac{A_2{}^3}{I_A} + \frac{A_2{}^2 B}{I_B}\right) - \frac{W}{3E}\left[\times \left(\frac{3\,A_2{}^2}{2\,I_A} + \frac{A_2 B}{I_B}\right) - \frac{x^3}{2\,I_A}\right]$$

This equation does not account for any support the shaft might receive from the hydrostatic stiffness at the impeller wear rings and shaft throttle bushing. The deflection calculated by this formula is conservative and should be used for relative comparisons; the actual value will be less.

The calculation of shaft deflection for a horizontal multi-stage pump is complicated by the fact that the distributed weight of the shaft as well as the concentrated weight of the impellers must be accounted for. A method of approximating the shaft deflection can be found in Stepanoff.

Key Stress

A simple but important part of shaft design is calculating key stress. In most centrifugal pumps, driver torque is transmitted across the major

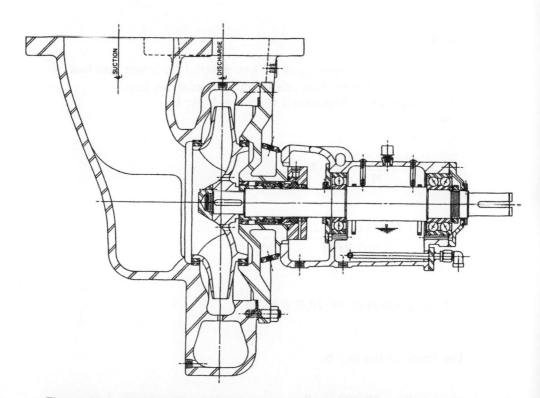

Figure 16-3. Single-stage overhung pump (courtesy BW/IP International, Inc. Pump Division, manufacturer of Byron Jackson/United™ Pumps).

component interfaces by means of a shear force applied to a key (e.g. driver shaft to driver half coupling hub; pump half coupling hub to pump shaft; and pump shaft to impeller). The general practice is to use a square key which has a height equal to one-fourth of the shaft diameter. (See ANSI-B17.1). The length of the key is determined by the strength requirements and the length of the hub. The strength of the key is evaluated by calculating the bearing and shear stress, assuming that the forces are uniformly distributed along the length of the key.

Example

For a pump with a $2^7/8$-in. diameter shaft at the coupling and a maximum horsepower of 1,580 at 3,560 RPM, determine the coupling key size. Shaft material is 4140 and the coupling key is 316 stainless steel.

Solution

The mechanical properties of the key material are lower than that of the shaft, so it will be adequate to only examine the key.
The torque to be transmitted from the driver is:

$$T = \frac{bhp \times 5,250}{N} \text{ ft-lb}$$

$$T = \frac{1,580 \text{ bhp} \times 5,250}{3,560 \text{ RPM}} \frac{\text{ft-lb RPM}}{\text{bhp}}$$

$$T = 2,330 \text{ ft-lb} = 28,000 \text{ in.-lb}$$

The force on the key is:

$$F = \frac{T}{r} = \frac{28,000 \text{ in.-lb}}{(2.875/2) \text{ in.}} = 19,478 \text{ lb}$$

Key section:

Using diameter divided by 4 as a rule of thumb, a 3/4-in. square key will be used.

Shear stress:

- 316 stainless steel.
- Yield Strength = 30,000 psi.
- For shear use .5 × 30,000 = 15,000 psi.
- With a safety factor of 1.5, the design stress should be 10,000 psi.
- Key shear area = Width × Length.

Shear stress is:

$$\tau_1 = \frac{F}{A_1}$$

Hence

Shaft Design, and Axial Thrust

$$L = \frac{F}{\tau_1 \times W_1} = \frac{19,478}{10,000 \times .75} = 2.597$$

Use 2⅝

See Figure 16-4 for a graphic illustration of the solution to this example.

Axial Thrust

In a centrifugal pump, axial thrust loads result from internal pressures acting on the exposed areas of the rotating element. These pressures can be calculated; however, such values should only be considered approximate, as they are affected by many variables. These include location of the impeller relative to the stationary walls, impeller shroud symmetry, surface roughness of the walls, wear ring clearance, and balance hole geometry. Calculated axial thrust is therefore based on a number of assumptions and should only be considered an approximation.

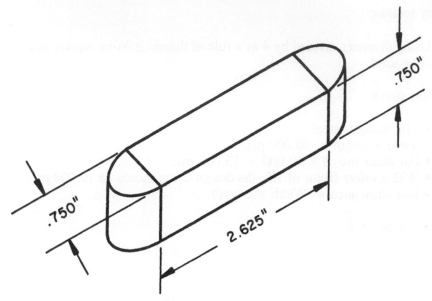

Figure 16-4. Coupling key.

In the methods of calculation described in this section, it is assumed that single- or multi-stage pumps will have 1/4-in. end play in either direction and that the impeller centerline will be located in the center of the volute, within reasonable limits (say ± 1/32 in.). It is also assumed that impeller profile is symmetrical and that the pressure on front and back shroud is equal and constant from the impeller outside diameter to the ring diameter. For simplification, parabolic reduction in pressure is ignored and the axial thrust calculations shown assume an average pressure acting on the impeller shrouds of 3/4 P_D.

On critical pump applications, it is recommended that performance testing include a thrust test. A thrust testing device mounted on the existing bearing housing will measure magnitude and direction of thrust by means of an inboard and outboard load cell. In addition, pressure sensing holes should be drilled through case and cover to record accurate pressure readings acting on the various parts of the rotor. Should the measured thrust be unacceptable, internal diameter changes can be analyzed and modified to reduce thrust to an acceptable level. This procedure is also useful for field observations on pumps experiencing thrust bearing problems and removes any doubt if in fact the problem is a result of excessive thrust.

Approximate methods of calculation for various pump configurations are now described.

Double-Suction Single-Stage Pumps

Theoretically double-suction pumps with double-suction impellers are in axial balance due to the symmetrical mechanical arrangement. If the flow approach to the impeller is also symmetrical, there is no need for thrust bearings. However, as minor mechanical or casting variations will cause axial unbalance, small capacity thrust bearings are normally supplied.

If symmetry of this type of pump is violated by relocating the impeller away from volute center, the developed pressure above the impeller rings on either side of the impeller will not be the same and substantial axial thrust will be developed. The amount of thrust will vary directly with axial displacement.

Single-Suction Single-Stage Overhung Pumps

Single-suction closed or open impellers have different geometries on either side of the impeller and thus develop substantial axial thrust. Calculating this thrust requires analysis of the different design variations normally used in commercial pumps.

Closed impeller with rings on both sides and balancing holes in back shroud. Balance holes through the impeller allow liquid leaking across the wear ring clearance to flow back to the suction side of the impeller. This aids in reducing thrust as the pressure behind the balance holes will be less than the pressure above the wear ring but greater than the suction pressure. Many simplified axial thrust calculations assume the pressure behind the balance holes to be the same as suction pressure; however, testing on small process pumps has shown this pressure to range from 2% to 10% pump differential pressure, depending on balance hole geometry. The calculation method shown in Figure 16-5 assumes liberal clearance holes with the hole area being a minimum of eight times the area between the wear rings. In the event that jetting action of leakage behind the back ring clearance creates a wear pattern on the back cover, it is often necessary to install a hardened ring on the pump cover.

Closed impeller with front ring only. All things being symmetrical, this impeller is axially balanced from the impeller outside diameter down to the suction ring diameter. Below this, there is an unbalance force acting on the area between the suction ring and the shaft sleeve. For the calculation method see Figure 16-6.

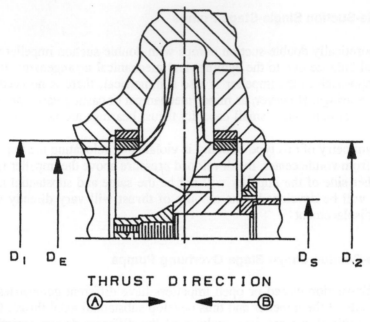

THRUST DIRECTION

$$\text{(A)} \longrightarrow \qquad \longleftarrow \text{(B)}$$

WHERE $D_1 = D_2$

NET THRUST $T = A_S P_S - [.03 P_D (A_1 - A_S)] + F_X$

WHERE $D_1 \neq D_2$

THRUST DIRECTION (A) $T_A = A_S P_S - [.03 P_D (A_1 - A_S)] + F_X$

THRUST DIRECTION (B) $T_B = 3/4\ P_D (A_1 - A_2)$

NET THRUST $T = T_A - T_B$

Figure 16-5. Axial thrust on single-suction closed impellers with balance holes.

Closed impeller with front ring and back balancing ribs. Where it is not desirable to design the impeller with a back ring, axial thrust can be balanced by designing integrally cast ribs on the back of the impeller shroud. A gap between rib and cover of 0.015 in. maximum, permits the liquid to rotate at approximately full impeller angular velocity, thus reducing the pressure on the impeller back shroud over the area of the rib. This method of balance is not very accurate or positive. If the gap is increased, the pressure acting on the back shroud increases, resulting in higher axial loads. Experience has shown that this type of impeller is less efficient than one with a back ring. Axial thrust can be estimated as shown in Figure 16-7.

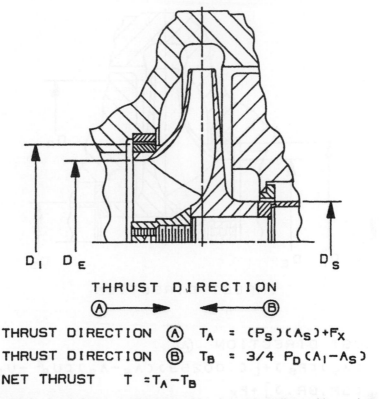

$$D_I \quad D_E \qquad\qquad\qquad\qquad D_S$$

$$\text{THRUST} \quad \text{DIRECTION}$$

$$\textcircled{A} \longrightarrow \qquad \longleftarrow \textcircled{B}$$

$$\text{THRUST DIRECTION} \quad \textcircled{A} \quad T_A = (P_S)(A_S) + F_X$$

$$\text{THRUST DIRECTION} \quad \textcircled{B} \quad T_B = 3/4 \; P_D (A_I - A_S)$$

$$\text{NET THRUST} \quad T = T_A - T_B$$

Figure 16-6. Axial thrust on single-suction closed impellers with no rings on back side.

Semi-open impellers. Depending on level of thrust, semi-open impellers can be designed with or without back ribs. Pumps of this type are quite often troublesome and inefficient. The gap between the open vane and liner must be carefully controlled at assembly, as any increase in this gap will affect the head developed by the pump. The problem is compounded when the back rib is added, as gaps on each side of the impeller must be controlled. Methods for calculating axial thrust are shown in Figures 16-8 and 16-9.

Multi-Stage Pumps

Axial thrust becomes more important with multi-stage pumps, because of the higher developed pressure and combined thrust of several stages. Impeller arrangement will normally be back to back for volute-type

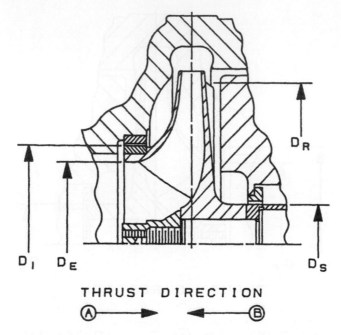

THRUST DIRECTION

Ⓐ ———▶ ◀——— Ⓑ

THRUST DIRECTION Ⓐ

$$T_A = (A_S)(P_S) + \left[(.00253)(A_R - A_S)(U_R^2 - U_S^2) (SP.GR.) \right] + F_X$$

THRUST DIRECTION Ⓑ

$$T_B = 3/4 \ P_D (A_I - A_S)$$

NET THRUST $\quad T = T_A - T_B$

Figure 16-7. Axial thrust on single-suction closed impeller with ribs on back side.

pumps and stacked (all impellers facing one direction) for diffuser pumps.

Back-to-back impellers. This arrangement simplifies axial balance for an odd or even number of stages. In this design there are two locations on the rotating element that break down half the total pump head. One is the center bushing between the center back-to-back impellers, and the other is the throttle bushing in the high-pressure stuffing box. Sleeve diameters at these locations can be sized to provide axial balance for any number of stages.

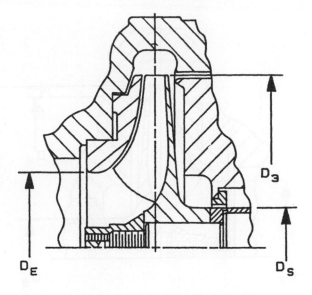

THRUST   DIRECTION

$$T_A = (A_S)(P_S) + \left[(A_3 - A_E) \frac{P_D}{2} \right] + F_X$$

THRUST   DIRECTION   Ⓑ

$$T_B = 3/4 \ P_D (A_3 - A_S)$$

NET   THRUST $T = T_A - T_B$

Figure 16-8. Axial thrust on single-suction open impellers without ribs on back side.

It is recommended that the element maintain a small unbalance (minimum 100 lb) in one direction to lightly load the bearing and prevent floating of the rotor. Normal wear of the original ring or bushing clearance will not affect balance. It is recommended that all impeller rings be the same size and on even-stage pumps that the center sleeve diameter be the same as the sleeve diameters in each stuffing box. To balance an odd number of stages, the simplest solution is to change only the diameter of the center sleeve.

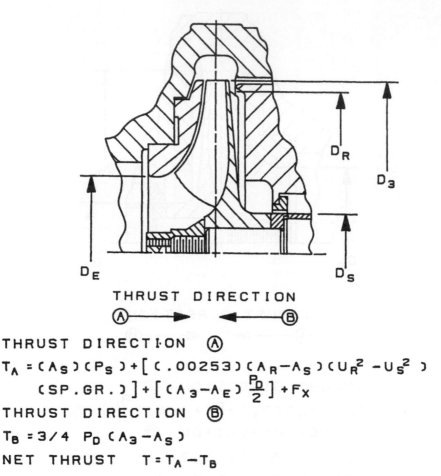

THRUST DIRECTION

THRUST DIRECTION (A)

$$T_A = (A_S)(P_S) + [(.00253)(A_R - A_S)(U_R^2 - U_S^2)$$
$$(SP.GR.)] + [(A_3 - A_E)\frac{P_D}{2}] + F_X$$

THRUST DIRECTION (B)

$$T_B = 3/4 \ P_D \ (A_3 - A_S)$$

NET THRUST $\quad T = T_A - T_B$

Figure 16-9. Axial thrust on single-suction open impeller with ribs on back side.

Stacked impellers. With all impellers facing in one direction, the area between the last impeller and the high-pressure stuffing box is under full discharge pressure. Pressure is broken down across a balance drum, with the diameter of the drum sized for axial balance. Wear on the balance drum will affect axial balance, therefore bearing rating should be sized accordingly. On critical applications, particularly where a high rate of erosion wear is anticipated, the permanent installation of load cells should be considered.

The low-pressure side of the balance drum is piped back to suction, as is the low-pressure side of the throttle bushing in the back-to-back ar-

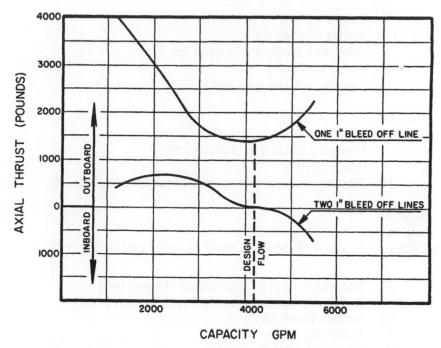

Figure 16-10. Increasing number of bleedoff lines reduces thrust.

rangement. This aids axial balance and subjects the inboard and outboard mechanical seal to only suction pressure. It is essential that the bleed-off line be adequately sized to transfer leakage back to suction without permitting a pressure build-up in the stuffing box. Any such build-up of pressure will have an adverse effect on axial thrust. If pressure build-up is detected, or suspected, the bleed-off line should be increased in size or additional lines added. This principle also applies to overhung process pumps with balance holes, where the back pressure can be reduced by increasing the size of the holes. The effect on thrust of a two-stage pump with one and two balance lines is illustrated in Figure 16-10. Note, that the thrust level and direction of thrust is not constant across the capacity range of the pump. This tendency is typical of most pumps.

Notation

Shaft Design

τ_1 Shear stress (psi)
τ Torsional stress (psi)

T	Torque (lb-in.)
r	Radius of shaft (in.)
J	Polar area moment of inertia (in.^{4})
P	Power (ft-lb/sec)
ω	Angular velocity (rad./sec)
D (with or without subscript)	Diameter of shaft as indicated (in.)
N	Speed (RPM)
M	Bending moment due to radial hydraulic load (in.-lb)
I (with or without subscript)	Moment of inertia of shaft as indicated (in.^{4}) = $\dfrac{\pi D^4}{64}$
Se	Endurance limit of mechanical element (psi)
S$'$e	Endurance limit of rotating beam specimen (psi)
Ka	Surface finish factor
Kc	Reliability factor
Ke	Stress concentration factor
A	Area of shaft cross section (sq in.)
W	Impeller weight plus radial thrust (lb)
E	Modulus of elasticity of shaft (psi)
A$_2$	Distance from centerline of impeller to centerline of inboard bearing (in.)
B	Bearing span centerline to centerline (in.)
X	Distance from centerline of impeller to location on shaft where deflection is to be calculated (in.)
bhp	Brake horsepower
A$_1$	Key shear area (sq in.)
Y	Deflection at any point on shaft (in.)
F	Force (lb)
L	Useful key length (in.)
W$_1$	Key width (in.)

Axial Thrust

A (with subscript)	Area of cross section as indicated by subscript (sq in.) = $\pi D^2/4$.

D	
(with subscript)	Diameter of cross section as indicated by subscript (in.).
P_D	Differential pressure (psi).
P_S	Suction pressure (psi).
T	Net axial thrust (lb).
T (with subscript)	Axial thrust in direction as indicated by subscript. (lb)
U (with subscript)	Peripheral velocity at diameter indicated by subscript (ft/sec) = D × RPM/229.
sp gr	Specific gravity.
g	32.2 ft/sec² (constant used in calculation for Figures 16-14 and 16-16)
γ	Fluid density of water lb/ft³ (constant used in calculation for Figures 16-14 and 16-16.
F_X	Thrust due to momentum change (lb) = (Q² × sp gr)/A_E × 722
Q	GPM

References

Peterson, R. E., *Stress Concentration Factors,* John Wiley and Sons, Inc., New York, 1974.

Roark, R. J., and Young, W. C., *Formula For Stress and Strain,* 5th edition, McGraw-Hill, Inc., 1975.

Shigley, J. E. *Mechanical Engineering Design,* 3rd edition, McGraw-Hill, Inc., 1977.

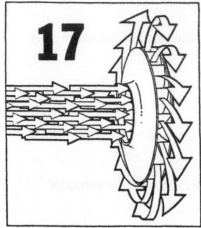

17

Mechanical Seals

by **James P. Netzel**
John Crane Inc.

The most common way to seal a centrifugal pump is with a mechanical seal. The basic components of a mechanical seal are the primary and mating rings. When in contact they form the dynamic sealing surfaces that are perpendicular to the shaft. The primary ring is flexibly mounted in the seal head assembly, which usually rotates with the shaft. The mating ring with a static seal, forms another assembly that is usually fixed to the pump gland plate (Figure 17-1). Each of the sealing planes on the primary and mating rings is lapped flat to eliminate any visible leakage. There are three points of sealing common to all mechanical seal installations:

- At the mating surfaces of the primary and mating rings.
- Between the rotating component and the shaft or sleeve.
- Between the stationary component and the gland plate.

When a seal is installed on a sleeve, there is an additional point of sealing between the shaft and sleeve. Certain mating ring designs may also require an additional seal between the gland plate and stuffing box.

Normally the mating surfaces of the seal are made of dissimilar materials and held in contact with a spring. Preload from the spring pressure holds the primary and mating rings together during shutdown or when there is a lack of liquid pressure.

The secondary seal between the shaft or sleeve must be partially dynamic. As the seal faces wear, the primary ring must move slightly forward. Because of vibration from the machinery, shaft runout, and thermal expansion of the shaft to the pump casing, the secondary seal must

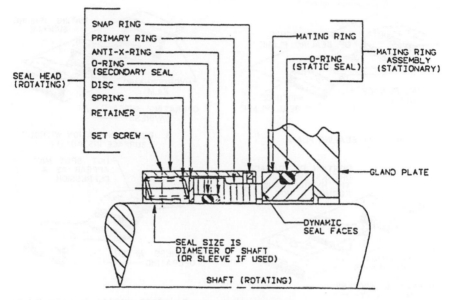

SNAP RING
PRIMARY RING
ANTI-X-RING
O-RING
(SECONDARY SEAL
SEAL HEAD
(ROTATING)
DISC
SPRING
RETAINER

SET SCREW

MATING RING
O-RING
(STATIC SEAL)
MATING RING
ASSEMBLY
(STATIONARY)

GLAND PLATE

DYNAMIC
SEAL FACES

SEAL SIZE IS
DIAMETER OF SHAFT
(OR SLEEVE IF USED)

SHAFT (ROTATING)

Figure 17-1. Mechanical seal.

move along the shaft. This is not a static seal in the assembly. Flexibility in sealing is achieved from such secondary seal forms as a bellows, O-ring, wedge, or V-ring. Most seal designs fix the seal head to the sleeve and provide for a positive drive to the primary ring.

Although mechanical seals may differ in various physical aspects, they are fundamentally the same in principle. The wide variation in design is a result of the many methods used to provide flexibility, ease of installation, and economy.

Theory of Operation

Successful operation of a seal depends upon developing a lubricating film and controlling the frictional heat developed at the seal faces. Whenever relative motion occurs between the primary and mating rings, frictional heat is generated. Normally, cooling of the seal faces is accomplished by a seal flush. Many theories have been developed to explain the formation of the lubricating film. These theories include the types of motion transferred to the seal, thermal distortion of the sealing plane, and surface waviness of both the primary and mating rings (Figure 17-2). The generation of a lubricating film is particularly important if the seal is run for a long period of time without exhibiting any appreciable wear at

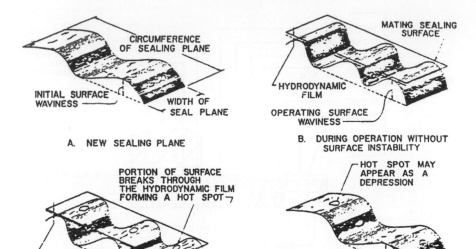

A. NEW SEALING PLANE

B. DURING OPERATION WITHOUT SURFACE INSTABILITY

C. DURING OPERATION WITH INSTABILITY

D. AFTER INSTABILITY WHEN SLIDING SYSTEM COMES TO REST

Figure 17-2. Mechanical seal waviness for different conditions of operation.

the faces or any seal leakage. Surface waviness is believed to be the primary reason for the generation of a lubricating film. Despite the fact that both seal faces are lapped flat, a very small amount of initial waviness remains (Figure 17-2A). For a hard seal face like tungsten carbide, the initial surface waviness will be 2 to 8 micro-inches. Softer seal face materials like carbon graphite would be within the range of 10 to 16 micro-inches. As sliding contact occurs between the primary and mating rings of the seal, frictional heat develops that increases the surface waviness of each seal face from its initial value to some operating waviness. One of the consequences of waviness is that the changes to the sealing surfaces are not uniform. However, continuous operation without solid contact between the seal faces is the result of an ideal wave formation for the operating conditions of the seal. This also results in the development of a lubricating film and a long life for the seal installation (Figure 17-2B).

Under certain conditions non-uniform heating of the seal faces may lead to even more concentration of friction. Portions of the sealing surfaces will break through the lubricating film, resulting in solid contact and increased frictional heat in localized areas. The formation of highly stressed hot spots on the seal face will reduce the effects of the lubricat-

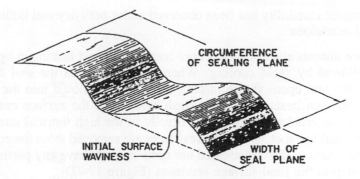

Figure 17-2A. New sealing plane.

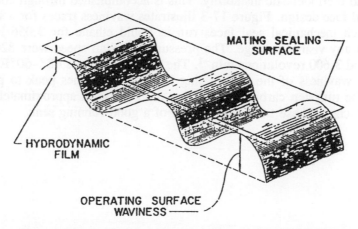

Figure 17-2B. During operation without surface instability.

ing film and these spots will grow in intensity, generating still more heat. This intense heat may cause the liquid being sealed to flash or vaporize, resulting in unstable operation of the seal. The transition from a normally flat condition to a highly deformed surface is called thermoelastic instability (Figure 17-2C).

During the mid-seventies, many experiments were conducted to determine if this could be present in mechanical seals. Burton et al. (1980) demonstrated this phenomenon by running a seal configuration against a glass plate. Hot patches or spots appeared on the surface of the metal primary ring. These spots moved through the entire circumference of the seal. Movement of these spots is believed to be the result of wear. Confirmation of hot patches or spots on conventional materials used in mechanical seals was determined by Kennedy (1984). Localized hot spots from thermoelastic instability has been observed under both dry and liquid lubricated conditions.

Surface distress at a hot spot occurs because of rapid heating in operation, followed by rapid cooling. When a liquid film at the seal faces flashes, the seal opens for an instant, allowing cool liquid into the seal faces. This then heats up only to flash again until the surface cracks. These thermo cracks are attributed directly to the high thermal stresses near the small hot spot. When the seal faces are removed from the equipment, the location of the distressed hot spots may not have any particular relationship to the final surface waviness (Figure 17-2D).

Stable operation of a seal is achieved through control on surface waviness and thermoelastic instability. This is accomplished through cooling and seal face design. Figure 17-3 illustrates waviness traces for a set of 3.94-inch mechanical seal faces run in liquid ethane for 3,056 hours without any visible leakage. The pressure and shaft speed were 820 lbs in.$^{-2}$ and 3,600 revolutions min^{-1}. The temperature was 48°–60°F. The surface waviness of the carbon ring is 75 micro-inches peak to peak, while the tungsten carbide surface has a waviness of approximately 12 micro-inches. These traces are typical of a good running seal.

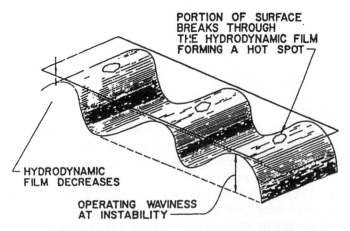

PORTION OF SURFACE
BREAKS THROUGH
THE HYDRODYNAMIC FILM
FORMING A HOT SPOT

HYDRODYNAMIC
FILM DECREASES

OPERATING WAVINESS
AT INSTABILITY

Figure 17-2C. During operation with instability.

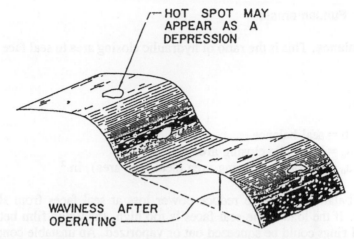

Figure 17-2D. After instability when sliding system comes to rest.

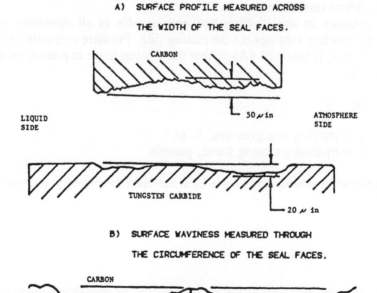

Figure 17-3. Wear of mechanical seal faces after 3056 hours of operation.

Design Fundamentals

Seal Balance. This is the ratio of hydraulic closing area to seal face area:

$$b = \frac{a_c}{a_o}$$

where b = seal balance
       a_c = hydraulic closing area in.^{2}
       a_o = hydraulic opening area (seal face area), in.^{2}

Seal balance is used to reduce power loss at seal faces from sliding contact. If the load at the seal faces is too high, the liquid film between the seal rings could be squeezed out or vaporized. An unstable condition from thermoelastic instability could result in a high wear rate of the sealing surfaces. Seal face materials also have a bearing limit that should not be exceeded. Seal balancing can avoid these conditions and lead to a more efficient installation.

The pressure in any stuffing box acts equally in all directions and forces the primary ring against the mating ring. Pressure acts only on the annular area a_c (Figure 17-4A) so that the closing force in pounds on the seal face is:

$$F_c = pa_c$$

where p = stuffing box pressure, lb/in.^{2}
       F_c = hydraulic closing force, pounds

The pressure in pounds per square inch between the primary and mating rings is:

$$\frac{F_c}{a_o} = \frac{pa_c}{a_o}$$

To relieve the pressure at the seal faces, the relationship between the opening and closing force can be controlled. If a_o is held constant and a_c decreased by a shoulder on a sleeve or on the seal hardware, the seal face pressure can be lowered (Figure 17-4B). This is called seal balancing. A seal without a shoulder in the design is an unbalanced seal. A balanced seal is designed to operate with a shoulder. Only a metal bellows seal is a balanced seal that does not require a shoulder. The balance diameter is the mean area diameter of the bellows.

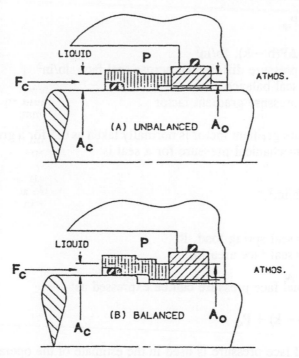

Figure 17-4. Hydraulic pressure acting on the primary ring.

Seals can be balanced for pressures at the outside diameter of the seal faces as shown in Figure 17-4B. This is typical for a seal mounted inside the stuffing box. Seals mounted outside the stuffing box can be balanced for pressure at the inside diameter of the seal faces. In special cases, seals can be double balanced for pressure at both the outside and inside diameters of the seal. Seal balance can range from 0.65 to 1.35, depending on operating conditions.

Face Pressure. This is an important factor in the success or failure of a mechanical seal. Hydraulic pressure develops within the seal faces that tend to separate the primary and mating rings. This pressure distribution is referred to as a pressure wedge (Figure 17-5). For most calculations it may be considered as linear. The actual face pressure P_f in pounds per square inch is the sum of the hydraulic pressure P_h and the spring pressure P_{sp} designed into the mechanical seal. The face pressure P_f is a further refinement of P which does not take into account the liquid film pressure or the load of the mechanical seal:

$$P_f = P_h + P_{sp}$$

where $P_h = \Delta P(b - k)$, lb/in^2
$\quad\quad$ P = pressure differential across seal face, lb/in^2
$\quad\quad$ b = seal balance
$\quad\quad$ k = pressure gradient factor

The pressure gradient factor is normally taken as 0.5 for a given installation. The mechanical pressure for a seal is:

$$P_{sp} = \frac{F_{sp}}{a_o} \ lb/in.^2$$

where F_{sp} = seal spring load, lb
$\quad\quad$ a_o = seal face area in.^{2}

then the actual face pressure can be expressed as:

$$P_f = \Delta P(b - k) + P_{sp}$$

The actual face pressure is used in the estimate of the operating pressure-velocity for a given seal installation.

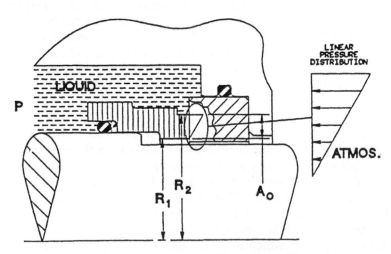

Figure 17-5. Pressure wedge opening force at the seal faces.

Pressure-Velocity. The value for a seal installation may be compared with values developed by seal manufacturers as a measure of adhesive wear. As the primary and mating rings move relative to each other, they are affected by the actual face pressure and rotational speed. The product of the two, pressure times velocity, is referred to as PV and is defined as the power N per unit area with a coefficient of friction of unity:

$$PV = \frac{N_f}{a_o}$$

For seals, the equation for PV can be written as follows:

$$PV = P \, V_m = [\Delta P(b - k) + P_{sp}]V_m$$

where V_m = velocity at the mean face diameter d_m, ft/min

Power Consumption. The power consumption of a seal system can be estimated using the PV value and the following equation:

$$N_f = (PV) \, f \cdot a_o \text{ ft. lb/min}$$

where f is the coefficient of friction.

As a general rule, the power to start a seal is five times the running value.

The coefficients of friction for various common seal face materials are given in Table 17-1. These coefficients were developed with water as a lubricant at an operating PV value of 100,000 lb/in.^{2} ft/min. Values in oil would be slightly higher as a result of viscous shear of the fluid film at the seal faces.

Example:

A pump having a 3-inch diameter sleeve at the stuffing box is fitted with a balanced seal of the same size and mean diameter. The seal operates in water at 400 lb/in.^{2}, 3,600 rpm and ambient temperatures. The material of construction for the primary and mating rings are carbon and tungsten carbide. Determine the PV value and power loss of the seal.

Given: $P = 400 \text{ lb/in.}^2$
$\quad b = 0.75$
$\quad k = 0.5$
$\quad d_m = 3 \text{ in.}$
$\quad P_{sp} = 25 \text{ lb/in.}^2$

$$V_m = \frac{\pi}{12} \times 3 \times 3,600 = 2,826 \text{ ft/min}$$

$\quad a_o = 1.88 \text{ in.}^2$
$\quad f = 0.07$

$$PV = [400 \, (.75 - 0.5) + 25] \, (2826) = 353,250 \, \frac{\text{lb.}}{\text{in}^2} \cdot \frac{\text{ft}}{\text{min}}$$

$$N = (353,250) \, (0.07) \, (1.88) = 46487 \, \frac{\text{ft. lb}}{\text{min}} = 1.4 \text{ hp.}$$

Temperature Control. At the seal faces, temperature control is desirable because wear is a function of temperature. Heat at the seal faces will cause the liquid film to vaporize, generating still more heat. Thermal distortion may also result, which will contribute to seal leakage. Therefore, many applications will require some type of cooling.

Table 17-1
Coefficient of Friction for
Various Seal Face Materials
(Water Lubrication, PV—100,000 Lb/In. W × Ft/min. John Crane Inc.)

SLIDING MATERIALS		COEFFICIENT OF FRICTION
PRIMARY RING	MATING RING	
CARBON-GRAPHITE (RESIN FILLED)	CAST IRON	0.07
	CERAMIC	0.07
	TUNGSTEN CARBIDE	0.07
	SILICON CARBIDE	0.02
	SILICONIZED CARBON	0.015
SILICON CARBIDE	TUNGSTEN CARBIDE	0.02
	SILICONIZED CARBON	0.05
	SILICON CARBIDE	0.02
	TUNGSTEN CARBIDE	0.08

The temperature of the seal faces is a function of the heat generated by the seal plus the heat gained or lost to the pumpage. The heat generated at the faces from sliding contact is the mechanical power consumption of the seal being transferred into heat. Therefore,

$$Q_s = C_1 N_f = C_1 \ (PV_f \ a_o)$$

where Q_s = heat input from the seal Btu/h
$C_1 = 0.077$

If the heat is removed at the same rate it is produced, the temperature will not increase. If the amount of heat removed is less than that generated, the seal face temperature will increase to a point where damage will occur. Estimated values for heat input are given in Figure 17-6 and Figure 17-7.

Heat removal from a single seal is generally accomplished by a seal flush. The seal flush is usually a bypass from the discharge line on the pump or an injection from an external source. The flow rate for cooling can be found by:

$$Gpm = \frac{Q_s}{C_2 \ (sp.ht.) \ (SpGr) \ \Delta T}$$

where Q_s = seal heat, Btu/h(W)
$C_2 = 500$
SpHt = specific heat of coolant, Btu/lb°F
SpGr = specific gravity of coolant
ΔT = temperature rise, °F

When handling liquids at elevated temperatures, the heat input from the process must be considered in the calculation of coolant flow. Then

$$Q_{net} = Q_p + Q_s$$

The heat load Q_p from the process can be determined from Figure 17-8.

Example:

Determine the net heat input for a 3-inch diameter balanced seal in water at 3600 rpm. Pressure and temperature are 400 lb/in² and 170°F. From Figure 17-7.

$$Q_s = (2750 \ Btu/h/1000 \ rpm) \ 3600 = 9900 \ Btu/h$$

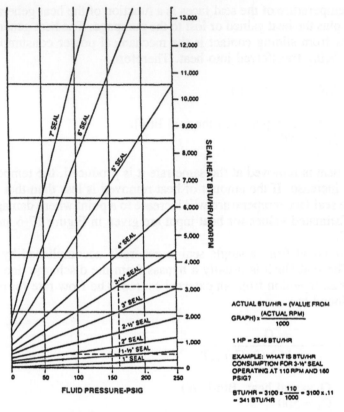

ACTUAL BTU/HR = (VALUE FROM

GRAPH) x (ACTUAL RPM) / 1000

1 HP = 2545 BTU/HR

EXAMPLE: WHAT IS BTU/HR
CONSUMPTION FOR 3-½" SEAL
OPERATING AT 110 RPM AND 160
PSIG?

BTU/HR = 3100 x $\frac{110}{1000}$ = 3100 x .11
= 341 BTU/HR

Figure 17-6. Unbalanced seal heat generation (courtesy of John Crane).

From Figure 17-8 assuming that the stuffing box will be cooled to 70°F and that the temperature difference between the stuffing box and pumpage is 100°F

Q_p = 255 Btu/h
Q_{net} = 10,155 Btu/hr

The total heat input is used to estimate the required flow of coolant to the seal. For single seal installations, the flow of coolant or seal flush may be achieved through the use of a bypass line from pump discharge. Some installations may require an injection from an external source of liquid. A bypass line can be built internal to the pump or may be constructed externally on the unit. External piping will allow for the installa-

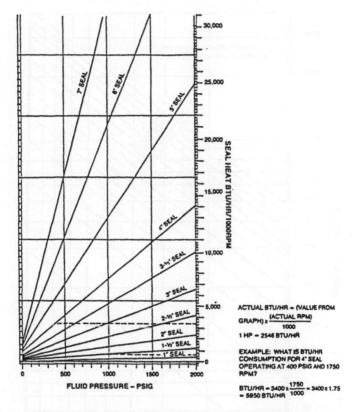

Figure 17-7. Balanced seal heat generation (courtesy of John Crane).

tion of a heat exchanger or abrasive separator when required (Figure 17-9). Another method of supplying additional cooling to a stuffing box is shown in Figure 17-10. This seal is equipped with a pumping ring and heat exchanger. The pumping ring acts like a pump, causing liquid to flow through the stuffing box. Liquid is moved out the top of the stuffing box and through the heat exchanger, returning directly to the faces at an inlet at the bottom of the end plate. As liquid is circulated, heat is removed at the seal and stuffing box. This is an extremely efficient method of removing heat because the coolant circulated is only in the stuffing box and does not reduce the temperature of the liquid in the pump. A closed loop system such as this is commonly used on hot water pumps.

For double seal installations, the heat load for each seal must be considered, as well as the heat soak from the process. The flow of coolant can be achieved through the use of an external circuit with a small pump.

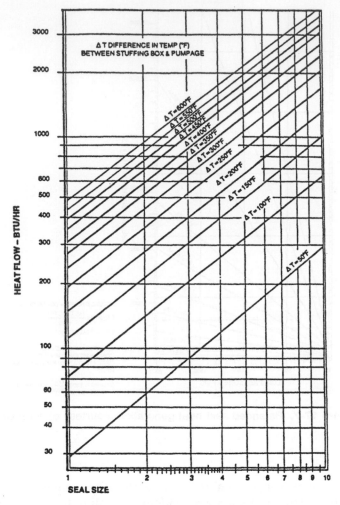

Figure 17-8. Heat soak from the process fluid (courtesy of John Crane).

Pressures in the stuffing box should be kept above the pressure at the inboard seal. As a rule of thumb, the stuffing box is normally 10% or a minimum of 25 psig above the product pressure against the inboard seal. For low pressure installations, a pressurized loop and axial flow pumping ring shown in Figure 17-11 can be considered. Here, as the shaft rotates, the pumping ring moves liquid to a reservoir tank where heat is removed from the system. Liquid returns to the bottom of the stuffing box and continues to remove heat from the seal area.

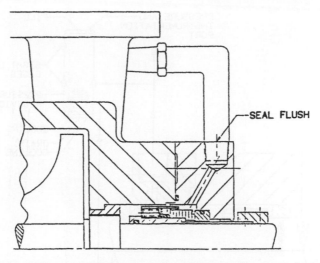

Figure 17-9. Package mechanical seal installation with external piping.

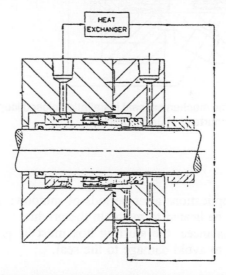

Figure 17-10. Package mechanical seal with pumping ring and heat exchanger closed looped for cooling (courtesy of John Crane).

On tandem seal installations, the inboard seal may be treated as a single seal installation for cooling. However, the outboard seal is normally a closed circuit with a pumping ring (Figure 17-16).

Positive methods for providing clean cool liquid to the seal must be considered at all times. The use of a thermal syphon system for removing heat on a pump application should be avoided.

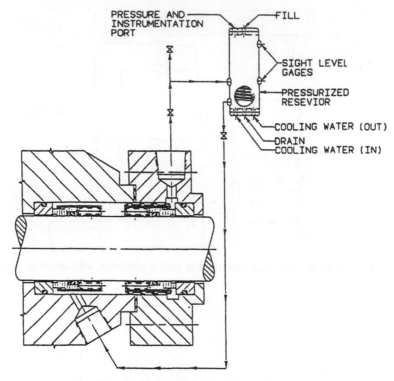

Figure 17-11. Double mechanical seal installation closed loop system with axial flow pumping ring (courtesy of John Crane).

Some special applications after periods of shutdown may require that the process liquid be heated to maintain the product being pumped as a liquid. In these instances, the stuffing box must be preheated prior to starting the pump to avoid damage to the seal.

Seal Leakage

An estimate for seal leakage in cubic centimeters per minute can be made from the following equation:

$$Q = -C_3 \frac{h^3 (P_2 - P_1)}{\mu \ln (R_2/R_1)}$$

where $C_3 = 532$

h = face gap, in

P_2 = pressure at face ID, lb/in.^{2}

P_1 = pressure at face OD, lb/in.^{2}

μ = dynamic viscosity, CP

R_2 = outer face radius, in.

R_1 = inner face radius, in.

Negative flow indicates flow from the face outer diameter to the inner diameter. The effect of centrifugal force from one of the rotating sealing planes is very small and can be neglected in normal pump applications. The gap between the seal faces is a function of the materials of construction, flatness, and the liquid being sealed. The face gap can range from 20×10^{-6} to 50×10^{-6} inches. Leakage from a seal is also affected by the parallelism of the sealing planes, angular misalignment, coning (negative face rotation), thermal distortion (positive face rotation), shaft run-out, axial vibration, and fluctuating pressure. Further discussions of these problem areas will be covered later in the chapter on seal installation and troubleshooting.

Seal Wear

The types of wear found in mechanical seals are adhesive, abrasive, corrosive, pitting or fatigue, blistering, impact, and fretting. Adhesive wear is the dominant type of wear in well designed seals. The other types of wear are related to problems with the entire sealing application and will be discussed later in seal installation and troubleshooting.

Even though the successful operation of a seal depends on the development of a lubricating film, solid contact and wear occur during startup, shutdown, and during periods of changing service conditions for the equipment. For most applications, seals are designed with a carbon graphite primary ring considered to be the wearing part in the assembly. The primary ring runs against the mating ring that is made of a harder material and wears to a lesser extent. In mechanical seal designs, face loads are sufficiently low so that only a mild adhesive wear process occurs. If the application is extremely abrasive, hard seal faces for both the primary and mating rings must be considered. Abrasive applications include sand type slurries, paint, and abrasives generated at elevated temperatures from carbonized products.

The PV criterion used in the design of seals is also used in the expression of the limit of mild adhesive wear. Table 17-2 gives the PV limitations for frequently used seal face materials. Each limiting value has been

Table 17-2
Frequently Used Seal Face Materials
and Their PV Limitations

SLIDING MATERIALS		PV LIMIT LB/IN2 FT.MIN	COMMENTS
PRIMARY RING	MATING RING		
CARBON GRAPHITE	NI-RESIST	100,000	BETTER THERMAL SHOCK RESISTANCE THAN CERAMIC
	CERAMIC (85% AL2O3)	100,000	POOR THERMAL SHOCK RESISTANCE. BETTER CORROSION RESISTANCE THAN NI-RESIST
	CERAMIC (99% AL2O3)	100,000	BETTER CORROSION RESISTANCE THAN 85% (AL2O3) CERAMIC
	TUNGSTEN CARBIDE (6% CO)	500,000	WITH BRONZE FILLED CARBON GRAPHITE. PV IS UP TO 800,000
	TUNGSTEN CARBIDE (6% NI)	500,000	NI BINDER FOR BETER CORROSION RESISTANCE
	SILICONIZED CARBON	500,000	GOOD WEAR RESISTANCE THIN LAYER OF SI-C MAKES RELAPPING QUESTIONABLE
	SILICON CARBIDE (SOLID)	500,000	BETTER CORROSION RESISTANCE THAN TUNGSTEN CARBIDE
CARBON GRAPHITE		50,000	LOW PV, BUT VERY GOOD AGAINST FACE BLISTERING
CERAMIC		10,000	GOOD SERVICE ON SEALING PAINT PIGMENTS
TUNGSTEN CARBIDE		120,000	PV IS UP TO 185,000 WITH GRADES THAT HAVE DIFFERENT BINDERS
SILICONIZED CARBON		500,000	EXCELLENT ABRASION RESISTANCE. MORE ECONOMICAL THAN SOLID SILICON CARBIDE
SILICON CARBIDE		500,000	EXCELLENT ABRASION AND GOOD CORROSION RESISTANCE MODERATE THERMAL SHOCK RESISTANCE

developed for a wear rate that provides an equivalent seal life of two years. A PV value for a given application may be compared with the limiting value to determine satisfactory service. These values apply to aqueous solutions at 120°F. For lubricating liquids such as oil, values 60% or higher can be used.

Three separate tests are performed by seal manufacturers to establish performance and acceptability of seal face materials. PV testing is a measure of adhesive wear. Abrasive testing establishes the relative ranking of materials in a controlled environment. Operating temperatures have a significant influence on wear. A hot water test is used to evaluate the behavior of seal face materials at temperatures above the atmospheric boiling point of the liquid.

Classification of Seals by Arrangement

Seal arrangement is used to describe the design of a particular seal installation and number of seals used on a pump. Sealing arrangements may be classified into two groups:

- Single seal installations
  · Internally mounted
  · Externally mounted
- Multiple seal installations
  · Double seals
    · Back to back
    · Opposed
  · Tandem

Single seals are commonly used on most applications. This is the simplest seal arrangement with the least number of parts. An installation mounted inside the stuffing box chamber is referred to as an inside mounted seal (Figure 17-12). Here, liquid in the stuffing box and under pressure acts with the spring load to keep the faces in contact.

An outside mounted seal refers to a seal mounted outside the stuffing box (Figure 17-13). Here, if the seal is not balanced for pressure at the inside diameter of the seal face, the pressure will try to open the seal. Outside mounted seals are considered only for low pressure applications. The purpose of an external seal installation is to minimize the effects of corrosion that might occur if the metal parts of the seal were directly exposed to the liquid being sealed.

Multiple seals are used in applications requiring:

- A neutral liquid for lubrication
- Improved corrosion resistance
- A buffered area for plant safety

Double seals consist of two seal heads mounted back to back with the carbon primary rings facing in opposite directions in the stuffing box chamber (Figure 17-14). A neutral liquid with good lubricating properties, at higher pressures than the pumpage, is used to cool and lubricate the seal faces. The inboard seal keeps the liquid being pumped from entering the stuffing box. Both the inboard and outboard seals prevent the loss of the neutral lubricating liquid. Circulation for cooling is normally achieved with an external circulation system and smaller pump. An axial flow pumping ring and closed system may be considered on low pressure systems.

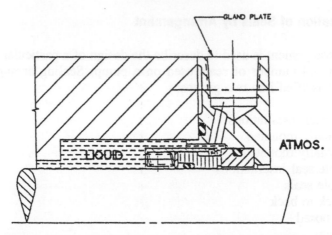

Figure 17-12. Single seal mounted inside the stuffing box.

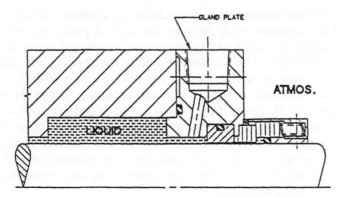

Figure 17-13. Single seal mounted outside the stuffing box. For low pressure service.

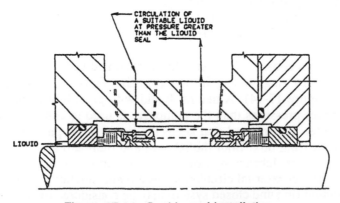

Figure 17-14. Double seal installation.

Double seals may be used in an opposed arrangement (Figure 17-15). Two seals are mounted face to face with the primary sealing rings running on mating rings supported by a common end plate. The neutral liquid is circulated between the seals at the inside diameter of the seal faces. The circulation pressure is normally less than the process liquid being sealed. The inboard seal is similar to a single inside mounted seal and carries the full differential pressure of the pump stuffing box to the neutral liquid. The outboard seal carries only the pressure of the neutral liquid to atmosphere. Both inboard and outboard primary rings are balanced to handle pressure at either the outside or inside diameter of the seal faces without opening. The purpose of this arrangement is to fit a stuffing box having a smaller confined space than what is possible with back-to-back double seals. Double seals are normally applied to toxic liquids for plant safety.

Tandem seals are an arrangement of two single seals mounted in the same direction (Figure 17-16). The inboard seal carries the full pressure differential of the process liquid to atmosphere. The outboard seal contains a neutral liquid and creates a buffered zone between the inboard seal and plant atmosphere. Normally the neutral lubricating liquid is maintained at atmospheric pressure. Developed heat at the inboard seal is removed with a seal flush similar to a single seal installation. The liquid in the outboard seal chamber should be circulated with a pumping ring to remove unwanted seal heat. Tandem seals are used on toxic or flammable liquids that require a buffered area or safety zone in the seal installation.

Package seals are an extension of other seal arrangements. A package seal requires no special measurements prior to seal installation. A package seal assembly consists of a seal head assembly, mating ring assembly, and adaptive hardware. Adaptive hardware are those structural parts that fit a specific seal to a pump. These hardware items are a gland plate, sleeve, and drive collar. A spacer or spacer clips are provided on most package seals to properly set the seal faces. The spacer or spacer clips are removed after the drive collar has been locked to the shaft and the gland plate bolted to the pump. Examples of package seals are shown in Figures 17-9, 17-10, and 17-15.

Classification of Seals by Design

Certain design features are considered important and may be used to describe a seal. These descriptions also form four classification groups. A seal may be referred to as:

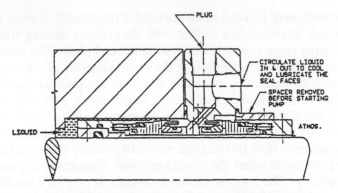

Figure 17-15. Package opposed double seals.

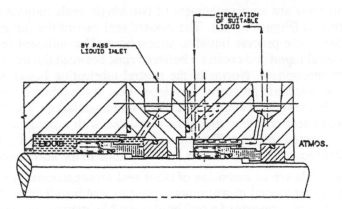

Figure 17-16. Tandem seal installation.

- Unbalanced or balanced
- Rotating or stationary seal head
- Single spring or multiple spring construction
- Pusher or non-pusher secondary seal design

The selection of an unbalanced or balanced seal is determined by the pressure in the pump stuffing box and the type of liquid to be sealed. Balance is a way of controlling the contact pressure between the seal faces and power loss at the seal. When the percentage of balance b (ratio of hydraulic closing area to seal face area) is 100 or greater, the seal is referred to as unbalanced. When the percentage of balance for a seal is less than 100, the seal is referred to as balanced. Figure 17-17 illustrates common balanced and unbalanced seals.

The selection of a rotating or stationary seal is determined by the speed of the pump shaft. A seal that rotates with a pump with the shaft is a rotating seal assembly. Examples are shown in Figures 17-15 and 17-16. When the mating ring rotates with the shaft the seal is stationary (Figure 17-18). Rotating seal heads are common in industry for normal pump shaft speeds and where stuffing box space is limited. As a rule of thumb, when the shaft speed exceeds 5,000 ft/min, stationary seals are required. Higher speed applications require a rotating mating ring to keep unbalanced forces that may result in seal vibration to a minimum. Also, for certain applications like vacuum tower bottom pumps, a stationary seal allows a steam quench to be applied to the entire inside diameter of the seal to prevent hangup. Stationary seals require a large cross section stuffing box.

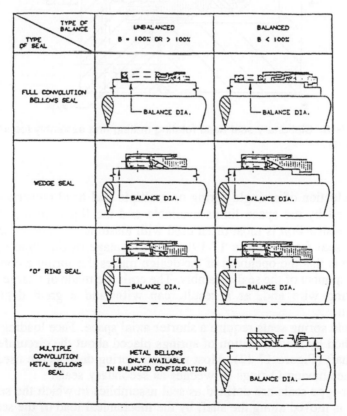

Figure 17-17. Common unbalanced and balanced seals.

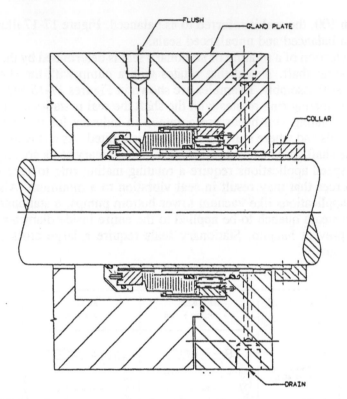

Figure 17-18. Stationary seal head rotating mating ring assembly (courtesy of John Crane).

The selection of a single spring or multiple seal head construction is determined by the space limits available and the liquid sealed. Single spring construction is most often used with elastomeric bellows seals to load the seal faces (Figure 17-19). The advantage of this type of construction is that the openness of the design makes the spring a non-clogging component of the seal assembly. The coils are made of a large diameter spring wire and, as a result, can withstand a great degree of corrosion.

Multiple spring seals require a shorter axial space. Face loading is accomplished by a combination of springs placed about the circumference of the shaft (Figure 17-12). Most multiple spring designs are used with assemblies having O-rings or wedges as secondary seals.

Pusher-type seals are defined as seal assemblies in which the secondary seal is moved along the shaft by the mechanical load of the seal and the hydraulic pressure in the stuffing box. This designation applies to

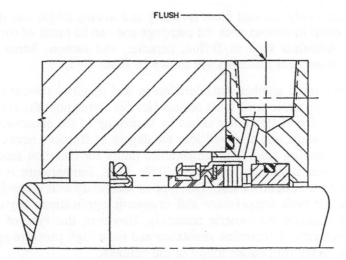

FLUSH

Figure 17-19. Full convolution elastomeric bellows with single spring construction (courtesy of John Crane).

seals that use an O-ring, wedge, or V-ring. Typical construction is illustrated in Figure 17-12.

Non-pusher seals are defined as seal assemblies in which the secondary seal is not forced along the shaft by the mechanical load or hydraulic pressure in the stuffing box. Instead, all movement is taken up by the bellows convolution. A non-pusher design is shown in Figure 17-19. This definition applies to those seals that use half, full, and multiple convolution bellows as a secondary seal.

Materials of Construction

The selection of materials of construction must be based on the operating environment for the seal. The effects of corrosion, temperature, deformation from pressure, and wear from sliding contact must be considered for good life. Each seal must be broken down into component parts for material selection. The effects of corrosion must be known for the secondary seal, primary and mating rings, as well as hardware items. The NACE (National Association of Corrosion Engineers) corrosion handbook is an excellent source of information for corrosion rates of various materials in a wide variety of liquids and gases. When hardware items of a seal are exposed to a liquid and the corrosion rate is greater than 2 mils per year, a double seal arrangement should be considered. The metal parts in this seal arrangement can be kept out of the corrosive

liquid. Then only the seal faces (primary and mating rings) and secondary seal come in contact with the pumpage and can be made of corrosion
resistant materials such as Teflon, ceramic, and carbon. Some of the
more common seal materials are shown in Table 17-3.

Temperature is another consideration in seal material selection. Both
secondary and static seals must remain flexible throughout the entire life
of the seal. If seal flexibility is lost by hardening of the elastomer from
high temperature, the seal will lose the degree of freedom necessary to
follow the mating ring. The temperature limits for common secondary
and static seal materials is given in Table 17-4. Temperature is not an
issue for a metal bellows seal. This type of secondary seal is specifically
designed for high temperature and cryogenic applications beyond the
limits of common elastomeric materials. However, this type of seal is
limited in terms of corrosion resistance and most high pressure applications due to the thin construction of the bellows.

Table 17-3
Most Common Materials of Construction
for Mechanical Seals

COMPONENT	MATERIALS OF CONSTRUCTION
SECONDARY SEALS O-RING	NITRILE, ETHYLENE PROPYLENE, FLUOROELASTOMER, PERFLUOROELASTOMER, CHLOROPRENE
BELLOWS	NITRILE, ETHYLENE PROPYLENE, FLUOROELASTOMER, CHLOROPRENE
WEDGE	FLUOROCARBON RESIN, GRAPHITE FOIL
METAL BELLOWS	STAINLESS STEEL, NICKEL BASE ALLOY
PRIMARY RING	CARBON, METAL FILLED CARBON, TUNGSTEN CARBIDE, SILICON CARBIDE, SILICONIZED CARBON
HARDWARE (RETAINER, DISC, SNAP RING, SET SCREWS, SPRINGS)	18-8 STAINLESS STEEL, 316 STAINLESS STEEL, NICKEL BASE ALLOYS, TITANIUM
MATING RING	CERAMIC, CAST IRON, TUNGSTEN CARBIDE, SILICON CARBIDE

Table 17-4
Temperature Limitations for Common Secondary
and Static Seals

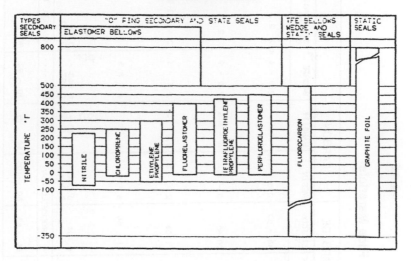

An additional consideration in the selection of the primary and mating rings is their PV limitations. As discussed in the section on seal wear, a PV value can be determined for each application and compared to those given in Table 17-2. A value of less than the value given in the table will result in a seal life greater than two years. Also shown in Table 17-5 are the physical properties of common seal face materials.

Mating Ring Designs

A complete seal installation consists of the seal head assembly and a mating ring assembly. The mating ring assembly consists of the mating ring and a static seal. There are five common mating ring types: a) groove O-ring, b) square section, c) cup mounted, d) floating, and e) clamped in (Figure 17-20). Other designs may exist but they are all variations of these common types. The purpose of a mating ring is to provide a hard surface for the softer member of the seal face to run against. Because wear will occur at the seal faces, the mating ring must be designed as a replaceable part in the seal installation. This is accomplished through the use of a static seal, as shown. The material of construction of the static seal will determine the temperature limit of the assembly.

Table 17-5
Properties of Common Seal Face Materials

PROPERTY	CAST IRON	NI RESIST	CERAMIC		CARBIDE		CARBON			
			85% (AL2O3)	99% (AL2O3)	TUNGSTEN (6% NI)	SILICON (SI-C)	RESIN	BABBITT	BRONZE	SILICONIZED CARBON
MODULUS OF ELASTICITY X 10E6 LB/IN2	13 - 16	10.5 - 16.9	32	50	90	48 - 57	2.5 - 4.0	1.04 - 4.1	2.9 - 4.4	2 - 2.3
TENSILE STRENGTH X 10E3 LB/IN2	65 - 120	20 - 45	20	39	123.25	20.65	8 - 8.6	8 - 8.6	7.5 - 9	2
CO-EFFICIENT OF THERMAL EXPANSION X 10E-6 IN/IN F	6.6	6.5 - 6.8	3.9	4.3	2.66	1.08	2.1 - 2.7	2.1 - 2.7	2.4 - 3.1	2.4 - 3.2
THERMAL CONDUCTIVITY BTU.FT/H X FT2 F	23 - 29	25 - 28	8.5	14.5	41 - 48	41 - 60	6 - 9	6 - 9	8 - 8.5	30
DENSITY LB/IN3	0.259 - 0.268	0.264 - 0.268	0.123	0.137	0.59	0.104	0.064 - 0.069	0.083 - 0.112	0.083 - 0.097	0.06 - 0.070
HARDNESS	----BRINELL---- 217 - 269	131 - 183	----ROCKWELL A---- 87	87	92	----ROCKWELL 45N--- 86 - 88	80 - 105	----SHORE---- 60 - 95	70 - 92	ROCKWELL 151 90

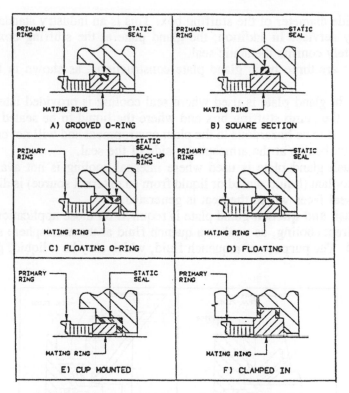

Figure 17-20. Common mating ring assemblies.

Adaptive Hardware

Structural parts of a seal installation such as shaft sleeve, gland plates, bushings, and collars used to fit a seal to a pump are referred to as adaptive hardware. These items may be furnished by the pump company or the seal manufacturer.

The success or failure of a seal installation can often be traced to the proper selection of the gland plate and associated piping arrangement. The purpose of the gland plate or end plate is to hold either the mating assembly or the seal assembly, depending on the type of seal construction. The gland plate is also a pressure containing component of the assembly. The alignment of one of the sealing surfaces, and possibly a bushing, is dependent on the fit of the gland plate to the stuffing box.

For a pump manufactured to API (American Petroleum Institute) specifications, a registered fit of the gland plate is required between the inside or outside diameter of the stuffing box. This is an industry standard for refinery service. In addition, the gland plate at the stuffing box must completely confine the static seal.

There are three basic gland plate constructions, as shown in Figure 17-21.

A **plain** gland plate is used where seal cooling is provided internally through the pump stuffing box and where the liquid to be sealed is not considered to be hazardous to the plant environment and will not crystallize or carbonize at the atmospheric side of the seal.

A **flush** gland plate is used where internal cooling is not available. Here, coolant (liquid sealed or liquid from an external source) is directed to the seal faces where the heat is generated.

A **flush and quench** gland plate is required on those applications that need direct cooling, as well as a quench fluid at the atmosphere side of the seal. The purpose of a quench fluid, which may be a liquid, gas, or

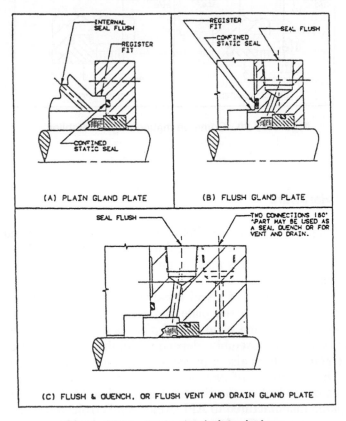

Figure 17-21. Basic gland plate designs.

steam, is to prevent the buildup of any carbonized or crystallized material along the shaft. When properly applied, a seal quench can increase the life of a seal installation by eliminating the loss of seal flexibility due to hangup.

A **flush, vent and drain** gland plate is used where seal leakage needs to be controlled. Flammable vapor from a seal can be vented to a flair and burned off, while non-flammable liquid leakage can be directed to a sump.

Figure 17-22 illustrates some common restrictive devices used in the gland when quench or vent and drain connections are used. These bushings may be pressed in place, as shown in Figure 17-22A, or allowed to float, as in figures 17-22B, C and D. Floating bushings allow for closer running fits with the shaft because such bushings are not restricted at their outside diameter. Floating bushings in Figures 17-22C and D, are

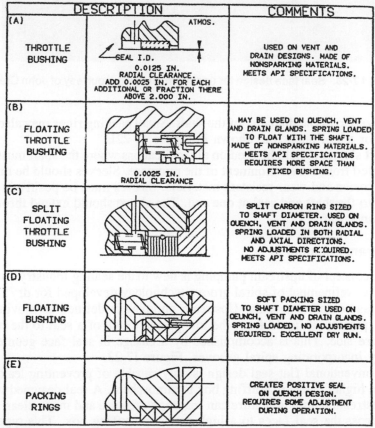

DESCRIPTION		COMMENTS
(A) THROTTLE BUSHING	ATMOS. SEAL I.D. 0.0125 IN. RADIAL CLEARANCE. ADD 0.0025 IN. FOR EACH ADDITIONAL OR FRACTION THERE ABOVE 2.000 IN.	USED ON VENT AND DRAIN DESIGNS. MADE OF NONSPARKING MATERIALS. MEETS API SPECIFICATIONS.
(B) FLOATING THROTTLE BUSHING	0.0025 IN. RADIAL CLEARANCE	MAY BE USED ON QUENCH, VENT AND DRAIN GLANDS. SPRING LOADED TO FLOAT WITH THE SHAFT. MADE OF NONSPARKING MATERIALS. MEETS API SPECIFICATIONS REQUIRES MORE SPACE THAN FIXED BUSHING.
(C) SPLIT FLOATING THROTTLE BUSHING		SPLIT CARBON RING SIZED TO SHAFT DIAMETER. USED ON QUENCH, VENT AND DRAIN GLANDS. SPRING LOADED IN BOTH RADIAL AND AXIAL DIRECTIONS. NO ADJUSTMENTS REQUIRED. MEETS API SPECIFICATIONS.
(D) FLOATING BUSHING		SOFT PACKING SIZED TO SHAFT DIAMETER USED ON QEUNCH, VENT AND DRAIN GLANDS. SPRING LOADED, NO ADJUSTMENTS REQUIRED. EXCELLENT DRY RUN.
(E) PACKING RINGS		CREATES POSITIVE SEAL ON QUENCH DESIGN. REQUIRES SOME ADJUSTMENT DURING OPERATION.

Figure 17-22. Common restrictive devices used with quench or vent-and-drain gland plate.

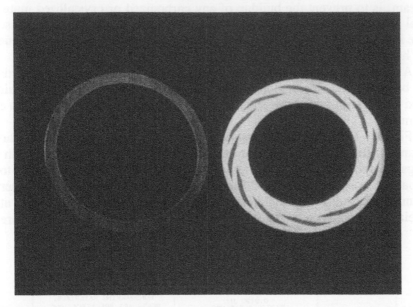

Figure 17-23. Seal face design for upstream pumping (courtesy of John Crane).

also sized to fit the diameter of the shaft. Small packing rings may also be used for a seal quench as shown in Figure 17-22E.

Shaft sleeves should be used on all applications where the shaft must be protected from the environment of the pumpage. Sleeves should be made of corrosion and wear resistant materials. Sleeves must be positively secured to the shaft and seal at one end. The sleeve should extend through the gland plate.

Upstream Pumping

The concept of upstream pumping is new to the sealing industry and is a further refinement of spiral groove technology developed for dry running gas compressor seals. Upstream pumping is defined as moving a small quantity of liquid from the low pressure side of a seal to the high pressure side. This is accomplished by a change in seal face geometry and by incorporating spiral grooves, Figure 17-23.

A conventional flat seal design is only capable of preventing leakage from a higher pressure stuffing box to atmosphere. A seal designed with the upstream pumping feature can seal high pressure and move clean liquid across the seal faces to a high pressure stuffing box. This type of design creates a full liquid film at the faces and reduces horsepower loss. This is also a new way to flush a seal face with just a few cc/minute rather than a few gpm with a conventional flush. Figure 17-24 illustrates a seal with the upstream pumping concept at the inboard seal. Early experiments demonstrated that this 1.750-inch diameter seal operating in water at 1800 rpm can move 0.6 cc/min to a stuffing box pressure of 800 psig. This concept when applied to multiple seals would allow a plant operator to bring a supply of water or a neutral liquid directly to the space between the seals. The low pressure liquid can be made to flow in small quantities to higher pressures through the pumping action of the seal faces. This improvement in seal design has resulted in a new way to flush a seal, which can be applied to abrasive slurry service, hot water, hazardous liquids, heat transfer, and liquids with poor lubricating properties. This is a significant development resulting in a non-contacting, non-leaking seal for liquid service. The upstream pumping concept is a patented development by John Crane Inc.

Mechanical Seals for Chemical Service

Traditionally, mechanical seals for this type of service are fitted to pumps that meet American National Standards Institute (ANSI) Specification B73.1. This specification defined the stuffing box cross section by the size of the packing that could be used on a given application. Standard sizes of cross section were $5/16$, $3/8$, and $7/16$ inches. The idea behind the specification was that if seals were used and had to be replaced they could be replaced by packing. However, in practice when a seal is worn out or

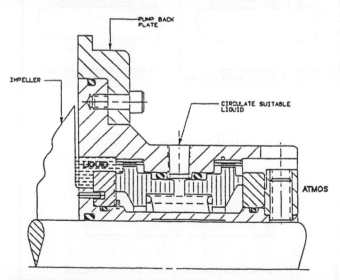

Figure 17-24. Package symmetrical seal with upstream pumping feature (courtesy of John Crane).

failed, it was replaced by another seal. Fitting a seal to packing cross sections was sometimes difficult, particularly when multiple seal arrangements were required. In addition, the smaller the space around the seal, the warmer it would run.

Figure 17-25 shows the original stuffing box and the new revised box that allows more than two times the clearance over the pump shaft. The new box dimensions allow a current ANSI seal to run 25°F cooler than the original design. By opening up the back of the stuffing box and providing a slight taper to the bore, the stuffing box will run approximately 40°F cooler than the original design. This data was developed testing conventional seals that are considered standard in the original stuffing box. These include elastomeric bellows, wedge, and O-ring seals. Stuffing box design plays an important part in providing a good environment for the mechanical seal.

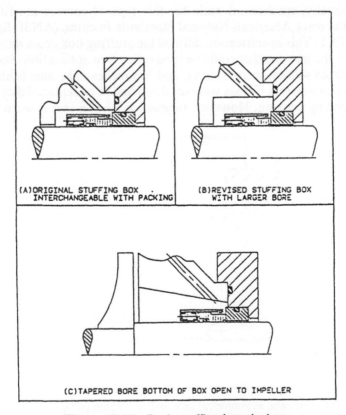

Figure 17-25. Basic stuffing box designs.

Mechanical Seals for Refinery Service

Refinery services can be broken down into two primary classes based on temperature:

- Applications below 350°F
- Applications above 350°F

For applications below 350°F elastomeric seals should be used. These applications may also involve the sealing of high pressure as well. A typical package seal for this class of service is shown in Figure 17-26.

For applications above 350°F, a metal bellows seal should be used. Pressure on most high temperature applications will be low and would not present a problem for most metal bellows available to industry. A typical package seal for this class of service is shown in Figure 17-27. Both seals illustrated in Figures 17-26 and 17-27 meet the requirements of American Petroleum Institute's (API) 610 Standard. This standard does not cover the design of individual seal components. That is the responsibility of the seal manufacturer. The standard does cover the minimum requirements of equipment for use in refinery services. These requirements would include:

- Seal chamber and shaft sleeve considerations.
- Hardness of seal materials to be used on a pump shaft.
- Preferred use of balanced seals unless otherwise specified.
- Containment of seal face leakage by one of the following arrangements:
    Single seal with an auxiliary sealing device such as a floating close clearance bushing or other suitable device.
    Tandem seal arrangement with a barrier fluid maintained at a pressure lower than seal chamber pressure.
    Double seal arrangement with a barrier fluid maintained at a pressure greater than seal chamber pressure.
- Gland plate design including register fit to the seal chamber and confinement of static seals.

Materials of construction are identified as well as piping plans for seal installations. Piping plans for seal installations are so important that they are shown in Figures 17-28A and B for easy reference. API Specification 610 represents the collective knowledge and experience of end users, pump and seal manufacturers. This specification is updated periodically, and anyone involved in refinery equipment should have an updated copy of the specification from the American Petroleum Institute.

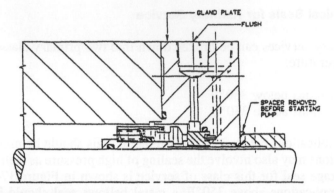

Figure 17-26. Package seal for refinery service for temperatures below 350°F (courtesy of John Crane, Inc.).

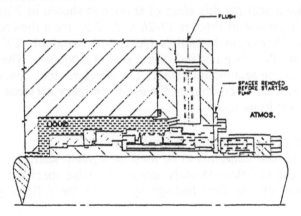

Figure 17-27. Package metal bellows seal (courtesy of John Crane).

Typical Applications

Mechanical seals are used to seal a wide range of liquids and gases at various conditions of pressure, temperature, and speed. As seal technology advances, progress is being made to move from the traditional contacting seal to what is becoming known as a class of seals referred to as non-contacting, non-leaking. Certain types of liquids will require the seal to operate without any appreciable power loss or wear at the seal faces.

Light Hydrocarbon Service

When sealing a light hydrocarbon as a liquid, conventional flat sealing faces can be considered when the operating conditions do not require the seal to work in an environment where the liquid in the seal chamber is near its vapor pressure. As a rule of thumb, when the stuffing box pressure is less than one-third of the vapor pressure, a seal face design with hydropads should be considered. Hydropads or lubrication recesses are

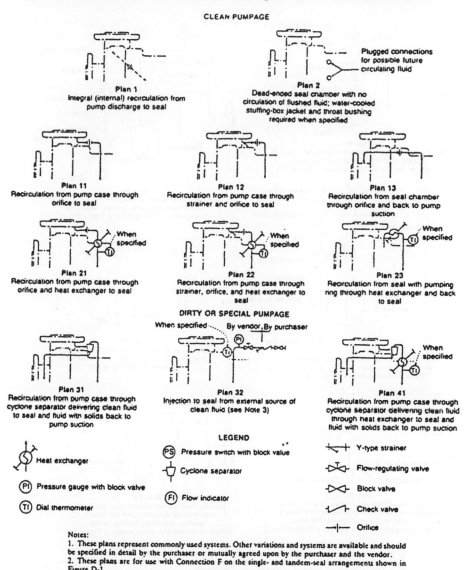

CLEAN PUMPAGE

Plan 1
Integral (internal) recirculation from pump discharge to seal

Plan 2
Dead-ended seal chamber with no circulation of flushed fluid; water-cooled stuffing-box jacket and throat bushing required when specified

Plugged connections for possible future circulating fluid

Plan 11
Recirculation from pump case through orifice to seal

Plan 12
Recirculation from pump case through strainer and orifice to seal

Plan 13
Recirculation from seal chamber through orifice and back to pump suction

Plan 21
Recirculation from pump case through orifice and heat exchanger to seal

Plan 22
Recirculation from pump case through strainer, orifice, and heat exchanger to seal

Plan 23
Recirculation from seal with pumping ring through heat exchanger and back to seal

When specified

DIRTY OR SPECIAL PUMPAGE

When specified   By vendor, By purchaser

Plan 31
Recirculation from pump case through cyclone separator delivering clean fluid to seal and fluid with solids back to pump suction

Plan 32
Injection to seal from external source of clean fluid (see Note 3)

Plan 41
Recirculation from pump case through cyclone separator delivering clean fluid through heat exchanger to seal and fluid with solids back to pump suction

When specified

LEGEND

Heat exchanger

(PS) Pressure switch with block value

(PI) Pressure gauge with block valve

Cyclone separator

(TI) Dial thermometer

(FI) Flow indicator

Y-type strainer

Flow-regulating valve

Block valve

Check valve

Orifice

Notes:
1. These plans represent commonly used systems. Other variations and systems are available and should be specified in detail by the purchaser or mutually agreed upon by the purchaser and the vendor.
2. These plans are for use with Connection F on the single- and tandem-seal arrangements shown in Figure D-1.
3. For Plan 32, the purchaser will specify the fluid characteristics, and the vendor shall specify the volume, pressure, and temperature required.

Figure 17-28A. Piping plans for mechanical seals from A.P.I. Standard 610, 6th Edition. (Reproduced courtesy American Petroleum Institute.)

an effective way of reducing the power loss of a flat sealing face by as much as 30%. The hydropad design feature is shown in Figure 17-29. In most applications this will permit the seal to operate in a very stable condition with very low leakage to the atmosphere. Leakage rates from these

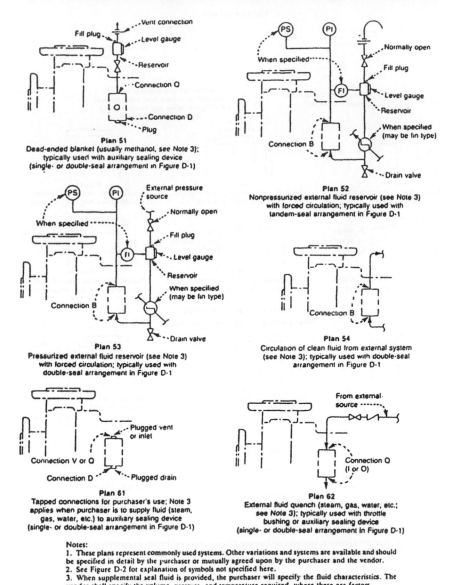

Plan 51
Dead-ended blanket (usually methanol, see Note 3); typically used with auxiliary sealing device (single- or double-seal arrangement in Figure D-1)

Plan 52
Nonpressurized external fluid reservoir (see Note 3) with forced circulation; typically used with tandem-seal arrangement in Figure D-1

Plan 53
Pressurized external fluid reservoir (see Note 3) with forced circulation; typically used with double-seal arrangement in Figure D-1

Plan 54
Circulation of clean fluid from external system (see Note 3); typically used with double-seal arrangement in Figure D-1

Plan 61
Tapped connections for purchaser's use; Note 3 applies when purchaser is to supply fluid (steam, gas, water, etc.) to auxiliary sealing device (single- or double-seal arrangement in Figure D-1)

Plan 62
External fluid quench (steam, gas, water, etc.; see Note 3); typically used with throttle bushing or auxiliary sealing device (single- or double-seal arrangement in Figure D-1)

Notes:
1. These plans represent commonly used systems. Other variations and systems are available and should be specified in detail by the purchaser or mutually agreed upon by the purchaser and the vendor.
2. See Figure D-2 for explanation of symbols not specified here.
3. When supplemental seal fluid is provided, the purchaser will specify the fluid characteristics. The vendor shall specify the volume, pressure, and temperature required, where these are factors.

Figure 17-28B. Piping plans for mechanical seals from A.P.I. Standard 610, 6th Edition. (Reproduced courtesy American Petroleum Institute.)

Figure 17-29. Carbon primary ring with hydropad face design (courtesy of John Crane).

types of seals are measured in parts per million. The seal illustrated in Figure 17-30 has been successfully applied to sealing light hydrocarbon at the following conditions:

- Liquid sealed: ethylene, ethane, or propane
- Stuffing box pressure: 500 to 1200 psig
- Temperature: 30 to 60°F
- Specific gravity: 0.3 to 0.9
- Speed: 1500 to 3480 rpm
- Seal size: 5.375 inches

Many installations have been made and thousands of hours of seal life have been realized. The materials of construction for the seal faces are carbon versus tungsten carbide.

Hydropads have also been successfully applied to high temperature applications requiring a metal bellows seal. The seal shown in Figure 17-31 is a typical installation that has been in service for more than three years. This was a difficult application for the original seals that were specified.

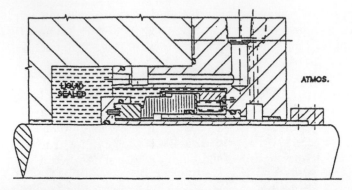

Figure 17-30. Package mechanical seal installation for a pipeline pump (courtesy of John Crane).

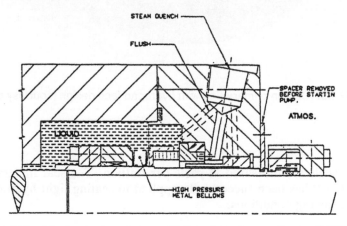

Figure 17-31. Mechanical seal for a high temperature refinery charge pump (courtesy of John Crane).

Failure occurred due to vaporization at the seal faces. The seal shown has been through many startups with no adverse effects. The operating conditions for this application are:

- Liquid sealed: gas oil
- Stuffing box pressure: 278 psig
- Temperature: 615 °F
- Specific gravity: 0.665
- Speed: 3550 rpm
- Seal size: 3.125 inches

The measured leakage from each end of this horizontal pump was 10 to 120 ppm. This installation is a rotating seal head with a tungsten carbide insert running on a silicon carbide mating ring. A steam quench is being used on the atmospheric side of the seal.

Hydropads are also used to aid the lubrication process on the refinery application shown in Figure 17-32. The pump is used in a hydrofluoric alkylation unit to move hydrofluoric acid. To protect the sealing system, a clean injection of isobutane is flushed in at the seal faces. The original seals experienced short seal life due to chemical attack from the hydrofluoric acid. The operating conditions for this application are:

• Liquid sealed: isobutane/hydrofluoric acid
• Stuffing box pressure: 240 psig
• Temperature: 89°F
• Specific gravity: 0.945
• Speed: 1180 rpm
• Seal size: 3.5 inches

This hydropadded seal has added years to the service life of the unit. The materials of construction for the seal faces are solid silicon carbide running on solid silicon carbide. This material combination was selected for its excellent corrosion resistance to hydrofluoric acid, because corrosion is a common cause for seal failure on these types of applications.

Hydropads or lubrication recesses when properly applied will result in improved seal life. This design feature is the first step in the process of advancement to non-contacting seals.

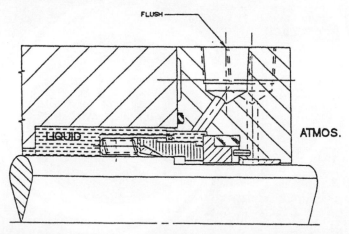

Figure 17-32. Mechanical seal installation for refinery HF Alkylation Unit (courtesy John Crane).

True non-contacting seals are being applied to pumping equipment and will change the way a seal is viewed for a given application. When pumping a light hydrocarbon near its vapor pressure, it is more efficient to allow the liquid to flash to a gas in the stuffing box and use a dry running seal to seal the gas. This is done with a non-contacting seal design. This most efficient method in achieving non-contact in operation is through the use of the spiral groove face seal. This patented concept was developed by John Crane Inc. for use in sealing high performance gas compressors and is illustrated in Figure 17-33. The spiral groove pattern is a series of logarithmic spirals recessed into the hard mating ring. The ungrooved portion of the seal face below the spiral is called the sealing dam. The spiral groove pattern is designed to rotate only in one direction. As the seal face begins to rotate from the outside diameter of the seal faces to the groove diameter, gas is compressed and then expanded across the sealing dam. This will generate sufficient opening force to separate the seal faces by a few micro-inches. Seal balance, face width and spiral groove diameters are all critical and determine the static and dynamic operation of the seal during periods of startup and shutdown when the seal faces contact. A comparison of hydropadded and spiral groove faces is made in Figure 17-34.

Figure 17-33. Dry running face design (courtesy of John Crane).

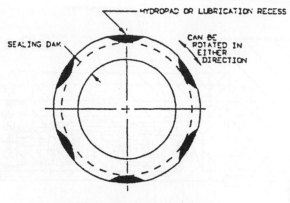

(A) HYDROPADDED SEALING SURFACE

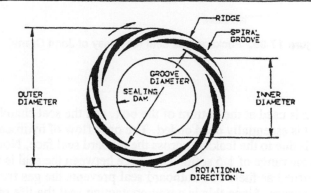

(B) SPIRAL GROOVE SEALING SURFACE

Figure 17-34. Comparison of seal face geometry.

A typical pump application is shown in Figure 17-35. The operating conditions for this seal are:

- Liquid sealed: light hydrocarbon flashed to gas in the stuffing box
- Stuffing box pressure: 387 psig
- Vapor pressure at pumping temperature: 375 psig
- Temperature: 31°F
- Specific gravity: 0.4
- Speed: 1500 rpm
- Seal size: 4.687 inches

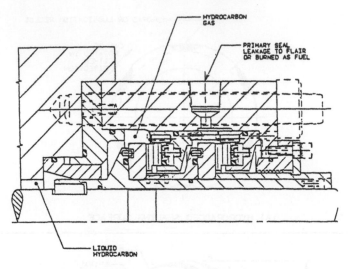

Figure 17-35. Package gas seal (courtesy of John Crane).

A bushing is used at the bottom of the box, and the seal chamber at the inboard seal is essentially dead ended. The only flow of hydrocarbon gas in this area is due to the leakage across the inboard seal face. Normal leak rates are in the range of 1.5 scfm. The space between the seal is vented to a flair or burned as fuel. The outboard seal prevents the gas from escaping to atmosphere. Since this is a non-contacting seal the life of the seal will not be affected by wear.

Some applications still require double seals to provide a good running environment for the seal. Figure 17-36 shows a solution to sealing high pressure liquid carbon dioxide service. These seals are used on pumping equipment used to inject the liquid carbon dioxide into the ground for crude oil extraction. On this application, oil is circulated through the seal chamber at a pressure 10% greater than the pressure at the inboard seal from the liquid carbon dioxide. This seal is operating at the following conditions.

- Liquid sealed: carbon dioxide
- Buffer liquid: oil
- Stuffing box pressure: 2400 psig
- Temperature: 65°F
- Specific gravity: 0.76 (oil)
- Speed: 3550 rpm
- Seal size: 3.750 inches

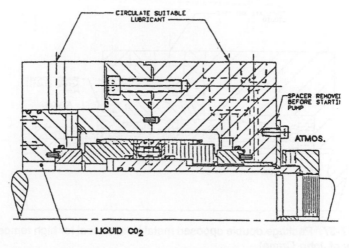

Figure 17-36. Package mechanical seal for CO_2 service (courtesy of John Crane).

The inboard seal faces are a metal alloy running against tungsten carbide. The choice of this material combination was to avoid blistering in a carbon material that would be exposed to the liquid carbon dioxide. The seal faces at the outboard position are running at a higher level of PV and here, carbon versus tungsten carbide is being used.

Double seals are also considered and used for harsh environments involving high temperature applications. Figure 17-37 shows a double-opposed stationary seal design. This seal is operating at the following conditions:

- Liquid sealed: residual bottoms
- Vapor pressure at pumping temperature: 1.25 psia
- Viscosity: 0.6 cp
- Temperature: 700°
- Buffer liquid: gas oil
- Temperature: 150°
- Stuffing box pressure: 200 psig
- Speed: 3550 rpm
- Seal size: 3.000 inches

Seal faces are carbon versus tungsten carbide. Double seals on applications such as this have a seal life that can be 2½ times longer than a single seal installation.

Figure 17-38 shows another multiple installation. This seal can be used as a double or a tandem seal, depending on pressure between the seal

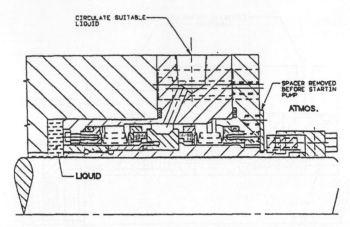

Figure 17-37. Package double opposed metal bellows seal for high temperature (courtesy of John Crane).

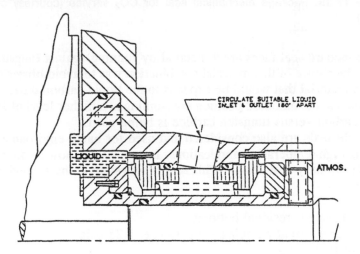

Figure 17-38. Package symmetrical seal (courtesy of John Crane).

heads. If the pressure is greater than the liquid being pumped, it is operating as a double seal. If the pressure is less than the liquid being pumped, it is operating as a tandem. This seal is designed to operate at the following operating conditions:

- Liquid sealed: various
- Buffered liquid: oil, water, or a non-hazardous liquid compatible with the process
- Temperature: 300°F (max)
- Speed: 3600 rpm

This design is a package seal that includes the stuffing box as an integral part of the seal assembly. This model is built for an end suction pump and can be removed from the pump without disassembling any bearings. This is referred to as a Symmetrical Seal and is a patented design by John Crane Inc. This seal design is ideally suited for incorporating the upstream pumping concept that was previously discussed.

High speed pump applications offer a unique opportunity for the use of a non-contacting, non-leaking seal. Through the use of the concept of upstream pumping, a stuffing box can be run dead ended. The only lubrication and cooling is achieved through the liquid being moved from the inside diameter of the seal face to its outside diameter. This design is illustrated in Figure 17-39. In this tandem seal arrangement, liquid between the tandem seals at 100 psig is being moved to a high pressure of 1200 psig in the stuffing box. Shaft speed is 15,600 rpm and seal size is 1.500 inches. Flow rate from low pressure to high pressure side of the seal is 17 cc/min. There is no measureable increase in surface temperatures at the inboard seal that is handling a differential pressure of 1100 psig. Statically and dynamically, there is no leakage of the liquid being pumped. Buffer liquid between the seals is water. Only a small amount of circulation is required to remove the seal heat generated at the outboard seal. Upstream pumping is a patented design by John Crane Inc. that will begin to change the way one views a seal application. A seal can also be used to do work in the system and reduce costly support equipment that is sometimes required. It also allows a totally different way to flush a seal with extremely small quantities of liquid.

Split seals are sometimes required on certain applications. Shown in Figure 17-40 is a patented split seal by John Crane. In this case all components are split, allowing the changing of a seal without dismantling the entire pump. Split seals have been applied to pumps with large shafts operating at speeds of 1800 rpm and pressures to 80 psig. The unique feature of the floating seal faces allows this type of seal to handle large amounts of shaft runout.

Many standard and unique seal designs are available from seal manufacturers. The most efficient design for a given application can be determined by making all of the operating conditions available to the seal manufacturer.

Mechanical Seal Installation and Troubleshooting

Long trouble-free operation of equipment includes proper seal design, selection of materials of construction, and correct seal installation. Any shortcuts taken during seal installation can result in the equipment being taken out of service in a few hours, days, or even weeks from startup. A

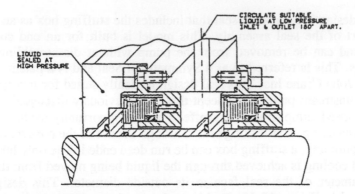

Figure 17-39. High speed application of patented upstream pump concept (courtesy of John Crane).

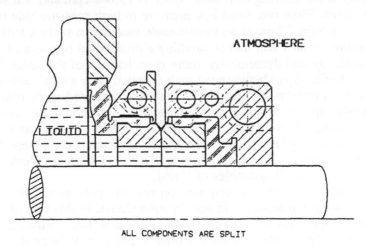

Figure 17-40. Split face seal design (courtesy of John Crane).

pump must operate within the pump manufacturer's specification and should be checked each time a seal is replaced. When a seal is removed due to excessive leakage, a study should be made to identify the cause for leakage. This study can yield results that can extend the life of the next installation.

When a pump shaft begins to rotate, the seal parts also begin to move relative to each other. Not only does sliding occur at the primary and mating rings, but other seal parts move as well. The key to successful seal life is to keep the types of motion transferred to a seal to a minimum. The common types of motion that influence seal performance are angular and axial movement. These types of motion are the result of angular misalignment, parallel misalignment, end play, and radial shaft runout (Figure 17-41). If wear has occurred between seal parts or between seal parts and a shaft sleeve or shaft, the pump should be checked for the type of condition illustrated and corrective measures should be taken. A pump must operate within the pump manufacturer's tolerances.

Angular misalignment results when the mating ring is not square with the shaft. As shaft rotation begins, the primary ring will try to follow the surface of the mating ring. This often results in uneven wear of the softer carbon primary ring, possible internal wear of seal parts, fretting of seal hardware or shaft sleeve. Fretting on a shaft or shaft sleeve occurs when the protective oxide coating is removed from the metal by rubbing motion from a secondary seal in a pusher-type seal design. A hard coating on the shaft or shaft sleeve can be used to eliminate this type of wear. A full convolution elastomeric bellows is designed to protect the shaft or shaft sleeve by taking all the movement in the bellows convolution instead of along the shaft. Angular misalignment may also occur from a stuffing box that has been distorted by piping strain developed at operating temperatures. In this case, damage to the wear rings in the pump may also be found if the stuffing box has been distorted.

Parallel misalignment results when the stuffing box is not properly aligned with the rest of the pump. Generally, with small amounts of parallel misalignment, the seal will operate properly. However, with a large amount of parallel misalignment, the shaft may strike the inside diameter of the mating ring. If damage has occurred, there will also be damage to any bushing located at the bottom of the stuffing box. The point of damage will be in the same location as that on the mating ring.

Excessive axial end play can damage the sealing surfaces and can also cause fretting. If the seal is continually being loaded and unloaded, abrasives can penetrate the seal faces and cause premature wear of the primary and mating rings. Thermal damage in the form of heat checking on the seal faces can result if excessive end play occurs and the seal is operated below its working height.

Radial runout in excess of limits established by the pump manufacturer will cause excessive vibration at the seal. This vibration, coupled with small amounts of other types of motion which have been defined, can shorten seal life. Radial runout from a bent shaft can damage the inside diameter of a mating ring through 360°.

Seal installation begins with a review of the assembly instructions and any seal layout drawings. The proper seal installation dimension or spacing is required to ensure that the seal is at its proper working height (Figure 17-42). The installation reference can be determined by locating the face of the box on the surface of the sleeve and then measuring along the sleeve after it has been removed from the unit. It is not necessary to use

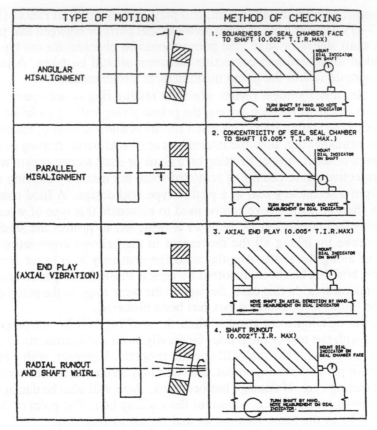

Figure 17-41. Common types of motion that influence seal performance and method of checking the equipment.

this procedure if a step in the sleeve or collar has been designed into the sleeve assembly to provide for proper seal setting. Assembly of other parts of the seal will bring the unit to its correct working height. A single seal mounted as shown in Figure 17-42 is the most economical seal installation available to industry.

Package seals as shown in Figure 17-26 can be assembled with relative ease because only the bolts on the gland plate and set screws in the collar need to be fastened to the stuffing box and shaft. After the spacer is removed, the pump is ready to operate, provided the piping has been replaced.

Because a seal has precision lapped faces and secondary seal surfaces are critical in the assembly, installation should be kept as clean as possible. All lead edges on sleeves and gland plates should have sufficient chamfers to facilitate installation.

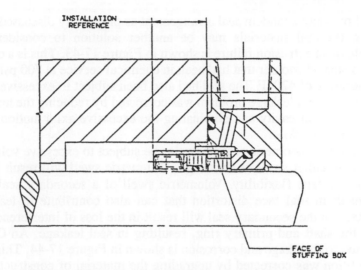

INSTALLATION
REFERENCE

FACE OF
STUFFING BOX

Figure 17-42. Typical installation reference dimensions.

When mechanical seals are properly applied, there should be no static leakage under normal conditions. Dynamic leakage can range from none to several drops per minute, depending on the operating conditions for the unit. For estimates of static and dynamic leakage, the user should consult the seal manufacturer. When conditions call for a vacuum, a mechanical seal is used to prevent air from leaking into the pump.

Seal leakage can occur in many ways. The cause of leakage may be the result of an improper seal selection, installation, environment, or operation of the pump. Leakage may occur past static and secondary seals or past the seal faces. Leakage may occur while the pump shaft is rotating and stationary or just when the pump shaft is rotating. Each observation made about the pump and its operation, as well as the condition of the failed seal parts, plays an important role in identifying the cause for seal leakage. Analyzing the causes for seal leakage can best be done with pictures of actual seal parts.

A leakage path past secondary and static seals are usually confined to elastomeric seals. These types of seals are more vulnerable to nicks, scratches, and cuts. Poor handling and installation practices are the most common causes for surface damage to these types of seal parts. Care should always be taken to remove all sharp burrs and edges from steps and keyways as well as previous set screw identations from prior seal installations. Extrusion failures of secondary elastomeric seals such as O-rings are caused by excessive pressure and can be prevented by reducing the pressure and/or temperature on the seal. This can be accomplished by using a tandem seal arrangement as previously discussed. Upgrading the seal materials may be another solution to consider. An example of an extrusion failure is shown in Figure 17-43. This is a corrosion resistant elastomer that had been in chemical service at 700 psig and a temperature of 420°F. The seal had also been subject to excessive axial motion. The seal leakage problem was corrected by reducing the temperature in the seal cavity and eliminating the excessive axial motion from the equipment.

Secondary elastomeric seals may also be subject to excessive volumetric change, either swell or shrinkage. Volumetric swell will result in the loss of seal face flexibility. Volumetric swell of a secondary seal may also result in seal face distortion that can also contribute to leakage. Shrinkage of the secondary seal will result in the loss of interference between the shaft and primary ring, resulting in seal leakage. An O-ring subjected to shrinkage and corrosion is shown in Figure 17-44. This type of condition was corrected by upgrading the material of construction.

Elastomeric O-rings are also subject to a condition referred to as explosive decompression. This can occur when sealing some liquefied gases such as carbon dioxide. Small amounts of gas can penetrate the O-ring at high pressure. As long as pressure is kept on the seal, nothing will happen to the O-ring. Once the pressure in the seal cavity is reduced, the O-ring will expand and burst at the point where the gas has been absorbed into the O-ring. This blistered condition can be avoided by upgrading the material used for the O-ring.

The advantages of using an O-ring as a secondary seal outweigh the disadvantages. This type of secondary seal offers an excellent solution to high pressure applications.

Metal bellows seals offer an excellent solution for temperatures greater than 400°F and pressures to 500 psig. A seal fitted with a metal bellows can be applied to those applications that are essentially non-corrosive and where angular misalignment is not a problem. A bellows used in an environment with a large amount of angular misalignment will fail from fatigue. Cracking of a metal bellows will occur above the weld near the inside diameter of the bellows, as shown in Figure 17-45. This type of seal problem can be eliminated by correcting the motion from angular misalignment.

Motion from angular misalignment, parallel misalignment, shaft runout or whirl, and shaft end play have an effect on seal operation. Motion from parallel misalignment is believed to aid the lubrication process. Motion from other types of conditions will result in larger than normal forces on a primary seal ring. This will lead to unstable operation for the mechanical seal. The increase in load from angular misalignment can be many times larger than the design load of the seal. This increase in load

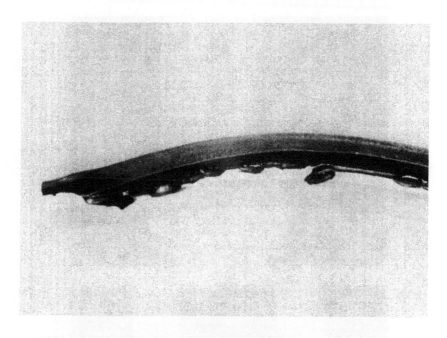

Figure 17-43. Extruded O-ring from high temperature and pressure.

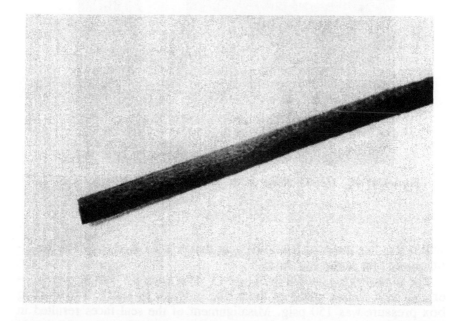

Figure 17-44. Condition of O-ring subject to corrosion, abrasion, and shrinkage.

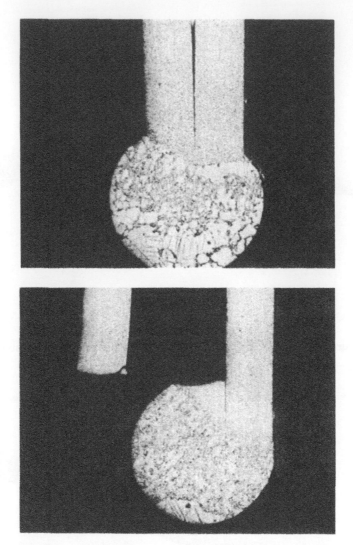

Figure 17-45. Metal bellows welds. A. Typical weld. B. Fatigue crack.

will result in additional frictional heat that will be detrimental to the lubricating film at the seal faces.

The primary ring shown in Figure 17-46 is from a 5.375-inch seal operating at a surface speed of 5000 fpm in water at 180°F. The stuffing box pressure was 150 psig. Misalignment of the seal faces resulted in wear at only one anti-rotation notch. At this point of greatest friction in

Figure 17-46. Damage to a carbon primary ring from angular misalignment.

the drive system, edge chipping has occurred at the seal face. This impact type of failure occurred after two months of operation. The surface waviness of this part measured 2550 micro-inches through 360°. This is an extremely large value of surface waviness for a sealing plane and would account for a large volume of leakage. The surface profile of this part is irregular, alternating from convex to concave at approximately every 90°. At the chipped edge, the profile surface is 350 micro-inches convex. The maximum concave surface is 200 micro-inches. The secondary seal has been severely damaged due to the motion transmitted to this seal from the misalignment problem. High surface waviness and irregular surface profile readings appear to be typical for seals subjected to impact type of failure from misalignment. In cases where the misalignment is very small, cracking and edge chipping of the carbon ring would not occur; however, uneven wear of the softer seal face materials would be experienced. An alignment problem on this application was corrected to achieve long seal life.

Leakage past the seal faces may be the result of distortion, heat checking, environmental causes, and also equipment motion. Leakage past the seal faces due to distortion can be the result of the pressure in the seal cavity, thermal effects from the environment and mechanical causes from gland plate loading. Leakage from pressure is the result of deflection at the seal faces. At higher pressure, the seal faces may deflect in a positive direction to the shaft centerline. As the seal is operated at higher pressure, no leakage occurs. At lower pressure, however, the seal will begin to leak because the wear pattern has changed and the seal faces are now opening to the pressure at the outside diameter. This type of leakage from coning or positive seal face rotation can be eliminated by upgrading seal materials or the design of the seal rings.

Leakage due to thermal distortion is just the opposite to what is observed when the seal faces deflect from pressure. Here, wear occurs at the inside diameter of the seal face. The seal face continues to expand, opening up to the liquid in the stuffing box. As the seal opens up at higher temperatures, leakage increases and little or no contact is observed at the outside diameter of the seal faces. This condition is corrected by increasing the cooling to the seal faces. An illustration of damage from thermal distortion is given in Figure 17-47. Damage to this cast iron mating ring was eliminated by providing adequate flow to the seal faces.

Mechanical distortion may result from bolt loading from the gland plate on the pump. Illustrations of mechanical distortion are shown in Figures 17-48, 17-49, and 17-50. In Figure 17-48 a tungsten carbide mating ring had been installed into a unit where the bolts had been overtightened. This changed the complete waviness pattern at the seal faces and resulted in two high spots 180° apart. When this condition occurs, the separation between the mating and primary rings can result in erosion of the softer carbon primary ring. This erosion can take two forms. If the unit is idle with pressure in the stuffing box and the shaft is not rotated, wire drawing can occur across the seal face. This condition is shown in Figure 17-49. In addition to wire drawing shown on the seal faces, erosion of the outside carbon shoulder can be seen. Erosion of the carbon can be eliminated by redirecting the seal flush to the side of the faces, and where possible, an abrasive separator should be added to the seal flush line.

Wire brushing of a seal face can occur when abrasives are present in the stuffing box and the seal is operated with a mechanically distorted mating ring, Figure 17-50. In both cases, erosion to the seal faces can be eliminated by not overtightening the gland plate bolts or by supporting the gland plate outside the bolt circle. Distortion can also occur when the top and bottom halves of a split case pump are not properly lined up. In this case, a review of the application should be made to determine if an abrasive separator should be included in the seal flush line.

Heat checking of a seal face can occur in a pattern through 360° as shown in Figure 17-51. When this condition occurs a few minutes after startup, it would indicate that the seal was operated below its working height or that the seal was operated without any flush to the seal cavity.

Figure 17-47. Damage to a mating ring surface from thermal distortion.

Figure 17-48. A form of damage from mechanical distortion to a mating ring surface.

Figure 17-49. Damaged carbon primary ring showing wire drawing across the seal face and erosion on exterior surfaces.

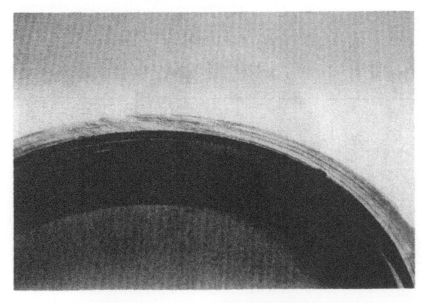

Figure 17-50. A carbon primary ring with surface damage referred to as wire brushing.

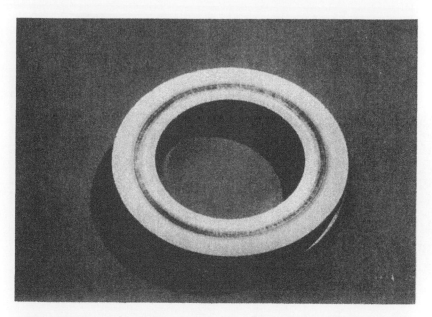

Figure 17-51. Heat checked mating ring surface damaged through 360°.

This type of failure may also occur after a longer period of time in service if the pump had developed excessive axial end play, causing the seal to run solid. In this case, checks should be made to determine if the seal had been properly set at its working height and had sufficient flow to the seal cavity.

Partial heat checking of the sealing surface through an arc less than 180° opposite the flush inlet is the result of not properly distributing the flow of coolant in the stuffing box. In this case, a circumferential groove may be added to the gland plate at the outside diameter of the mating ring or an additional inlet 180° from the existing inlet should be considered. In this particular case, it would also be a good idea to check the squareness of the face of the stuffing box with the shaft to be sure the alignment is correct and that the mating ring has not been pulled to one side when installed.

Hot localized patches of material have been previously discussed in the theory of operation for a mechanical seal. The development of a hot spot is illustrated in Figure 17-52. When a seal operates in an unstable condition for a short period of time, the localized hot spot will appear to be very small. The surface distress at the hot spot occurs because of rapid heating in operation followed by rapid cooling due to the liquid at the seal faces changing to a gas. When the liquid at the seal faces flashes or va-

Figure 17-52. Small hot spot on a seal face.

porizes, the seal will open, cooling the spot which results in heat checking of the surface. When removed from the equipment after running for a short period of time, the deformed surface may measure between 5 to 10 micro-inches larger than the surrounding surface. In operation, however, due to the developed heat at the hot spot, the deformed surface will appear to be larger than the lubricating film. If this seal is allowed to run with this type of surface condition, the liquid being sealed will continue to flash and vaporize. Carbonized debris from the liquid being sealed and/or carbon wear particles from the seal faces will begin to build up on exterior surfaces of the seal parts as illustrated in Figure 17-53. The flashing or vaporization of the liquid being sealed may be heard above the normal equipment sounds as a spitting or sputtering sound that may occur every few seconds to several minutes. This type of condition indicates abnormal wear occurring from the damaged mating ring. If a seal is believed to be operating in an unstable condition, it may be checked for sound as well as for the amount of carbonized debris on the exterior surfaces of the gland plate. Eliminating hot patches of thermally distressed material from instability can be achieved through increased cooling direct to the seal faces. If this is not possible, then the seal manufacturer should be contacted to determine whether or not hydropads could be incorporated into the seal design. Thermal distressed or heat checked seal faces commonly occur when a seal is run in liquids that are exceptionally

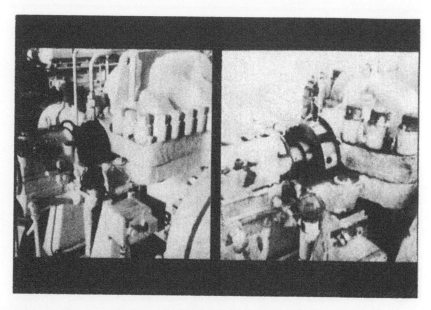

Figure 17-53. Condition of external seal parts at stuffing box. (A) Unstable operation with surface disturbance. (B) Stable operation without any surface disturbance.

poor lubricants. In this case, it is also a good idea to check for mechanical distortion as well.

Mechanical seals are applied to an infinite number of environmental conditions. Some of the liquids to be sealed are extremely poor lubricants for the seal faces and can result in severe wear problems. Figure 17-54 illustrates the faces of an inboard double seal from an agitator. The material of construction for both the primary and mating rings is tungsten carbide. This is the result of running in a liquid with a very low viscosity. The seal lubricant in this case was a silicone oil. The seal life was 11 months. During that time the primary ring wore into the mating ring by as much as 0.125 inch. This type of wear condition was overcome by replacing the lubricant in the seal chamber. In this case, water was selected as the replacement liquid. The result was a substantial increase in seal life.

Normally, in poor lubricating medias, the liquid to be sealed cannot develop a sufficient film to separate both the primary and mating rings. Figure 17-55 illustrates the results of running a silicon carbide primary ring against a tungsten carbide mating ring in a liquid with poor lubricating qualities. The tungsten carbide heat checked and the silicon carbide wore approximately 0.030 inch into the tungsten carbide. The liquid

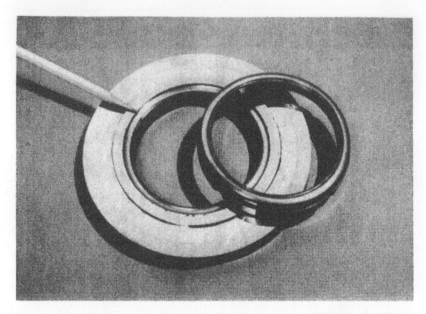

Figure 17-54. Wear of hard seal faces from an environment with poor lubricant.

sealed was ethane at 1200 psig and 3600 rpm. These hard seal faces offered no additional lubricant from the materials of construction. This problem was solved by using a mechanical grade of carbon with hydropads running against a tungsten carbide mating ring.

Poor lubrication can also be a problem on higher temperature applications. Figure 17-56 illustrates the results of operating a metal bellows seal on or near its atmospheric boiling point. Note the large amount of wear on the tungsten carbide surfaces. This also resulted in seal face vibration being transferred to the bellows, causing a leak at that point. This problem was corrected by lowering the temperature in the stuffing box to 50° below the liquid's atmospheric boiling point. Another solution is to consider the possible use of double opposed metal bellows seals with a good lubricant between the seals. Double seals, when properly applied, can last 2½ times longer than single seals on high temperature applications.

Corrosion of seal parts can be a serious problem that will appear over a longer period of time. Corrosion may not be isolated to just one structural part of the seal, but may be common to all parts of the assembly. Carbon materials can become gray, appear to be porous, and lose their hardness. Metals such as tungsten carbide, as shown in Figure 17-57, can have their binder etched out. In this case, once the corrosion covers the entire seal face, leakage occurs. Material selection should be made care-

Figure 17-55. Wear of hard faces in light hydrocarbon service at high pressure and high speed.

Figure 17-56. Wear of hard seal faces at high temperature.

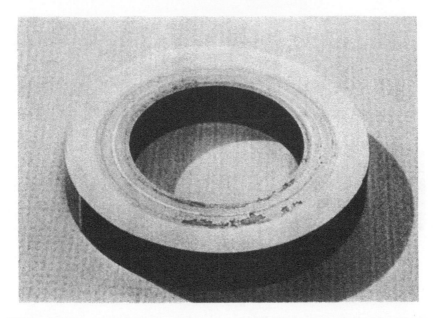

Figure 17-57. Corrosion of a mating ring surface exposed to the product being sealed.

fully. Most seal manufacturers have a wealth of information on materials for various services and should be contacted for their recommendations on the best materials to be used.

Finally, when a seal has been properly selected and installed, it will result in years of trouble-free operation. The seal shown in Figure 17-58 has been in service for 220,000 hours or just over 25 years. This 3-inch seal was used to seal gasoline, fuel oil, and occasionally, propane. Normal operating pressure and temperature were 800 psig and 60°F. The shaft speed was 3560 rpm. The condition of the carbon primary ring exhibits virtually no adverse wear of the sealing face or at the drive notches, which represent the drive system on this seal. Wear at the carbon seal face is only 0.032 inches for this period of time. The mating ring appears to be in good condition after running for this length of time. It is reported that this seal had not been leaking even though some minor heat checking had occurred. The seal aged in service; the O-ring hardness increased from 70 to 90 durometer. This would eventually lead to a loss of flexibility for the seal head.

A checklist for identifying causes of seal leakage is found in Table 17-6. Many other references on causes of seal leakage are available from seal manufacturers. One such source is found in *Identifying Causes of Seal Leakage, Chart and Bulletin S-2031*, available from John Crane Inc.

Figure 17-58. Three-inch diameter seal after approximately 220,000 hours of service in petroleum product.

Table 17-6
Checklist for Identifying Causes for Seal Leakage

SYMPTOM	POSSIBLE CAUSES	CORRECTIVE PROCEDURES
SEAL SPITS AND SPUTTERS ("FACE POPPING") IN OPERATION	SEAL FLUID VAPORIZING	O INCREASE COOLING OF SEAL FACES O ADD BY-PASS FLUSH LINE IF NOT IN USE O ENLARGE BY-PASS FLUSH LINE AND/OR ORIFICES IN GLAND PLATE O CHECK WITH SEAL MANUFACTURER FOR PROPER SEAL BALANCE O CHECK SEAL MANUFACTURER FOR ADDED COOLING BY USING HYDROPADS
SEAL DRIPS STEADILY	O FACES NOT FLAT O CARBON GRAPHITE FACES BLISTERED O SEAL FACES THERMALLY DISTORTED	O CHECK FOR INCORRECT INSTALLATION DIMENSION O CHECK FOR IMPROPER MATERIALS OR SEAL DESIGN FOR APPLICATION O IMPROVE COOLING FLUSH LINES O CHECK FOR GLAND PLATE DISTORTION DUE TO OVER TORQUING GLAND BOLTS O CHECK GLAND GASKET FOR PROPER COMPRESSION O CLEAN OUT FOREIGN PARTICLES BETWEEN SEAL FACES; RELAP FACES O CHECK FOR CRACKS AND CHIPS AT SEAL FACES; REPLACE IF NECESSARY

(table continued on next page)

Table 17-6 Continued
Checklist for Identifying Causes for Seal Leakage

SYMPTOM	POSSIBLE CAUSES	CORRECTIVE PROCEDURES
SEAL DRIPS STEADILY	O SECONDARY SEALS NICKED OR SCRATCHED DURING INSTALLATION O O-RINGS OVERAGED O SECONDARY SEALS HARD AND BRITTLE FROM COMPRESSION SET O SECONDARY SEALS SOFT AND STICKY FROM CHEMCIAL ATTACK	O REPLACE SECONDARY SEALS O CHECK FOR PROPER LEAD IN CHAMFER; REMOVE ANY BURRS O CHECK FOR PROPER SEALS WITH SEAL MANUFACTURER O CHECK WITH SEAL MANUFACTURER FOR OTHER MATERIALS
	O SPRING FAILURE O HARDWARE DAMAGED BY EROSION O DRIVE MECHANISMS CORRODED	O REPLACE PARTS O CHECK WITH SEAL MANUFACTURER FOR OTHER MATERIALS
SEAL SQUEALS DURING OPERATION	O AMOUNT OF LIQUID INADEQUATE TO LUBRICATE SEAL FACES	O ADD BY-PASS FLUSH LINE O ENLARGE BY-PASS LINE AND/OR ORIFICES IN GLAND PLATE
CARBON DUST ACCUMULATES ON OUTSIDE OF GLAND PLATE	O AMOUNT OF LIQUID INADEQUATE TO LUBRICATE SEAL FACES O LIQUID FILM EVAPORATING BETWEEN SEAL FACES	O ADD BY-PASS FLUSH LINE O ENLARGE BY-PASS FLUSH LINE AND/OR ORIFICES IN GLAND PLATE O IF CHAMBER PRESSURE IS HIGH, CHECK SEAL DESIGN WITH SEAL MANUFACTURER
SEAL LEAKS	O NOTHING APPEARS TO BE WRONG	O REFER TO LIST UNDER "SEAL DRIPS STEADILY" O CHECK FOR SQUARENESS OF CHAMBER FACE TO SHAFT O PROPERLY ALIGN SHAFT, IMPELLER, BEARING ETC.TO PREVENT SHAFT VIBRATION AND/OR DISTORTION AT GLAND PLATE AND MATING RING
SHORT SEAL LIFE	O ABRASIVE FLUID O ABRASIVE MATERIALS BUILDING UP AT ATMOSPHERIC SIDE OF SEAL O SEAL RUNNING TOO HOT O EQUIPMENT MECHANICALLY OUT-OF-LINE	O PREVENT ABRASIVES FROM ACCUMULATING AT SEAL FACES O ADD BY-PASS FLUSH LINE O USE ABRASIVE SEPARATOR O ADD A LIQUID OR STEAM QUENCH O INCREASE COOLING TO SEAL FACES O INCREASE BY-PASS FLUSH LINE FLOW O CHECK FOR OBSTRUCTED FLOW IN COOLING LINES O ALIGN--CHECK FOR RUBBING OF SEAL ON SHAFT

References

American Petroleum Institute: Centrifugal Pumps for General Refinery Services, API Standard 610, 7th Edition, Washington, DC, 1980.

ANSI/ASME B73.1 Specifications for Horizontal, End Suction Centrifugal Pumps for Chemical Pumps. Mechanical Engineering, June 1985, pp. 93.

Burton, R. A., "Thermal Deformation in Frictionally Heated Contact," *Wear,* 59, 1980, pp. 1–20.

Hamaker, J. B., "Mechanical Seals Lubrication Systems," 1977 ASLE Education Program-Fluid Sealing Course, American Society of Lubrication Engineers and John Crane Inc., Morton Grove, IL, May 1977.

John Crane Inc. *Engineered Fluid Sealing: Materials, Design, and Applications,* Morton Grove, IL, 1979.

John Crane Inc. *Identifying Causes of Seal Leakage,* S-2031 and Bulletin, Morton Grove, IL, 1979.

Kennedy, Jr., F. E. and Grim, J. N., "Observation of Contact Conditions in Mechanical Face Seals," ASLE Trans. 27, 1984, pp. 122–128.

Kilaparti, R. and Burton, R. A., "A Moving Hot Spot Configuration for a Seal-like Geometry, With Frictional Heating, Expansion and Wear," ASLE Trans. 20, 1977, pp. 64–70.

Netzel, J. P., "Identifying Causes of Seal Leakage," Pacific Energy Association, Pump Workshop, Anaheim, CA, October 13, 1982.

_____, "Mechanical Seals for Light Hydrocarbon Service," Fifth Annual Energy-Sources Technology Conference, New Orleans, Louisiana, March 8–10, 1982, The Petroleum Division, ASME, pp. 47–53.

_____, "Observation of Thermoelastic Instability in Mechanical Face Seals," *Wear,* 59, 1980, pp. 135–148.

_____, "Sealing Light Hydrocarbon as a Liquid or Gas," ASME, South Texas Section, Houston, Texas, October 29, 1985.

_____, "Symmetrical Seal Design: A Sealing Concept For Today," First International Pump Symposium, Houston, Texas, May 1984, Proceedings, pp. 109–112.

_____, "Wear of Mechanical Seals in Light Hydrocarbon Service," *Wear* 102, 1985, pp. 141–151.

Schoenherr, K., "Design Terminology for Mechanical Face Seals," SAE Trans 74, (65030), 1966.

Schoenherr, K. and Johnson, R. L., "Seal Wear," *Wear Control Handbook,* M. Peterson and W. Winer, Eds., American Society of Mechanical Engineers, New York, 1980.

Snapp, R. B., "Theoretical Analysis of Face Type Seals With Varying Radial Face Profiles," 64-WA/Lub 6, American Society of Mechanical Engineers, New York, 1964.

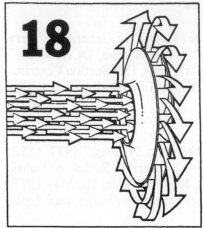

18

Vibration and Noise in Pumps

by **Fred R. Szenasi***
Engineering Dynamics Incorporated

Introduction

Although a certain amount of noise is to be expected from centrifugal pumps and their drivers, unusually high noise levels (in excess of 100 dB) or particularly high frequencies (whine or squeal) can be an early indicator of potential mechanical failures or vibration problems in centrifugal pumps. The purpose of this chapter is to concentrate on the mechanisms that may produce noise as a by-product; however, reduction of the noise, *per se*, is not the main concern. The main point of interest of this chapter is to study the mechanisms and their effect on the reliability of the pump system. Methods will be presented to reduce the vibration (and noise) or eliminate the basic causes by modifying the pump or piping system.

The occurrence of significant noise levels indicates that sufficient energy exists to be a potential cause of vibrations and possible damage to the pump or piping. Defining the source and cause of noise is the first step in determining whether noise is normal or whether problems may exist. Noise in pumping systems can be generated by the mechanical motion of pump components and by the liquid motion in the pump and piping systems. Noise from internal mechanical and liquid sources can be transmitted to the environment.

Effective diagnosis and treatment of noise sources to control pump noise require a knowledge of the liquid and mechanical noise-generation

* The author wishes to acknowledge the contributions by the engineering staff of Engineering Dynamics Inc., who performed many of the analyses and field tests.

422

mechanisms and common noise conduction paths by which noise can be transmitted. If noise itself is the major concern, it can be controlled by acoustic enclosures or other treatment [1, 2].

Sources of Pump Noise

Mechanical Noise Sources

Common mechanical sources that may produce noise include vibrating pump components or surfaces because of the pressure variations that are generated in the liquid or air. Impeller or seal rubs, defective or damaged bearings, vibrating pipe walls, and unbalanced rotors are examples of mechanical sources.

In centrifugal machines, improper installation of couplings often causes mechanical noise at twice pump speed (misalignment). If pump speed is near or passes through the lateral critical speed, noise can be generated by high vibrations resulting from imbalance or by the rubbing of bearings, seals, or impellers. If rubbing occurs, it may be characterized by a high-pitched squeal. Windage noises may be generated by motor fans, shaft keys, and coupling bolts. Damaged rolling element bearings produce high-frequency noise [3] related to the bearing geometry and speed.

Liquid Noise Sources

These are pressure fluctuations produced directly by liquid motion. Liquid noise can be produced by vortex formation in high-velocity flow (turbulence), pulsations, cavitation, flashing, water hammer, flow separation, and impeller interaction with the pump cutwater. The resulting pressure pulsations and flow modulations may produce either a discrete or broad-band frequency component. If the generated frequencies excite any part of the structure including the piping or the pump into mechanical vibration, then noise may be radiated into the environment. Four types of pulsation sources occur commonly in centrifugal pumps [2]:

- Discrete-frequency components generated by the pump impeller such as vane passing frequency and multiples.
- Flow-induced pulsation caused by turbulence such as flow past restrictions and side branches in the piping system.
- Broad-band turbulent energy resulting from high flow velocities.
- Intermittent bursts of broad-band energy caused by cavitation, flashing, and water hammer.

A variety of secondary flow patterns [4] that produce pressure fluctuations are possible in centrifugal pumps, as shown in Figure 18-1, particularly for operation at off-design flow. The numbers shown in the flow stream are the locations of the following flow mechanisms:

1. Stall
2. Recirculation (secondary flow)
3. Circulation
4. Leakage
5. Unsteady flow fluctuations
6. Wake (vortices)
7. Turbulence
8. Cavitation

Causes of Vibrations

Causes of vibrations are of major concern because of the damage to the pump and piping that generally results from excessive vibrations. Vibrations in pumps may be a result of improper installation or maintenance,

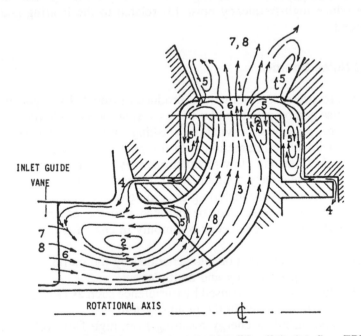

Figure 18-1. Secondary flow around pump impeller off-design flow EPRI Research Project 1266-18, Report CS-1445 [4].

incorrect application, hydraulic interaction with the piping system, or design and manufacturing flaws. Some of the common causes of excessive vibrations and failures are [5]:

Installation/Maintenance

Unbalance
Shaft-to-shaft misalignment
Seal rubs
Case distortion caused by piping loads
Piping dynamic response (supports and restraints)
Support structural response (foundation)
Anchor bolts/grout
Improper assembly

Application

Operating off of design point
Improper speed/flow
Inadequate NPSH
Entrained air

Hydraulic

Interaction of pump (head-flow curve) with piping resonances
Hydraulic instabilities
Acoustic resonances (pressure pulsations)
Water hammer
Flow distribution problems
Recirculation
Cavitation
Flow induced excitation (turbulence)
High flow velocity

Design/Manufacturing

Lateral critical speeds
Torsional critical speeds
Improper bearings or seals
Rotor instability
Shaft misalignment in journals
Impeller resonances
Bearing housing/pedestal resonances

Many of these causes are a result of an interaction of the pump (or its driver) with the fluid or the structure (including piping). This interactive relationship requires that the complete system be evaluated rather than investigating individual components when problems occur. Although prototype pumps or a new design may run the gambit of these problems, standard design or "off-the-shelf" pumps are not immune, particularly to system problems.

Installation/Maintenance Effects

Unbalance. Unbalance of a rotating shaft can cause large transverse vibrations at certain speeds, known as critical speeds, that coincide with the lateral natural frequencies of the shaft. Lateral vibration due to unbalance is probably the most common cause of downtime and failures in centrifugal pumps. Damage due to unbalance response may range from seal or bearing wipes to catastrophic failures of the rotor. Excessive unbalance can result from rotor bow, unbalanced couplings, thermal distortion, or loose parts. All too often, field balancing is required even after careful shop balancing has been performed.

Although a pump rotor may be adequately balanced at startup, after a period of operation the pump rotor may become unbalanced by erosion, corrosion, or wear. Unbalance could also be caused by non-uniform plating of the pumped product onto the impeller. In this instance, cleaning the impeller could restore the balance. Erosion of the impeller by cavitation or chemical reaction with the product may cause permanent unbalance requiring replacement of the impeller. Wear of the impeller or shaft caused by rubs will require the repair or replacement of the damaged component. Another cause of unbalance can occur if lubricated couplings have an uneven build-up of grease or sludge.

Assembly or manufacturing procedures may cause a new pump rotor to be unbalanced because of slight manufacturing imperfections or tolerance build-up resulting in the center of mass of the rotor not being exactly at the center of rotation. Forging or casting procedures can produce local variations in the density of the metal due to inclusions or voids. On large cast impellers, the bore for the shaft may not be exactly centered with the casting geometry. Stacking a rotor can result in thermal distortions of the shaft or impellers that can result in a cocked impeller. Nonsymmetries of just a few mils caused by these manufacturing or assembly methods can result in significant forces generated by a high speed rotor. Most of these nonsymmetries can be compensated for by balancing the rotor.

Misalignment. Angular misalignment between two shafts connected with a flexible coupling introduces an additional driving force that can produce torsional or lateral vibrations. The forces in a typical industrial coupling are similar to those in a universal joint (Figure 18-2). When a small angular misalignment occurs, the velocity ratio across the joint is not constant. If one shaft speed is assumed constant, then the other shaft has a faster rotational rate [6] for part of the revolution and a slower rotational rate for part of the revolution. This variation of rotating speed results in a second harmonic (twice shaft speed) vibrational component.

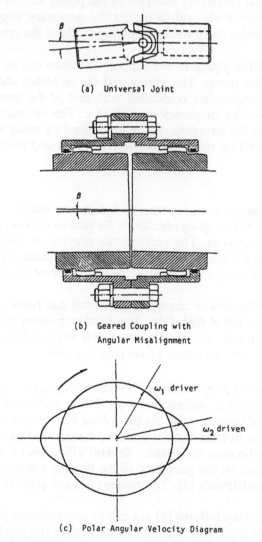

(a) Universal Joint

(b) Geared Coupling with
Angular Misalignment

ω_1 driver

ω_2 driven

(c) Polar Angular Velocity Diagram

Figure 18-2. Effects of angular misalignment in shaft couplings.

Piping and Structure. The pump should be relatively isolated from the piping. The weight and thermal loading on the suction and discharge connections should be minimized. The American Petroleum Institute (API) Standard 610 [7] specifies allowable external nozzle forces and moments. Most pump manufacturers specify allowable weight and thermal loads transferred from the pipe to the pump case. Static forces from the piping may misalign the pump from its driver, or for excessive loading, the pump case may become distorted and cause rubs or seal and bearing damage. Thermal flexibility analyses of the piping should be performed to evaluate piping loads and to design the necessary supports and restraints to minimize the transfer of piping loads to the operating equipment.

Vibrations of the piping or the support structure can be mechanically transferred to the pump. The piping and the structure should not have their resonant frequencies coincident with any of the pump excitations such as vane passing frequency or multiples. The vibrations transferred from the pipe to the structure can be minimized by using a visco-elastic material (i.e., belting material) between the pipe and the pipe clamp.

Application

The initial stage of pump system design should include the task of defining the range of operating conditions for pressure, flow, temperatures, and the fluid properties. The vendors can provide the correct pump geometry for these design conditions. Expected variations in operating conditions and fluid composition, if a significant percentage, may influence the design.

Improper application or changing conditions can result in a variety of problems. Operation at high-flow, low-head conditions can cause vibrations of the rotor and case. Inadequate NPSH can result in cavitation that will cause noise and vibration of varying degrees.

Bearings. General purpose, small horsepower pumps in process plants generally have rolling element bearings. Noise and vibrations are commonly a result of bearing wear. As the rolling elements or races wear, the worn surfaces or defects initially produce a noise and as wear increases vibrations may become noticeable. Several vibrational frequencies may occur that depend on the geometry of the bearing components and their relative rotational speeds [3]. The frequencies are generally above operating speed.

Many ball bearing failures [8] are due to contaminants in the lubricant that have found their way into the bearing after the machine has been placed in operation. Common contaminants include moisture, dirt, and other miscellaneous particles which, when trapped inside the bearing, may cause wear or permanently indent the balls and raceways under the tremendous stresses generated by the operating load.

Special purpose pumps and large boiler feed pumps commonly have oil film (hydrodynamic) bearings. The hydrodynamic bearing is superior to rolling element bearings for high speed or high load application. The hydrodynamic bearing supports the rotor on a film of oil as it rotates. The geometry of the hydrodynamic bearing and the oil properties play an important role in controlling the lateral critical speeds and consequently the vibrational characteristics of the pump.

Seals. The fluid dynamics of flow through seals have a dramatic effect on rotordynamics [9]. Hydrodynamic forces involved may contribute to the stabilization of rotating machinery or make it unstable. Seals with large axial flow in the turbulent range, such as in feed water pumps, tend to produce large stiffness and damping coefficients that are beneficial to rotor vibrations and stability. Wear of the seals will increase the clearance and cause greater leakage and possibly change the rotordynamic characteristics of the seal resulting in increased vibrations.

Hydraulic Effects

Hydraulic effects and pulsations can result in almost any frequency of vibration of the pump or piping from once per revolution up to the vane passing frequency and its harmonics. Frequencies below running speed can be caused by acoustical resonances. Generally, these effects are due to the impeller passing the discharge diffuser or some other discontinuity in the case. Any nonsymmetry of the internals of the pump may produce an uneven pressure distribution that can result in forces applied to the rotor.

Transients. Starting and stopping pumps with the attendant opening and closing of valves is a major cause of severe transients in piping systems. The resulting pressure surge, referred to as water hammer, can apply a sudden impact force to the pump, its internals, and the piping. Severe water hammer has caused cracks in concrete structures to which the pipe was anchored.

Rapid closure of conventional valves used in feedwater lines can cause severe water hammer. Increasing the closure time of the valve can reduce the severity of the surge pressure. Analytical methods are available to evaluate the severity of water hammer in a particular piping configuration for various closure rates [10].

Cavitation and Flashing. For many liquid pump piping systems, it is common to have some degree of flashing and cavitation associated with the pump or with the pressure control valves in the piping system. High flow rates produce more severe cavitation because of greater flow losses through restrictions.

Cavitation produces high local pressures that may be transmitted directly to the pump or piping and may also be transmitted through the fluid to other areas of the piping. Cavitation is one of the most commonly occurring and damaging problems in liquid pump systems. The term cavitation refers to the formation and subsequent collapse of vapor bubbles (or cavities) in a liquid caused by dynamic pressure variations near the vapor pressure. Cavitation can produce noise, vibration, loss of head and capacity as well as severe erosion of the impeller and casing surfaces.

Before the pressure of the liquid flowing through a centrifugal pump is increased, the liquid may experience a pressure drop inside the pump case. This is due in part to acceleration of the liquid into the eye of the impeller and flow separation from the impeller inlet vanes. If flow is in excess of design or the incident vane angle is incorrect, high-velocity, low-pressure eddies may form. If the liquid pressure is reduced to the vaporization pressure, the liquid will flash. Later in the flow path the pressure will increase. The implosion which follows causes what is usually referred to as cavitation noise. The collapse of the vapor pockets, usually on the nonpressure side of the impeller vanes, causes severe damage (vane erosion) in addition to noise.

When a centrifugal pump is operated at flows away from the point of best efficiency, noise is often heard around the pump casing. The magnitude and frequency of this noise may vary from pump to pump and are dependent on the magnitude of the pump head being generated, the ratio of NPSH required to NPSH available, and the amount by which actual flow deviates from ideal flow. Noise is often generated when the vane angles of the inlet guides, impeller, and diffuser are incorrect for the actual flow rate.

Cavitation can best be recognized by observing the complex wave or dynamic pressure variation using an oscilloscope and a pressure transducer. The pressure waveform will be non-sinusoidal with sharp maximum peaks (spikes) and rounded minimum peaks occurring at vapor pressure as shown in Figure 18-3. As the pressure drops, it cannot produce a vacuum less than the vapor pressure.

Cavitation-like noise can also be heard at flows less than design, even when available inlet NPSH is in excess of pump required NPSH, and this has been a puzzling problem. An explanation offered by Fraser [11, 12] suggests that noise of a very low, random frequency but very high inten-

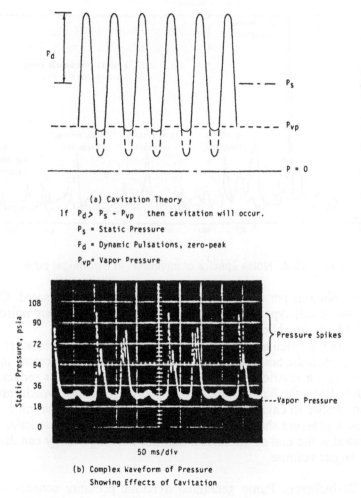

(a) Cavitation Theory

If $P_d > P_s - P_{vp}$ then cavitation will occur.

P_s = Static Pressure

P_d = Dynamic Pulsations, zero-peak

P_{vp} = Vapor Pressure

(b) Complex Waveform of Pressure
Showing Effects of Cavitation

Figure 18-3. Cavitation effects on the dynamic pressure.

sity results from backflow at the impeller eye or at the impeller discharge, or both. Every centrifugal pump has this recirculation under certain conditions of flow reduction. Operation in a recirculating condition can be damaging to the pressure side of the inlet and/or discharge impeller vanes (and also to casing vanes). Recirculation is evidenced by an increase in loudness of a banging type, random noise, and an increase in suction and/or discharge pressure pulsations as flow is decreased.

Sound levels measured at the casing of an 8000 hp pump and near the suction piping during cavitation [2] are shown in Figure 18-4. The cavitation produced a wide-band shock that excited many frequencies; however, in this case, the vane passing frequency (number of impeller vanes

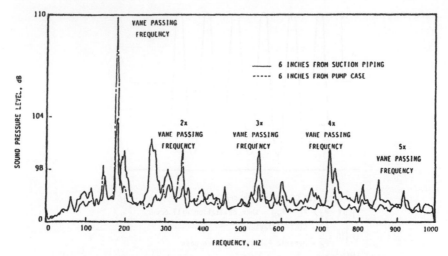

Figure 18-4. Noise spectra of cavitation in centrifugal pump.

times revolutions per second) and multiples of it predominated. Cavitation noise of this type usually produces very high frequency noise, best described as "crackling."

Flashing is particularly common in hot water systems (feedwater pump systems) when the hot, pressurized water experiences a decrease in pressure through a restriction (i.e., flow control valve). This reduction of pressure allows the liquid to suddenly vaporize, or flash, which results in a noise similar to cavitation. To avoid flashing after a restriction, sufficient back pressure should be provided. Alternately, the restriction could be located at the end of the line so that the flashing energy can dissipate into a larger volume.

Flow Turbulence. Pump generated dynamic pressure sources include turbulence (vortices or wakes) produced in the clearance space between impeller vane tips and the stationary diffuser or volute lips. Dynamic pressure fluctuations or pulsations produced in this manner can cause impeller vibrations or can result in shaft vibrations as the pressure pulses impinge on the impeller.

Flow past an obstruction or restriction in the piping may produce turbulence or flow-induced pulsations [2]. These pulsations may produce both noise and vibration over a wide-frequency band. The frequencies are related to the flow velocity and geometry of the obstruction. These pulsations may cause a resonant interaction with other parts of the acoustic piping system.

Most of these unstable flow patterns are produced by shearing at the boundary between a high-velocity and low-velocity region in a fluid field. Typical examples of this type of turbulence include flow around obstructions or past deadwater regions (i.e., a closed bypass line) or by bi-directional flow. The shearing action produces vortices, or eddies that are converted to pressure perturbations at the pipe wall that may result in localized vibration excitation of the piping or pump components. The acoustic natural response modes of the piping system and the location of the turbulence has a strong influence on the frequency and amplitude of this vortex shedding. Experimental measurements have shown that vortex flow is more severe when a system acoustic resonance coincides with the generation frequency of the source. The vortices produce broad-band turbulent energy centered around a frequency that can be determined with a dimensionless Strouhal number (S_n) from 0.2 to 0.5, where

$$f = \frac{S_n V}{D}$$

where   f = vortex frequency, Hz
S_n = Strouhal number, dimensionless (0.2 to 0.5)
V = flow velocity in the pipe, ft/sec
D = a characteristic dimension of the obstruction, ft

For flow past tubes, D is the tube diameter, and for excitation by flow past a branch pipe, D is the inside diameter of the branch pipe. The basic Strouhal equation is further defined in Table 18-1, items 2A and 2B. As an example, flow at 100 ft/sec past a 6-inch diameter stub line would produce broad-band turbulence at frequencies from 40 to 100 Hz. If the stub were acoustically resonant to a frequency in that range, large pulsation amplitudes could result.

Pressure regulators or flow control valves may produce noise associated with both turbulence and flow separation. These valves, when operating with a severe pressure drop, have high-flow velocities which generate significant turbulence. Although the generated noise spectrum is very broad-band, it is characteristically centered around a frequency corresponding to a Strouhal number of approximately 0.2.

Pulsations. Pumping systems produce dynamic pressure variations or pulsations through normal pumping action. Common sources of pulsations occur from mechanisms within the pump. The pulsation amplitudes in a centrifugal pump are generated by the turbulent energy that depends upon the clearance space between impeller vane tips and the stationary diffuser or volute lips, the installed clearances of seal, wear rings and the symmetry of the pump rotor and case. Because these dimensions are not accurately known, predicting the pulsation amplitudes is difficult. Even identical pumps often have different pressure pulsation amplitudes.

Table 18-1
Pulsation Sources

Generation Mechanism	Excitation Frequencies	
1. Centrifugal Compressors & Pumps	$f = \frac{nN}{60}$	
	$f = \frac{nBN}{60}$	B = Number of Blades
	$f = \frac{nvN}{60}$	v = Number of Volutes or Diffuser Vanes
2. Flow Excited		
A. Flow through Restrictions or Across Obstructions	$f = S\frac{V}{D}$	S = Strouhal Number = .2 to .5 V = Flow Velocity, ft/sec D = Restriction diameter, ft
B. Flow Past Stubs	$f = S\frac{V}{D}$	S = .2 to .5
C. Flow Turbulence Due to Quasi Steady Flow	$f = 0 - 30$ Hz (Typically)	
D. Cavitation and Flashing	Broad Band	

Even with the pump operating at its best efficiency point and proper conditions (NPSH, etc.) pulsations may be generated by high-flow velocities and turbulence at the vane tips or at the cutwater. As operating conditions deviate from the design conditions, more sources may come into play such as cavitation, recirculation, flow instabilities, etc.

These pulsations can interact with the hydraulic or acoustic natural frequencies of the piping system to amplify the pulsation. Acoustic natural frequencies in piping systems are a function of the fluid properties, the piping, and pump geometry. The acoustic interaction can be compared to the action of an organ pipe resonance where turbulence produced at the lip is amplified into an audible tone. Similarly, pulsations from the pump are amplified into pressure pulsations that react at elbows, restrictions, closed valves, and piping size changes to cause dynamic shaking forces. This conversion of hydraulic energy into mechanical forces can result in vibrations of the pump, piping, and their support structure.

In the design stage, the acoustical natural frequencies of piping systems can be calculated using either digital [13] or analog [14] modeling procedures. As an example, a model of a piping system analyzed by a digital acoustic analysis technique is given in Figure 18-5. The system was for chemical service with three pumps (3000 gpm, 250 psi, 3600 rpm) each 50% capacity (one spare). The predicted frequency response in the pump system at selected locations is given in Figure 18-6. The natural frequencies of the energy in the piping system can be compared to discrete frequencies generated by the pump (i.e., vane passing frequency, etc). It can immediately be seen that these 3600 rpm pumps (A and B operating) with a six-vane impeller could cause severe pulsations in the piping system because its vane passing frequency (6 × 60 rps) matches an acoustic response at 360 Hz. Based on the acoustic analysis, a seven-vane impeller should be used that would have its vane passing frequency at 420 Hz (7 × 60 rps) which has minimal response.

While in the design stage, changing the pump impeller for this system was a simple solution; however, the primary use of this acoustic analysis technique is to evaluate alternate piping configurations when the pump cannot be readily changed as in existing installations. Modifications to the piping system (i.e., lengths, routing) can be easily simulated to evaluate the effectiveness in attenuating a particular response mode.

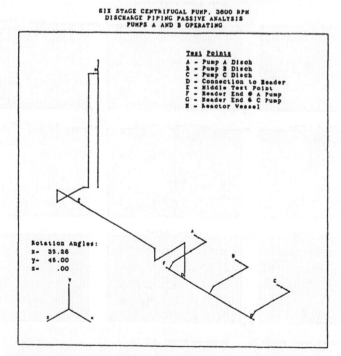

Figure 18-5. Simulation of centrifugal pump piping system.

The mechanical natural frequencies of the piping spans should not occur in the same range as the acoustic response frequencies. The analysis aids in determining the allowable frequency range that can be used to establish the proper design of the piping supports and span lengths to minimize the potential for exciting a piping resonance. The acoustic analysis can be used to redesign the piping to modify the frequency response and particular acoustic modes that may be predominant.

Acoustic Resonance. When a dynamic pressure pulse propagates down a pipe and reaches a restriction or pipe size change (flow area), the pulse is reflected [13, 15]. As the series of pressure pulses continue to be reflected, a standing wave is generated; that is, at a point in the pipe, the pressure periodically rises above and drops below the average line pressure (simple harmonic variation). The super-position of an incident pulse and a reflected pulse, being the sum of two pulses traveling in opposite directions, produce the standing wave.

If the timing (phasing) of a reflected pulse matches a new pulse, the two pulses will add, or amplify. The timing of the pulses are dependent

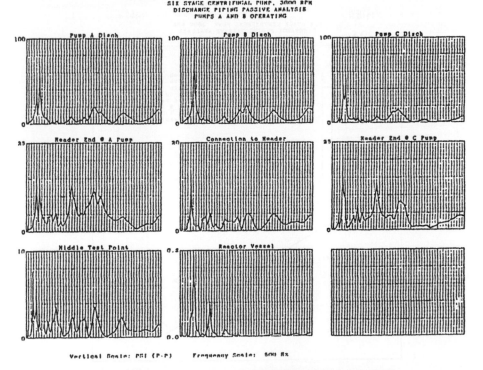

Figure 18-6. Passive acoustic response of piping system.

upon the pump speed (frequency) and pipe length (distance traveled) and the physical properties of the fluid.

The acoustic velocity, a function of the fluid density and bulk modulus, is an important factor in determining the resonant frequency of a pipe length. The API has published a comprehensive handbook for the physical properties of hydrocarbon gasses and liquids [16]. Calculation procedures for the acoustic velocity of water and other common liquids are presented in the Appendix at the end of this chapter.

A resonant condition [17] exists when the standing wave amplitude is reinforced so that the actual maximum dynamic pressure (pulsation) amplitude is substantially greater than the induced pulsation. Thus at frequencies (pump speeds) corresponding to resonance, there would be considerably higher amplitude levels generated from the same amount of energy than for frequencies off resonance.

If the wave frequencies are such that the incident and reflected waves are additive, the pulsations are amplified. If no damping is present, the pressure amplitudes at the anti-nodes would, theoretically, go to infinity. Actual piping systems have acoustic damping as a result of:

- Viscous fluid action (intermolecular shearing).
- Transmission, i.e., lack of total reflection, at a line termination.
- Piping resistance, i.e., pipe roughness, restrictions, orifices.

Therefore, damping of acoustic modes may be accomplished by placement of resistance elements, such as an orifice, that will work most effectively at velocity maxima.

Length Resonances in Distributed Acoustic Systems. The concepts of acoustic waves, reflections, and resonance can be applied to describe some of the classical length resonances [13].

The length resonances of certain piping elements are described in terms of a full-wave length. The acoustic wave length is the distance required for a complete cycle of dynamic pressure reversal. The wave length is related to the driving frequency and the acoustic velocity or speed of sound:

$$\lambda = \frac{c}{f}$$

where λ = wave length, ft/cycle
       c = acoustic velocity, ft/sec
       f = driving frequency, Hz

Half-Wave Resonance (Open-Open and Closed-Closed)

The first three modes for an open-open pipe are shown in Figure 18-7. Resonances may also occur at integer multiples of the half-wave frequency. For a closed-closed pipe, the formula also applies since both elements have a standing wave that is one-half of a sine wave even though the peaks occur at different locations. The pressure mode shapes of the first three modes are also shown for the open-closed configuration. The length should be corrected for entrance and exit effects (add approximately 80% of the pipe inside diameter) to calculate the half-wave resonance of open-open configurations. The end correction factor becomes very important in short pipes.

Quarter-Wave Resonance (Open-Closed)

The first three modes for an open-closed pipe, commonly referred to as a "quarter-wave stub" are depicted in Figure 18-7. The stub has its resonant frequencies at odd integer multiples of the fundamental quarter-wave frequency. Examples of a quarter-wave stub include a bypass line with a closed valve or a test connection with a pressure gauge.

A quarter-wave resonance can cause erroneous measurements [13] when obtaining dynamic pressure data. A typical test connection, depicted in Figure 18-8, with a short nipple and valve connected to a main line is an acoustical quarter-wave stub. This length can tune up to pulsations in the main line and cause the needle on a pressure gauge to wobble or indicate severe pressure variations that do not actually exist in the main line. Similarly, the data from a dynamic pressure transducer can be misinterpreted.

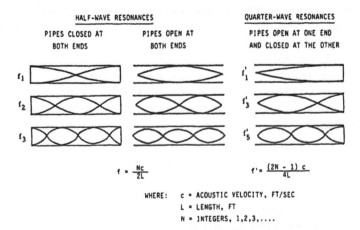

HALF-WAVE RESONANCES

QUARTER-WAVE RESONANCES

PIPES CLOSED AT BOTH ENDS

PIPES OPEN AT BOTH ENDS

PIPES OPEN AT ONE END AND CLOSED AT THE OTHER

$$f = \frac{Nc}{2L}$$

$$f' = \frac{(2N - 1) c}{4L}$$

WHERE: c = ACOUSTIC VELOCITY, FT/SEC
L = LENGTH, FT
N = INTEGERS, 1,2,3,....

Figure 18-7. Acoustic resonances of piping elements.

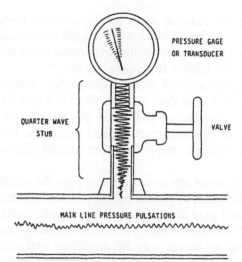

PRESSURE GAGE
OR TRANSDUCER

QUARTER WAVE
STUB

VALVE

MAIN LINE PRESSURE PULSATIONS

Figure 18-8. Erroneous pressure measurement caused by stub resonance.

For example, an installation where the speed of sound was 4500 ft/sec and the transducer was mounted one foot from the inside surface, the stub (quarter-wave) frequency would be 1125 Hz. A fictitious pressure pulsation component will be measured at this frequency and also at 3X, 5X, 7X, etc. The pulsation exists in the stub; however, it may not be present in the main pipe. If this stub frequency is close to the vane or blade passing frequency, the measured amplitude at the stub frequency will not be valid because it would be amplified by the acoustical resonance. The acoustical amplification factor can be as high as 200. To measure high-frequency pulsations, the transducer should be mounted flush to the inside surface of the pipe.

Measurements of peak-to-peak pulsations on an oscilloscope are often suspect. If the transducer signal frequency spectrum is analyzed and the quarter-wave resonance identified, then credible results can be obtained. An electronic filter may be used to eliminate the undesired frequency.

Coincidence of Driving Acoustic Frequencies and Length Resonances. The existence of quarter or half-wave natural frequencies alone does not constitute resonances. For resonance to occur, a dynamic pulse must be generated at a frequency equal to an acoustical natural frequency. The build-up in amplitude occurs because a reflected wave arrives at the proper time to reinforce the wave generated at the pump. The arrival of the reflected wave is dependent upon the path length of the piping elements. Therefore, the standing wave pattern amplitude is reinforced so that the actual maximum pulsating wave amplitude is substantially greater than the induced level. Multiples of the resonant modes can be excited; however, the multiple wave length resonances generally decrease in severity at the higher multiples because more acoustic energy is required to drive the higher frequency modes.

The acoustic resonances of piping systems for constant-speed pumps can usually be adjusted to detune them from the pump operating speed and vane-passing frequencies and avoid pulsation amplification. However, if the pump is operated over a speed range, the frequency band of the excitations is widened, requiring more careful placement of acoustic resonances.

Actual piping systems are more complex than the simple quarter-wave and half-wave elements. A typical piping system with tees, flow control valves (restrictions) pipe size changes, vessels, etc., will have a complicated pattern of pressure pulse reflection patterns (standing waves). Some of the standing waves may be amplified and others, attenuated. Each of the standing waves will have a particular acoustic length pertaining to a pipe segment between two end conditions. Calculations of the acoustic resonances of a complex piping system require the use of computer codes to consider the acoustic interaction between the pump and its piping system.

Instabilities. Hydraulic instabilities [14] can be a result of the dynamic interaction of a centrifugal pump (particularly the head-flow characteristics) and the acoustic response of the piping system. A centrifugal pump operating at constant speed in a piping system may amplify or attenuate pressure disturbances that pass through the pump. The action of the pump in causing amplification or attenuation of this energy is quite complex, but basically is dependent upon:

- The head curve slope and operating point
- System flow damping (in the piping)
- The existence of strong reactive resonances in the piping, particularly if they coincide with vortex frequencies
- The location of the pump in the standing wave field (i.e., at a velocity maximum rather than a pressure maximum)
- The compressibility (bulk modulus) of the liquid

Pulsations can be amplified by the piping system and cause a variety of problems such as damage to pump internals, torsional reactions, cavitation, vibrations at elbows, valves, or other restrictions. The amplitude of the pulsation is dependent upon operating conditions such as speed, flow rate, and losses (pressure drop) as well as fluid properties and acoustic natural frequencies. Consequently, pulsation amplitudes are usually affected by changes in the operating conditions or fluid composition.

Pulsations are commonly initiated by flow turbulence at changes in flow cross section, at restrictions (orifices, valves, etc.) or at the pump impeller. When the frequency of this turbulent energy excites one of the acoustic resonances of the piping system, and if the pump is situated near a velocity maximum in the resonant piping system, then high amplitude, self-sustaining pulsations can result. Pulsations can be minimized by moving the pump to a velocity minimum, but often a new pulsation frequency will be generated such that the pump is again situated near a velocity maximum for a higher mode oscillation. With proper care and detailed analysis of the relative strength of the various pulsation resonance modes of the piping, piping designs can be developed to avoid these strong resonances. Controlling piping stub lengths (i.e., in by-pass piping) so that their quarter-wave stub resonances are far removed from the Strouhal excitation frequencies, will also help in minimizing the potential for resonant pulsations.

This type of instability is more probable at low flows because acoustic damping that is generated by flow friction effects is greater at higher flow rates.

Design/Manufacturing

Dynamic response of the pump components to normal exciting forces within its operating frequency range can result in problems from excessive maintenance to catastrophic failure. Improper manufacture or assembly can cause unbalance resulting in damaging vibrations if the speed approaches natural frequencies of the rotor system. A rotordynamic audit [18] of the pump rotor design is crucial in avoiding speed-related vibrations.

The following section discusses the factors that are important in controlling the rotordynamic responses of a centrifugal pump.

Rotordynamic Analysis

A lateral critical speed is defined by API [7] as the speed at which a peak vibrational response occurs. At the critical speed, the rotor is more sensitive to unbalance than at any other speed. The critical speeds of a pump should be avoided to maintain acceptable vibration amplitudes. This section discusses the techniques involved in calculating the lateral critical speeds of centrifugal pumps.

Lateral Critical Speed Analysis

Pump rotordynamics are dependent on a greater number of design variables than are many other types of rotating equipment. Besides the journal bearing and shaft characteristics, the dynamic characteristics of the seals and the impeller-stationary lip interaction can have significant effects on the critical speed location, rotor unbalance sensitivity, and rotor stability [9, 19]. In this context, a seal is an element having a liquid film within a tight clearance. The liquid film has dynamic characteristics similar to a bearing. There are a variety of seal configurations including floating ring seals, grooved seals, and others. Several seal geometries will be discussed.

For modeling purposes, seals can be treated as bearings in the sense that direct and cross-coupled stiffness and damping properties can be calculated based on the seal's hydrostatic and hydrodynamic properties [20]. Seal clearances, geometry, pressure drop, fluid properties, inlet swirl, surface roughness, and shaft speed are all important in these calculations. The high pressure liquid being pumped also flows (or leaks) through the small annular spaces (clearances) separating the impellers under different pressures, such as wear rings and interstage bushings, and creates a hydrodynamic bearing effect that transforms the pump rotor from a two-bearing system to a multi-bearing system. The additional stiffness generated by the pumped liquid as it lubricates these internal bearings (seals, etc.) is referred to as the "Lomakin effect" [21].

The Lomakin stiffness effect minimizes the shaft deflections when the pump is running, and in some cases, the Lomakin effect can be of sufficient magnitude to prevent the critical speed of the rotor from ever being coincident with the synchronous speed. Since the pressure drop across seals increases approximately with the square of the pump speed, the seal stiffness also increases with the square of the speed.

The amount of support derived from the seals as bearings depends upon (a) the pressure differential, and therefore disappears completely when the pump is at rest, and (b) the clearance that increases significantly as the sealing surfaces wear. Consequently, contact between the rotor and stationary parts may take place each time the pump is started or stopped. In consideration of these facts, the rotordynamic analyses should include the effects of worn seals (loose clearances) as well as new seals (tight clearances).

Analytical techniques [22, 23, 24] have been developed whereby the seal geometry can be specified and the characteristics calculated for specific assumptions with regard to inlet swirl, groove design, etc. A series of grooved seal designs used in commercial pumps has been tested to verify the adequacy of the techniques.

A thorough lateral critical speed analysis is essential for developing a reliable, trouble-free pump system. The design audit [18] should include the following calculations:

- Critical speed map
- Undamped natural frequencies and mode shapes
- Bearing stiffness and damping properties
- Seal stiffness and damping properties
- Rotor response to unbalance
- Pedestal and foundation effects on response
- Rotor stability

The first step in performing a lateral critical speed analysis is to model the shaft with sufficient detail and number of masses to accurately simulate the rotor responses through its speed range. An accurate shaft drawing giving the dimensions, weights, and centers of gravity of all added masses is needed to develop the model. Generally, each significant shaft diameter change is represented by one or more stations. A station is generally located at each added mass or inertia, at each bearing and seal location, and at each potential unbalance location. A typical rotor shaft drawing and the computer model is given in Figure 18-9.

Rotating elements such as impellers are modeled as added masses and inertias at the appropriate locations on the shaft. The polar and transverse mass moments of inertia are included in the analyses to simulate the gyroscopic effects on the rotor. The gyroscopic effects are particularly significant on overhung rotors where the impeller produces a restoring moment when whirling in a deflected position.

Couplings are simulated as concentrated added weights and inertias. Normally half the coupling weight is placed at the center of gravity of the half coupling. When necessary, the entire train, including the driver and driven equipment, can be modeled by utilizing programs that can simulate the shear loading and moment transfer across the coupling. Once the shaft model is completed, the critical speed map can be calculated.

Critical Speed Map. The critical speed map is a logarithmic plot of the undamped lateral critical speeds versus the combined support stiffness, consisting of the bearing and support structure as springs in series. The critical speed map provides the information needed to understand the basic response behavior of rotors; therefore, it is important to understand how the map is developed [25].

For large values of support stiffness, the rotor critical speeds are called the rigid bearing critical speeds. If the bearing stiffness is infinity, the vibrations are zero at the bearings, and the first natural frequency for

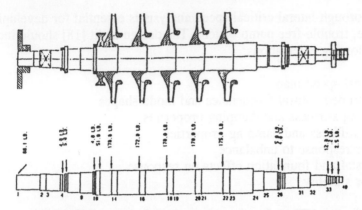

Figure 18-9. Typical shaft drawing and computer model.

shafts that do not have overhung impellers is analogous to a simply supported beam.

A critical speed map, normalized to the first rigid bearing critical speed is given in Figure 18-10 to illustrate the ratios of the various criticals for low and high support stiffness values and to illustrate the mode shapes that the rotor will have at different bearing and support stiffness values. For the rigid bearing critical speeds, the mode shape for the first mode would be a half-sine wave (one loop), the second critical speed would be a two-loop mode and would occur at a frequency of four times the first mode critical, the third critical speed would be a three-loop mode and would be nine times the first critical, etc. For most rotors, the bearing stiffnesses are less than rigid and the second critical will be less than four times (typically two–three times) the first critical.

For low values of support stiffness (shaft stiffness is large compared to support stiffness), the first critical speed is a function of the total rotor weight and the sum of the two support spring stiffnesses. For an ideal long slender beam, the second mode is similar to the rocking of a shaft on two springs and is equal to 1.73 times the first critical speed. Since both the first and second modes are a function of the support stiffness, the slope of the frequency lines for the first and second critical speeds versus support stiffness is proportional to the square root of the stiffness for low values of support stiffness compared to the shaft stiffness.

For a support stiffness of zero, the third and fourth modes would be analogous to the first and second free-free modes of a beam. For an ideal uniform beam, the ratio of the frequencies for these modes compared to the first critical speed for rigid bearings is 2.27 and 6.25.

To aid in the discussion, an example of the critical speed analysis [9] of an eight-stage pump will be presented. The critical speed map for the

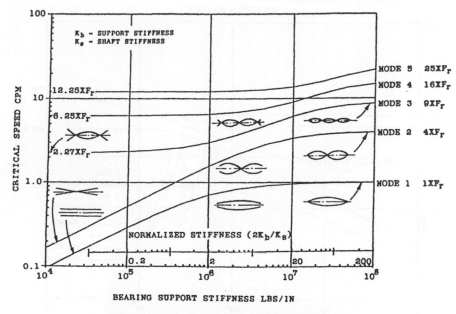

Figure 18-10. Normalized critical speed map.

"dry rotor" model (without the hydrodynamic effect of the seals) is shown in Figure 18-11. The first four undamped lateral critical speeds are plotted versus bearing support stiffness. Horizontal and vertical journal bearing stiffness curves (Kxx and Kyy) for both minimum and maximum assembled clearances are plotted to define the range of critical speeds. Intersections between the bearing stiffness curves and the mode curves represent undamped critical speeds (circled in Figure 18-11). Note that the "Mode 1" curve is fairly flat in the region where the intersections occur. The first two lateral mode shapes were calculated for a nominal bearing stiffness of 500,000 lb/in. and are shown in Figures 18-12 and 18-13. The dry rotor undamped natural frequencies as predicted by the critical speed map for the first, second, and third modes were 1710, 6650, and 8830 cpm, respectively.

Bearing Stiffness and Damping. The dynamic stiffness and damping coefficients of bearings [26] can be adequately simulated using eight linear coefficients (Kxx, Kyy, Kxy, Kyx, Cxx, Cyy, Cxy, Cyx) as shown in Figure 18-14. This information along with the lubricant minimum film thickness, flow, power loss, and temperature rise at operating conditions is needed to evaluate the bearing design. The bearing stiffness and damping coefficients are calculated as functions of the bearing type, length,

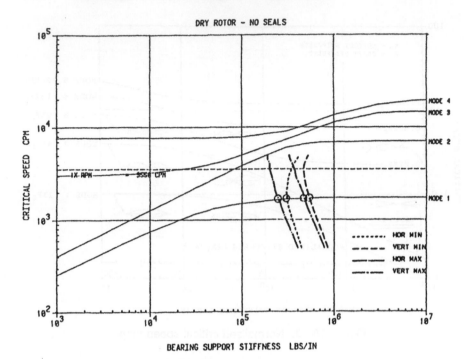

Figure 18-11. Eight-stage pump critical speed map—no seals.

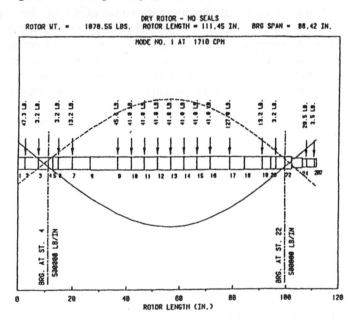

Figure 18-12. Eight-stage pump—first mode response—no seals.

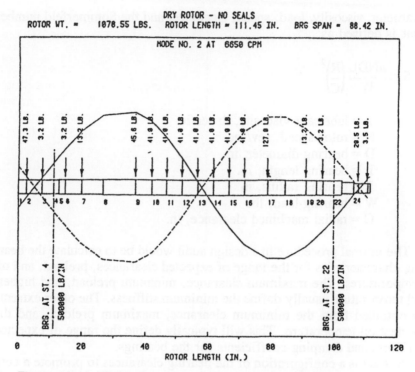

Figure 18-13. Eight-stage pump—second mode response—no seals.

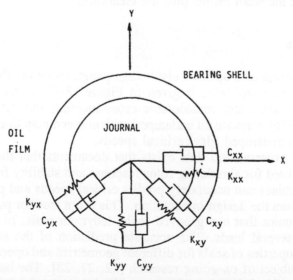

Figure 18-14. Hydrodynamic bearing stiffness and damping coefficients.

diameter, viscosity, load, speed, clearance, and the Sommerfeld number that is defined as:

$$S = \frac{\mu NDL}{W} \left(\frac{R}{C}\right)^2$$

> μ = lubricant viscosity, lb-sec/in.
> N = rotor speed, rev/sec
> D = bearing diameter, in.
> L = bearing length, in.
> R = bearing radius, in.
> W = bearing load, lbs
> C = radial machined clearance, in.

The normal procedure in a design audit would be to calculate the bearing characteristics for the range of expected clearances, preload, and oil temperatures. The maximum clearance, minimum preload, and highest oil temperature usually define the minimum stiffness. The other extreme is obtained from the minimum clearance, maximum preload, and the coldest oil temperature. This will typically define the range of expected stiffness and damping coefficients for the bearings.

Preload is a configuration of the bearing clearances to promote a converging wedge of oil that increases the oil pressure and consequently the bearing stiffnesses. A preloaded bearing has its radius of curvature greater than the shaft radius plus the clearance.

Seal Effects

The critical speed map for a eight-stage pump, including the effects of seal and bearing stiffness, is given in Figure 18-15. Even though the bearings and seals add considerable cross-coupling and damping, it is still desirable to generate an undamped critical speed map to establish the range of the undamped (dry) critical speeds.

Adequate experimental data exists that documents that the analytical procedures used for simulating rotor response and stability for compressors and turbines can accurately predict critical speeds and potential instabilities from the design information. This is not true for pumps, especially for pumps that use grooved seals, labyrinth seals, or screw type seals with several leads. The accurate prediction of the stiffness and damping properties of seals for different geometries and operating conditions is a subject of on-going research [22, 27, 28]. The basic theories presented by Black [29] have been modified to account for finite length

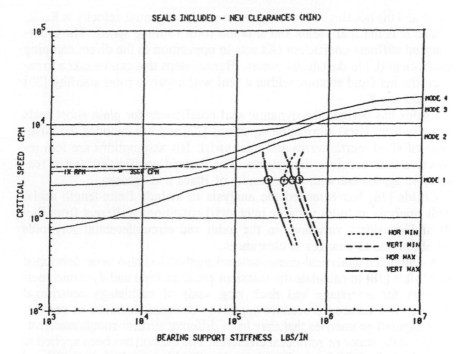

SEALS INCLUDED - NEW CLEARANCES (MIN)

Figure 18-15. Eight-stage pump critical speed map—seals included.

seals, inlet swirl, groove, and other important parameters. However, a universally accepted procedure to accurately predict seal properties is not available for all the seal types that are in use today. If the seal effects are not correctly modeled, calculated critical speeds can be significantly different from actual critical speeds.

The forces in annular pressure seals can have a significant effect on the vibration characteristics of a pump rotor. The hydrodynamic and hydrostatic forces involved can significantly affect unbalanced response characteristics. The fluid film interaction with the shaft and the pressure drop across the seal give rise to a load capacity and a set of dynamic stiffness and damping coefficients similar to those used to represent the oil film in journal bearings.

Unlike hydrodynamic bearings, seals develop significant direct stiffness in the centered, zero-eccentricity position due to the distribution of the axial pressure drop between the inlet losses and an axial pressure gradient due to friction losses. The cross-coupled stiffnesses arise due to fluid rotation (swirl) within the seal. As a fluid element proceeds axially along an annular seal, shear forces at the rotor accelerate or decelerate the fluid tangentially until an asymptotic value is reached. For a seal with the same directionally homogeneous surface-roughness treatment on the rotor and the housing, the average asymptotic tangential velocity is $R\omega/2$, where R is the seal radius and ω is the rotor running speed. The cross-coupled stiffness coefficient (K) acts in opposition to the direct damping coefficient (C) to destabilize rotors. Hence, steps that can be taken to reduce the net fluid rotation within a seal will improve rotor stability [30] by reducing K.

Childs has defined the dynamic seal coefficients for plain short seals directly from Hirs' lubrication equations [27] and has included the influence of fluid inertia terms and inlet swirl. His assumptions are less restrictive than previous derivations. The derived coefficients are in reasonable agreement with prior results of Black and Jenssen.

Childs [28] has extended the analysis to include finite-length seals. This analysis includes variable inlet swirl conditions (different from $R\omega/2$) and considers variations in the axial and circumferential Reynolds numbers due to changes in clearances.

A combined analytical-computational method has also been developed by Childs [28] to calculate the transient pressure field and dynamic coefficients for interstage and neck ring seals of multistage centrifugal pumps. The solution procedure applies to constant-clearance or convergent-tapered geometries that may have different surface-roughness treatments of the stator or rotor seal elements. The method has been applied to the calculation of "damper" seals as described by von Pragenau [31] and several roughened stator designs, such as knurled-indentation, diamond-grid post pattern, and round-hole pattern, have been tested. These procedures can be used to calculate serrated or grooved seals of various geometries.

Critical Speed Map—Considering Seals. The seal configurations at the balance piston, neck ring, and interstage bushings of the eight-stage pump are shown in Figure 18-16. The critical speed map (Figure 18-15) includes the support stiffnesses of the neck ring seals and interstage bushings combined at each impeller. The seal stiffness and damping coefficients are listed in Table 18-2 for nominal clearances. Note that a negative principal stiffness (K) value is predicted for the balance piston.

For this analysis the pump rotor was analyzed as if it had 11 bearings consisting of two tilted-pad bearings, the balance piston, and eight seals located at the impellers. For the purposes of developing the critical speed map, the seal stiffness values were held constant at their maximum levels (minimum clearances) that represent new seals.

The lateral mode shape of the first critical speed including these seal effects is shown in Figure 18-17. A bearing stiffness value of 500,000 lb/in. was again used for the mode shape calculations. Comparing Figure 18-17 with Figure 18-12, it is seen that the seals increase the frequency of

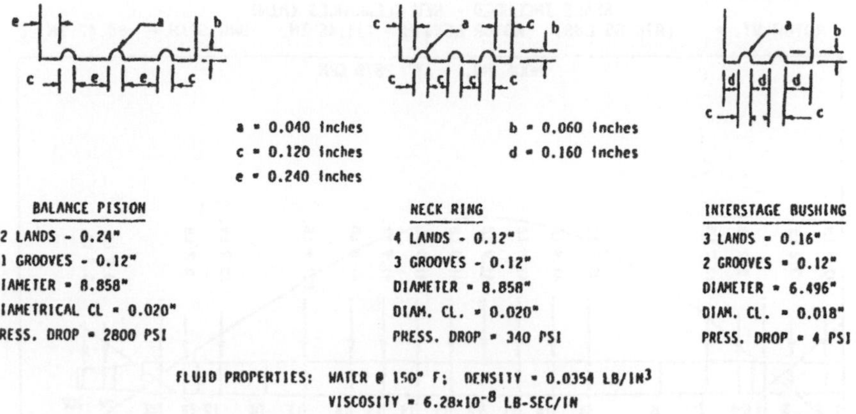

a = 0.040 inches
c = 0.120 inches
e = 0.240 inches

b = 0.060 inches
d = 0.160 inches

BALANCE PISTON	NECK RING	INTERSTAGE BUSHING
32 LANDS - 0.24"	4 LANDS - 0.12"	3 LANDS - 0.16"
31 GROOVES - 0.12"	3 GROOVES - 0.12"	2 GROOVES - 0.12"
DIAMETER - 8.858"	DIAMETER - 8.858"	DIAMETER - 6.496"
DIAMETRICAL CL - 0.020"	DIAM. CL. - 0.020"	DIAM. CL. - 0.018"
PRESS. DROP - 2800 PSI	PRESS. DROP - 340 PSI	PRESS. DROP - 4 PSI

FLUID PROPERTIES: WATER @ 150° F; DENSITY - 0.0354 LB/IN^3
VISCOSITY - 6.28x10^{-8} LB-SEC/IN

Figure 18-16. Eight-stage pump seal geometries and fluid properties.

Table 18-2
Summary of Seal Coefficients

	Stiffness		Damping	
Seal Type	Principal K-lb/in	Cross-Coupled K-lb/in	Principal C-lb sec/in	Cross-Coupled C-lb sec/in
Neck-Ring	13,400	4900	23	0
Int-Stg Bush	-8	370	3	0
Balance Piston	-271,000	627,000	25,000	3000

the first mode without altering the mode shape significantly. For an assumed bearing stiffness of 500,000 lb/in., the first, second, and third critical speeds from Figure 18-15 are 2570, 7120, and 8830 cpm.

Evaluation of Critical Speed Calculations. To summarize, in the evaluation of the adequacy of the rotor from the critical speed map and the mode shapes, the following items should be examined [18]:

• *The proximity of the critical speed to running speed or speed range.* The undamped lateral critical speeds should not coincide with the running speed. Various codes [7] address the allowable margin between lateral critical speeds and exciting frequencies. To determine if the actual critical speed will cause excessive vibrations, a rotor response to unbalance analysis should be performed.

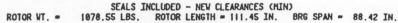

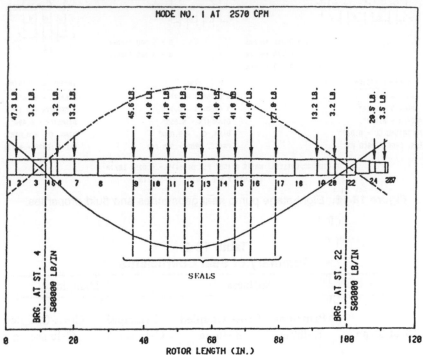

Figure 18-17. Eight-stage pump—first mode response—seals included.

- *The location of the critical speed relative to the support stiffness.* If the critical speed is near the rigid bearing criticals (flexible shaft region), increasing the bearing stiffness will not increase the critical speed because the weaker spring controls the resonant frequency. Vibration amplitudes may be low at the bearings (first mode), and therefore, low damping will be available. This can contribute to rotordynamic instabilities that will be discussed later. If the critical speeds are in the area of low support stiffness (stiff shaft region), the critical speeds are strongly dependent upon the bearing stiffness and damping characteristics and the critical speeds will be dependent upon bearing clearance. Bearing wear could be a significant problem.
- *The mode shape of the critical speed.* The mode shapes are used to assess the response of the rotor to potential unbalances. For example, a rotor that has a conical whirl mode (second critical) would be sensitive to coupling unbalance, but not strongly influenced by midspan unbalance.

Response To Unbalance

The location of a pump critical speed is defined by its response to unbalance. It is important to recognize the difference between critical speeds excited by unbalance and damped eigenvalues that are frequently also called critical speeds [32]. Generally, the effect of damping is to raise the frequency of the critical speed response due to unbalance; however, the effect of damping on the damped eigen values is to lower the frequency. The damped eigenvalues are primarily used for evaluating the stability of the rotor system. For compressors and turbines with tilting pad bearings, the damped eigenvalues are usually comparable to the unbalanced response criticals. However, in a pump with a large number of seals, the added damping to the system can be considerable, resulting in large differences in the unbalanced response critical speeds and the damped eigenvalues.

Rotor unbalance response calculations are the key analysis in the design stage for determining if a pump rotor will be acceptable from a dynamics standpoint. An accurate prediction of rotor unbalanced response is difficult for centrifugal pumps because of the sensitivity to bearing and seal clearances that may be at the tight or loose end of the tolerance range.

Computer programs are available that can calculate the elliptical shaft orbit at any location along the length of a rotor for various types of bearings, pedestal stiffnesses, pedestal masses, seals, labyrinths, unbalance combinations, etc. These programs can be used to determine the response of the installed rotor to unbalance and accurately predict the critical speeds over the entire range of variables. The actual critical speed locations as determined from response peaks caused by unbalance are strongly influenced by the following factors [33]:

- bearing direct stiffness and damping values

- bearing cross-coupled stiffness and damping values

- location of the unbalance

- location of measurement point

- bearing support flexibility

The normal unbalance used in the analysis would produce a force equal to 10% of the rotor weight at operating speed. Usually the rotor response to unbalance calculations are independently made for midspan unbalance, coupling unbalance, and moment unbalance. An unbalance equal to a force of 5% of the rotor weight is usually applied at the coupling to excite the rotor. For moment unbalance, an unbalance equal to 5% of the rotor weight is used at the coupling and another equal unbalance is used out-of-phase on the impeller furthest from the coupling.

The unbalance response of a pump should be analyzed for several cases to bracket the expected range of critical speeds. The first analysis should include minimum seal and bearing clearances that represent the maximum expected support stiffness and therefore, the highest critical speed. The second analysis should consider the maximum bearing clearances and seal clearances of twice the design clearance to simulate the pump condition after long periods of service. The third analysis should simulate the worn condition with no seal effects and maximum bearing clearances that represent the overall minimum expected support stiffness for the rotor (lowest critical speed).

The unbalanced response of the eight-stage pump with maximum bearing clearances and no seals is shown in Figure 18-18. The peak response at 1700 rpm was the minimum calculated critical speed. The unbalance was applied at the rotor midspan to excite the first mode. The response for the intermediate analysis (worn seals) is plotted in Figure 18-19. The worn seals increased the predicted response peak to approximately 1800 cpm. With minimum clearances at the bearings and seals, the response was lower and the frequency increased to 2200 cpm, as shown in Figure 18-20 (note scale changes).

Shop acceptance test data was available for the eight-stage pump which was analyzed. The calculated unbalanced response is compared with the measured vibration data from the test stand in Figure 18-21. For these calculations, maximum bearing clearance and the design values of seal clearance were used. Based on these results, Childs' finite length method [28] provides favorable results compared with measured data. The "shape" of the response curve using Childs' seal values compares closely with the measured results, indicating that the damping contribution of the seals is of the right magnitude.

The anticipated range of rotor response should be calculated with the range of bearing values and various combinations of unbalance. This is important because it is not possible in the design stage to know the exact installed configuration with regard to bearings (clearance, preload) and balance (location of unbalance). Usually a mechanical test will be limited to one configuration (clearance, preload, unbalance) that may not show any problem. Changes introduced later by spare parts during maintenance turnarounds may change sensitive dimensions that may result in a higher response. For this reason, the vibration characteristics of some satisfactorily operating machines may change after an overhaul.

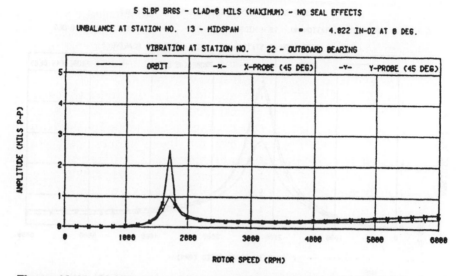

Figure 18-18. Unbalance response at outboard bearing with API unbalance at midspan—no seals.

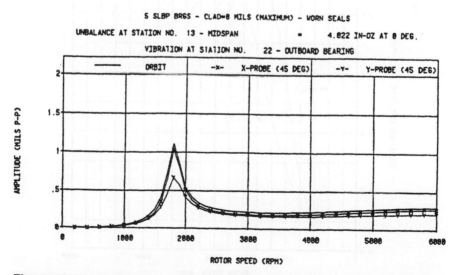

Figure 18-19. Unbalance response at outboard bearing with API unbalance at midspan—maximum clearance seals.

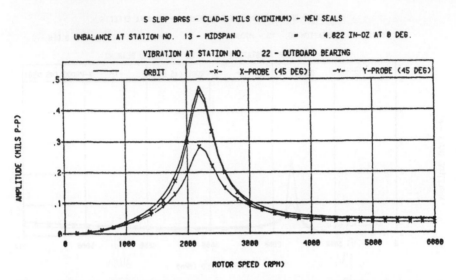

Figure 18-20. Unbalance response at outboard bearing with API unbalance at midspan—minimum clearance seals.

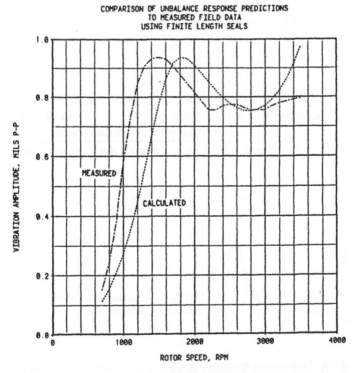

Figure 18-21. Comparison of calculated and measured response with seals.

Acceptable Unbalance Levels

Various engineering organizations have set forth criteria for allowable residual unbalance. The Acoustical Society of America defines balance quality grades for various types of rotors as described in Table 18-3. Pumps may have a range of balance quality grade from G2.5 to G6.3 depending upon size. The ASA Standard 2-1975 defines maximum residual unbalance [34] that is dependent upon speed as shown in Figure 18-22. For example, a 3,600 rpm pump with a rotor weight of 1,000 lbs for balance grade G2.5 would have an allowable residual unbalance of 4.5 in.-oz.

The revised API-610 (seventh edition, 1989) specifies an allowable residual unbalance for centrifugal pumps. The maximum allowable residual unbalance per plane (journal) may be calculated by the following formula:

$$U_b = \frac{4W}{N_{mc}}$$

where U_b = allowable unbalance, inch-ounces
W = journal static weight load, lbs
N_{mc} = maximum continuous speed, rpm

The total allowable unbalance (two planes near the journals) for a 3,600 rpm pump with a 1,000 lb rotor would be 1.1 in.-oz. This calculated residual unbalance from the current edition of the API code is significantly less than the allowable unbalance from earlier editions of the API codes for speeds less than 10,000 rpm. The previous edition of API-610 simply specified dynamic balance for all major rotating components with no specific value for an allowable unbalance. The various balance criteria are compared in Figure 18-23 for a 1,000 lb rotor.

Allowable Vibration Criteria

It is difficult to define the absolute maximum vibration level that can be tolerated without damage to the rotor. Some allowable vibration criteria are based on bearing housing vibrations. With rolling element bearings, the ratio of shaft vibrations to case vibrations is close to unity. The API-610 [7] specifies that the unfiltered vibration measured on the bearing housing should not exceed a velocity of 0.30 ips (inches per second) nor exceed a displacement of 2.5 mils peak-peak at rated speed and capacity ± 10%.

Balance quality grades G	$S\omega^{a,b}$ (mm/sec)	Rotor types—General examples
G 4 000	4 000	Crankshaft-drives[c] of rigidly mounted slow marine diesel engines with uneven number of cylinders.[d]
G 1 600	1 600	Crankshaft-drives of rigidly mounted large two-cycle engines.
G 630	630	Crankshaft-drives of rigidly mounted large four-cycle engines. Crankshaft-drives of elastically mounted marine diesel engines.
G 250	250	Crankshaft-drives of rigidly mounted fast four-cylinder diesel engines.[d]
G 100	100	Crankshaft-drives of fast diesel engines with six or more cylinders.[d] Complete engines (gasoline or diesel) for cars, trucks, and locomotives.[e]
G 40	40	Car wheels, wheel rims, wheel sets, drive shafts Crankshaft-drives of elastically mounted fast four-cycle engines (gasoline or diesel) with six or more cylinders.[d] Crankshaft-drives for engines of cars, trucks, and locomotives.
G 16	16	Drive shafts (propeller shafts, cardan shafts) with special requirements. Parts of crushing machinery. Parts of agricultural machinery. Individual components of engines (gasoline or diesel) for cars, trucks, and locomotives. Crankshaft-drives of engines with six or more cylinders under special requirements. Slurry or dredge pump impeller.
G 6.3	6.3	Parts or process plant machines. Marine main turbine gears (merchant service). Centrifuge drums. Fans. Assembled aircraft gas turbine rotors. Fly wheels. Pump impellers. Machine-tool and general machinery parts. Normal electrical armatures. Individual components of engines under special requirements.
G 2.5	2.5	Gas and steam turbines, including marine main turbines (merchant service). Rigid turbo-generator rotors. Rotors. Turbo-compressors. Machine-tool drives. Medium and large electrical armatures with special requirements. Small electrical armatures. Turbine-driven pumps.
G 1	1	Tape recorder and phonograph (gramophone) drives. Grinding-machine drives. Small electrical armatures with special requirements.
G 0.4	0.4	Spindles, disks, and armatures of precision grinders. Gyroscopes.

[a] $\omega = 2\pi n/60 \approx n/10$, if n is measured in revolutions per minute and ω in radians per second.
[b] In general, for rigid rotors with two correction planes, one-half of the recommended residual unbalance is to be taken for each plane, these values apply usually for any two arbitrarily chosen planes, but the state of unbalance may be improved upon at the bearings. (See Ses. 3.2 and 3.4.) For disk-shaped rotors the full recommended value holds for one plane. (See Sec. 3.)
[c] A crankshaft-drive is an assembly which includes the crankshaft, a flywheel, clutch, pulley, vibration damper, rotating portion of connecting rod, etc. (See Sec. 3.5.)
[d] For purposes of this Standard, slow diesel engines are those with a piston velocity of less than 9 m/sec; fast diesel engines are those with a piston velocity of greater than 9 m/sec.
[e] In complete engines, the rotor mass comprises the sum of all masses belonging to the crankshaft-drive described in Note c above.

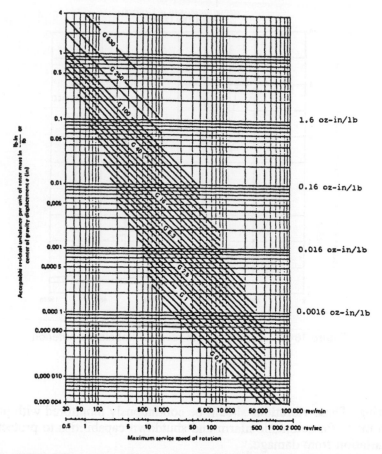

Figure 18-22. Allowable residual unbalance for industrial machinery ASA Standard 2-1975 [34].

However, with oil film bearings, the clearance between the shaft and bearings and the damping of the oil reduces the vibrational force transmitted to the case. For rotors with oil film bearings, shaft vibrations are a better indicator of unbalance conditions. The API-610 allowable vibrations for pumps with sleeve bearings (oil film) are based on shaft vibration measured at rated speed and capacity. The API-610 specifies the allowable unfiltered vibration shall not exceed a velocity of 0.40 ips nor exceed a displacement of 2.0 mils peak-peak including shaft runout.

For critical installations, vibration monitoring equipment should include non-contacting proximity probes (vertical and horizontal) at each

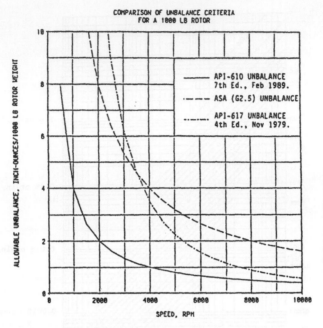

Figure 18-23. Comparison of residual unbalance criterion.

bearing. These vibrations should be continuously monitored with provision made for automatic alarm and shutdown capabilities to protect the installation from damage.

For oil film bearings, the following guidelines for defining maximum acceptable vibration levels are sometimes used when code allowables do not apply. If the vibrations are less than one-fourth of the diametrical bearing clearance, then the vibrations may be considered acceptable. Vibration amplitudes (A, peak-to-peak) greater than one-half the diametrical clearance (C_d) are unacceptable, and steps should be taken to reduce them.

$$A < \frac{C_d}{4} \text{ acceptable}$$

$$\frac{C_d}{4} < A < \frac{C_d}{2} \text{ marginal}$$

$$A > \frac{C_d}{2} \text{ unacceptable}$$

As with most experience-based criteria, these allowable amplitudes are based upon the synchronous vibration component only. Many manufacturers however, still assess acceptable vibrations on their equipment by case or bearing housing vibrations. In plant preventative maintenance programs, hand-held velocity pickups are commonly used to monitor vibrations.

Rotor Stability Analyses

Stability continues to be of major concern for rotors with hydrodynamic bearings, especially for high pressure pumps [36, 37]. In high performance pumps with vaned diffusers, large hydraulic forces can be exerted on the rotor at partial loads. The close internal clearances of pressure retaining seals create hybrid bearings that can produce a destabilizing effect.

Rotor instability occurs when the rotor destabilizing forces are greater than the rotor stabilizing forces. The destabilizing forces can be caused by: the bearings, seals, rotor unbalance, friction in shrink fits, or by loading effects such as inlet flow mismatching the impeller vane angle, turbulence at impeller tips, pressure pulsations, and acoustical resonances.

Instabilities in rotors can cause high vibrations with several different characteristics. They generally can be classified as bearing-related and self-excited. Oil whirl and half-speed whirl are bearing-related instabilities and are caused by the cross-coupling from the bearing stiffness and damping in fixed geometry bearings. Half-speed whirl will result in rotor vibrations at approximately one-half of the running speed frequency. Oil whirl describes a special type of subsynchronous vibration that tracks approximately half-speed up to the point where the speed is two times the first critical. As the speed increases, the subsynchronous vibration will remain near the first critical speed. These types of instabilities can generally be solved by changing the bearing design to a pressure dam, elliptical, offset-half bearing, or a tilting pad bearing.

Self-excited instability vibrations can occur on any rotor, including those with tilted pad bearings. The vibrations will usually occur near the rotor first critical speed or may track running speed at some fractional speed. These types of instability vibrations are sometimes called self-excited vibrations because the motion of the rotor creates the forcing mechanism that causes the instability.

The predominant method used in performing a stability analysis [38] is to calculate the damped (complex) eigenvalues and logarithmic decrement (log dec) of the rotor system including the bearings and seals. The log dec is a measure of the damping capability of the system to reduce vibrations by absorbing some of the vibrational energy. A positive log dec indicates that a rotor system can damp the vibrations and remain stable, whereas a negative log dec indicates that the vibration may actually increase and become unstable. Experience has shown that due to uncertainties in the calculations, the calculated log dec should be greater than + 0.3 to ensure stability. The damped eigenvalue and log dec are sometimes plotted in a synchronous stability map. The frequency of damped eigenvalues is generally near the shaft critical speeds; however, in some heavily damped rotors it can be significantly different from the unbalanced response.

Rotor stability programs are available that can model the rotor stability for most of the destabilizing mechanisms; however, some of the mechanisms that influence it are not clearly understood. It has been well documented that increased horsepower, speed, discharge pressure, and density can cause a decrease in the rotor stability. Many rotors that are stable at low speed and low pressure become unstable at higher values. To predict the stability of a rotor at the design operating conditions, the rotor system is modeled and the log dec is calculated as a function of aerodynamic loading.

Torsional Critical Speed Analysis

All rotating shaft systems have torsional vibrations to some degree. Operation on a torsional natural frequency can cause shaft failures without noticeable noise or an obvious increase in the lateral vibrations. In geared systems, however, gear noise may occur that can be a warning of large torsional oscillations. Therefore, it is important to ensure that all torsional natural frequencies are sufficiently removed from excitation frequencies.

A torsional analysis should include the following:

- Calculation of the torsional natural frequencies and associated mode shapes.
- Development of an interference diagram that shows the torsional natural frequencies and the excitation components as a function of speed.
- Calculation of the coupling torques to ensure that the coupling can handle the dynamic loads.
- Calculation of shaft stresses, even if allowable margins are satisfied.
- Calculation of transient torsional stresses [39] and allowable number of starts for synchronous motor drives.

Torsional natural frequencies are a function of the torsional mass inertia and the torsional stiffness between the masses. The natural frequencies and mode shapes are generally calculated by the Holzer method or by eigenvalue-eigenvector procedures [40]. Either of the methods can give accurate results. A good design practice would be to locate the torsional natural frequencies a minimum margin of 10% from all potential excitation frequencies.

An example of the mass-elastic diagram of a torsional system of a 3,600 rpm motor-driven, six-stage pipeline pump is given in Figure 18-24. The natural frequencies and mode shapes associated with the first four natural frequencies are given in Figure 18-25. The mode shapes can be used to determine the most influential springs and masses in the system. This information is important if a resonance is found near the operating speed and system changes must be made to detune the frequencies.

SIX STAGE CENTRIFUGAL PUMP
1750 HP, 3600 RPM
GEAR TYPE COUPLING

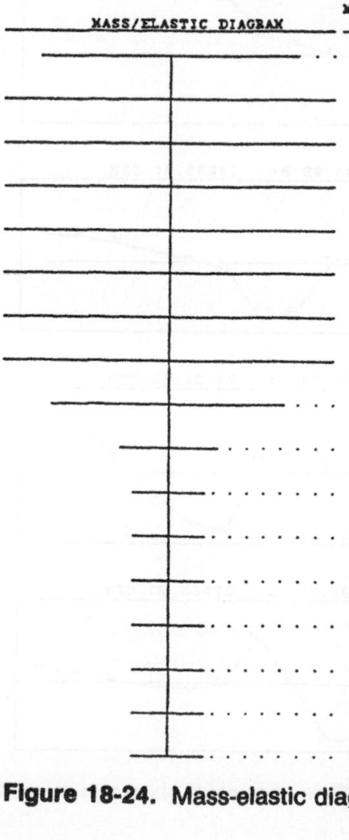

MASS/ELASTIC DIAGRAM	MASS NO.	WR2 in-lb-s2	K(1E-6) in-lb/rad	STATION DESCRIPTION
	1	4.30	652.68	Motor
	2	6.55	632.68	Motor
	3	6.55	632.68	Motor
	4	6.55	632.68	Motor
	5	6.55	632.68	Motor
	6	6.55	632.68	Motor
	7	6.55	632.68	Motor
	8	6.55	632.68	Motor
	9	3.79	9.36	Motor
	10	1.73	13.90	Cplg
	11	1.28	2.62	Cplg
	12	.74	19.42	Stage 1
	13	.70	19.39	Stage 2
	14	.71	19.39	Stage 3
	15	.71	19.19	Stage 4
	16	.73	19.19	Stage 5
	17	.71	.00	Stage 6

Figure 18-24. Mass-elastic diagram of six-stage pump train.

Parametric variations of the coupling stiffness should be made if changes are necessary, because most torsional problems can be solved by coupling changes.

An interference diagram for the six-stage pipeline pump is given in Figure 18-26. In this system, excitation by several orders is possible as the pump is started; however, operation at 3,600 rpm has an adequate margin from the critical speeds. Once the system has been modeled and the natural frequencies have been determined, potential forcing functions should be identified. The forcing functions represent dynamic torques applied at locations in the system that are likely to generate torque varia-

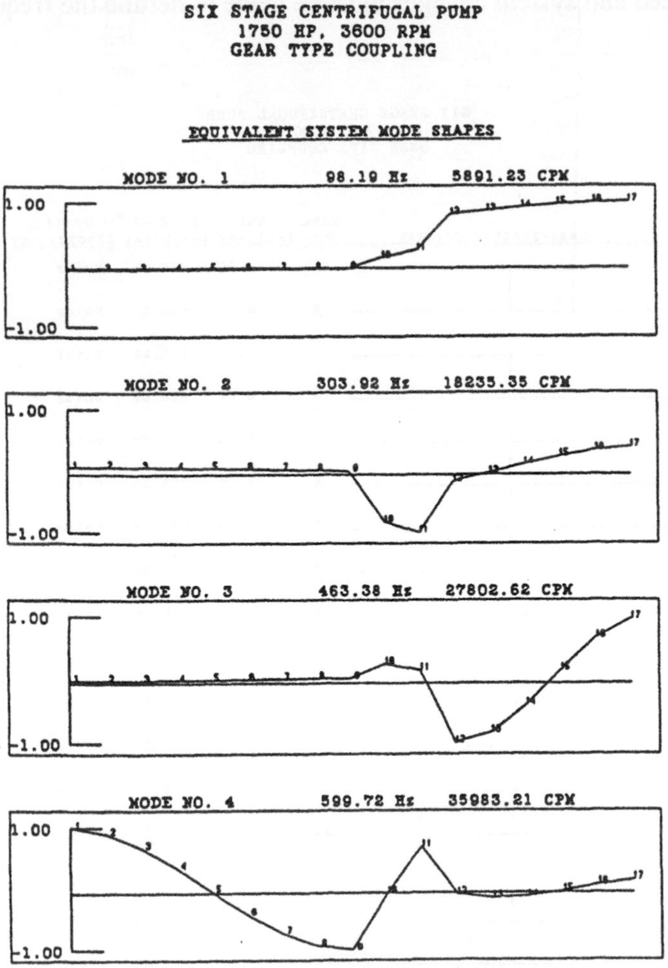

SIX STAGE CENTRIFUGAL PUMP
1750 HP, 3600 RPM
GEAR TYPE COUPLING

EQUIVALENT SYSTEM MODE SHAPES

MODE NO. 1   98.19 Hz   5891.23 CPM

MODE NO. 2   303.92 Hz   18235.35 CPM

MODE NO. 3   463.38 Hz   27802.62 CPM

MODE NO. 4   599.72 Hz   35983.21 CPM

Figure 18-25. Torsional resonant mode shapes of six-stage pump train.

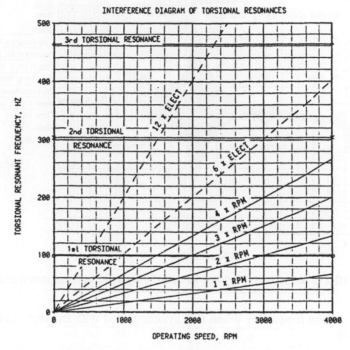

Figure 18-26. Interference diagram of torsional resonances of six-stage pump train.

tions. Identification of all possible sources of dynamic energy is an important step in diagnosing an existing vibration problem or avoiding problems at the design stage.

The most likely sources of dynamic torques include the following:

- Pumps, turbines, and compressors
- Motors (synchronous and induction)
- Couplings
- Gears
- Fluid interaction (pulsations)
- Load variations

The transient torques of a synchronous motor were measured during startup (Figure 18-27) by attaching a strain gauge to the motor shaft and obtaining the signal with an FM telemetry system. A synchronous motor produces a pulsating torque [39] that varies from twice line frequency as

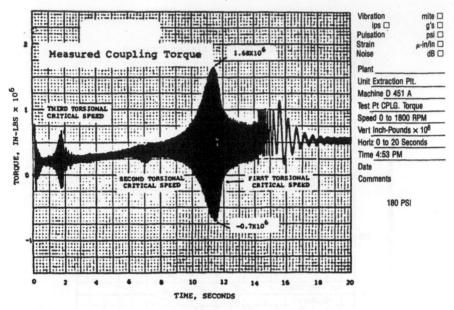

Figure 18-27. Measured coupling torque during synchronous motor startup.

it starts to zero frequency when it is synchronized with the line at operating speed.

Reliability Criteria. The overall system reliability depends upon the location of the torsional natural frequencies with regard to the potential excitation frequencies. An interference diagram generated for each system helps to identify coincidences between expected excitation frequencies and torsional natural frequencies within the operating speed range. Whenever practical, the coupling torsional stiffness and inertia properties should be selected to avoid any interferences within the desired speed range. If resonances cannot be avoided, coupling selection can be optimized based on torsional shaft stress calculations as well as location of critical speeds.

When coupling changes are implemented that have different weights than the vendor originally specified, the lateral critical speeds may be affected. Generally, heavier couplings lower the lateral critical speeds, while lighter couplings raise them. The lateral critical speeds should be reviewed to evaluate the possibility of operating near a lateral response that could result in high radial vibrations.

The acceptability of the torsional system is determined by comparison to typical engineering criteria. Common criteria used for industrial machinery (API Standards) recommend separation of the torsional critical speed and the frequency of all driving energy by a margin of 10%. The effect of the torque modulation on coupling life should be compared to the manufacturer's vibratory torque criterion.

Allowable Torsional Stresses. For long-term reliability, torsional shaft stresses should be compared to applicable criteria. The allowable shaft stress values given by Military Standard 167 [41] are appropriate for most rotating equipment. The Military Standard 167 defines an allowable endurance limit stress of 4,000 psi zero-peak or 8,000 psi peak-peak, based on 100,000 psi ultimate tensile strength shaft material. The general equation for allowable zero-peak endurance limit is the ultimate tensile strength divided by 25.

When comparing calculated stresses to allowable stress values, the appropriate stress concentration factor and a safety factor must be used. Generally a safety factor of 2 is used for fatigue analysis. The standard keyway (USA Standard ANSI B17.1) has a stress concentration factor of 3 [42]. When these factors are used, it can be shown that fairly low levels of torsional stress can cause failures. A typical torsional stress allowable thus becomes the ultimate tensile strength divided by 150.

To evaluate the stresses at resonance, the expected torsional excitation must be applied to the system. For systems with gear boxes, a torque modulation of 1%, zero-peak is a representative torque value that has proven to be appropriate for most industrial machinery trains. As a rule of thumb, excitations at the higher orders for gears are inversely proportional to the order numbers: the second order excitation is 0.5%, the third is 0.33%, etc.

The torque excitation should be applied at the appropriate location and the torsional stresses calculated on the resonant frequencies and at the running speed. An example of the stress calculations of the first natural frequency resonance for the six-stage pump are given in Table 18-4. The maximum stresses occurred at 2,946 rpm on startup as the second order matched the first critical speed. The stresses at 3,600 rpm were low because there was a margin of approximately 17% from the nearest natural frequency. The torque excitation at the second order would cause a maximum torsional stress of 395 psi peak-peak in shaft 11, which is the pump input shaft between the coupling and the first stage impeller. The dynamic torque modulation across the couplings was calculated for the applied input modulation. For this mode, the maximum torsional vibrations occur across the coupling and the dynamic torque modulation was 48 ft-lbs.

Table 18-4
Six-Stage Centrifugal Pump
1750 HP, 3600 RPM
Gear Type Coupling

| | Dynamic Torques (1 Percent Zero-Peak) Applied at Motor | | |
| | Maximum Resultant Torsional Stresses at 2945.62 RPM | | |
Shaft	Stress PSI P-P	SCF	Stress PSI P-P
1	.99	2.00	1.98
2	2.47	2.00	4.93
3	3.87	2.00	7.74
4	5.21	2.00	10.41
5	6.51	2.00	13.02
6	7.82	2.00	15.65
7	9.16	2.00	18.32
8	10.52	2.00	21.03
9	130.49	3.00	391.47
10	*DYNAMIC TORQUE VARIATION*		48.32*
11	131.77	3.00	395.30
12	72.65	2.00	145.30
13	59.57	2.00	119.13
14	45.51	2.00	91.03
15	30.84	2.00	61.68
16	15.31	2.00	30.63

* —*Values are dynamic torque variation across coupling, ft-lbs*

Variable Speed Drives

Systems that incorporate variable frequency drives [43] require additional considerations in the design stage over conventional constant speed equipment. The wide speed range increases the likelihood that at some operating speed a coincidence between a torsional natural frequency and an expected excitation frequency will exist. In addition to the fundamental mechanical frequency (motor speed), excitation frequencies include the fundamental electrical frequency (number of pole pairs times motor speed) and the sixth and twelfth orders of electrical frequency [44]. The variable frequency power supply introduces a pulsating torque with strong sixth and twelfth order harmonics.
As an example, the same six-stage pump system (described in Figure 18-24) was analyzed for a variable frequency drive motor. The dashed line in Figure 18-26 represents the sixth harmonic speed line that shows that the sixth order would excite the first torsional natural frequency at 982 rpm.

Because of the strong sixth order electrical torques produced by the variable frequency drive, the stresses (Table 18-5) are significantly greater than the stresses for the same system with a constant speed motor (compare with Table 18-4). Even though the natural frequencies of the rotor system remain constant, the exciting torques are greater and occur at a different speed.

It is dffficult to remove all coincidence of resonances with the excitation sources over a wide speed range; therefore, stress calculations must be made to evaluate the adequacy of the system response. The input shaft stresses for the six-stage pump with a variable frequency drive are shown in Figure 18-28. They are significantly different stresses than for the constant speed motor. This example demonstrates that converting a constant speed system (motor or turbine) to a variable frequency drive motor must be considered carefully. Although the variable speed provides greater efficiency; adjustments may be required (i.e., proper choice of couplings) to obtain acceptable torsional response.

Some variable speed systems use couplings with flexible elements to increase the damping as critical speeds are passed. Several couplings in common use have rubber elements to add damping that increases stiffness with higher transmitted torque. Other couplings use flexible grids or springs that again have an increased stiffness with load. These couplings are sometimes necessary when several torsional frequencies occur within the speed range. The torsional stiffness of these nonlinear couplings changes with load (transmitted torque) and speed. This nonlinearity adds complexity to the analysis. The pump load must be considered which also varies as a function of speed squared.

Diagnosis of Pump Vibration Problems

Large plants that handle liquids (i.e., chemical plants) may have hundreds or more small pumps, making it almost impossible to measure each pump in detail on a regular basis. The cost of a detailed analysis with several transducers, spectrum analyses, and the necessary study of the data can quickly exceed the cost of repair or even replacement of some of the small pumps. Only the pumps in a critical system, which could affect production if a failure occurred, can afford the expense of a detailed analysis or a permanent monitoring system. However, the hundreds of small pumps need regular attention to keep plant efficiency (and flow rates) from dropping.

An effective preventative maintenance program should periodically measure and record the vibrations on each bearing of the pump and its driver. Measurements with a hand-held velocity transducer or data collector system are usually adequate to obtain periodic data to evaluate the

Table 18-5
Six-Stage Centrifugal Pump
1750 HP, 3600 RPM Variable Frequency Drive Motor
Gear Type Coupling

Dynamic Torques (1 Percent Zero-Peak) Applied at Motor
Maximum Resultant Torsional Stresses at 981.87 RPM

Shaft	Stress PSI P-P	SCF	Stress PSI P-P
1	48.12	2.00	96.25
2	121.31	2.00	242.62
3	193.97	2.00	387.93
4	265.92	2.00	531.84
5	336.94	2.00	673.89
6	406.63	2.00	813.26
7	474.66	2.00	949.32
8	540.76	2.00	1081.52
9	6676.68	3.00	20030.04
10	*DYNAMIC TORQUE VARIATION*		2489.59*
11	6822.28	3.00	20466.84
12	3763.91	2.00	7527.81
13	3087.43	2.00	6174.85
14	2359.91	2.00	4719.81
15	1599.41	2.00	3198.81
16	794.30	2.00	1588.60

* —*Values are dynamic torque variation across coupling, ft-lbs*

vibrational trend. If the vibrations show an increasing trend, the unit should be monitored more frequently. Vibrational guidelines in common usage can aid in determining the severity or extent of damage so that maintenance can be scheduled. A vibrational velocity of less than 0.3 ips (inches per second) is generally accepted as satisfactory operation for pumps [7] and motors. Velocity levels above 0.3 ips are warnings of potential problems. Velocity levels of 0.5 ips and above may be indicative of significant damage to the bearings, seals, or rotor.

The basis of troubleshooting is obtaining test data on troublesome pumps during normal operation and comparing the data to the purchase specifications, vendor guarantees, or applicable vibration criteria. It is crucial to the success of the troubleshooting effort to have adequate instrumentation on the pump system. In many cases, the instrumentation can be temporarily added to obtain the required measurements.

Measurement Techniques

Typically pumps are installed with a minimum of vibration monitoring equipment. Some pumps may have a velocity pickup on each bearing

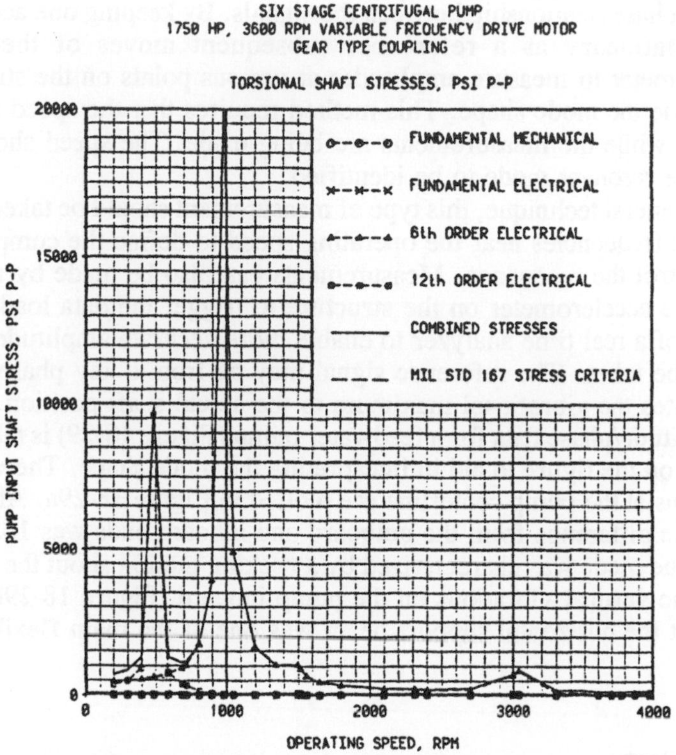

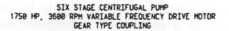

SIX STAGE CENTRIFUGAL PUMP
1750 HP, 3600 RPM VARIABLE FREQUENCY DRIVE MOTOR
GEAR TYPE COUPLING

TORSIONAL SHAFT STRESSES, PSI P-P

Figure 18-28. Pump shaft stress with variable frequency drive motor.

housing. These systems are satisfactory for pumps with rolling element bearings, because the bearings transmit the rotor forces directly to the case. However, additional instrumentation is often required to define difficult problems. For pumps with oil film bearings, it is desirable to have two proximity probes 90° apart near each bearing and an axial probe. Accelerometers and velocity probes attached to the bearing housings or case are often used to measure pump vibration; however, if the proximity probes are not installed, the data will be limited and may be a detriment to the diagnosis of some types of problems. A pressure transducer is vital to diagnostic work for measuring dynamic pulsations in the piping, impeller eye, diffuser, and across flow meters. Accurate flow measurements are necessary to define the pump vibration characteristics as a function of its location on the head-flow performance map.

Accelerometers or velocity transducers can be used to determine frequencies and amplitudes of the pump case or support structure during normal operation [45]. Two accelerometers and a dual channel oscilloscope displaying their complex vibration waveforms can be used to define the phase relationship between the signals. By keeping one accelerometer stationary as a reference, subsequent moves of the other accelerometer to measure amplitudes at various points on the structure can define the mode shape. This method requires that the speed remain constant while the measurements are being made. The speed should be set at the resonant mode to be identified.

As a general technique, this type of measurement should be taken at the resonant frequencies near the operating speed to define the components that control the resonance. Measurements can also be made by using a reference accelerometer on the structure to trigger the data loading sequence of a real time analyzer to enable more accurate amplitude/phase data to be taken. The reference signal may be from a key phase signal that relates the vibrational maximum to the actual shaft position.

Amplitude/phase data for a feedwater pump (Figure 18-29) is tabulated directly on the figure to aid in interpreting the mode shape. The vertical vibrations at the pump centerline are plotted in Figure 18-29a. Although the phase difference from the outboard and inboard ends was 165° (not 180°), the mode was characterized by a rocking motion about the pivotal axis. The vertical vibrations on the pump inboard (Figure 18-29b) indicate that the horizontal support beam was one of the main flexible elements.

Impact Tests

An impact test is a simple technique that can be used to excite resonances of flexible substructures. A significant advantage of the method is that mechanical natural frequencies can be measured while units are down. Typical instrumentation includes an accelerometer attached to the head of a rubber-tipped hammer to measure the impact force and a second accelerometer used to measure the response of the test structure. In modal analysis testing, the accelerometer remains at one location and an impact hammer is used to excite the structure at selected locations. The response signal can be automatically divided by the impact signal using a fast fourier transform (FFT) spectrum analyzer with transfer function (XFR) capabilities.

For best results, the impact velocity should not be greatly different from the vibrational velocity of the vibrating member. For example, the lowest beam mode of a piping span can be excited with a rubber mallet by applying the impact with a forceful, medium-speed swing. However, a sharp rap with a steel-faced hammer could produce higher modes of beam vibration or possibly a pipe shell wall resonance. These higher energy modes will be quickly damped and difficult to identify. Using modal

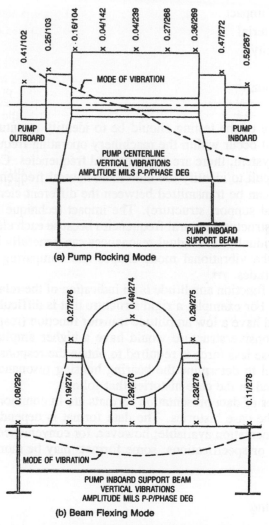

Figure 18-29. Pump vibrational mode shapes during operation.

analysis techniques, the input spectrum can be evaluated and the impact can be adjusted to obtain the best results for the structure being tested.

In nonsymmetrical members or structures, a resonant mode has a preferred direction of motion. Impact testing should be approached on an experimental basis by trying variations of impact velocity, direction, etc. Numerous frequencies may occur that make the mode shapes difficult to identify. Generally, the lower modes of vibration will be predominate. The mode of vibration is strongly influenced by the following:

- Direction of impact
- Interface material
- Impact velocity
- Contact time
- Impact location

The objective of the testing should be to identify structural resonant frequencies that occur within the machinery operating frequency range. In a complex system, there are many natural frequencies. Consequently, it may be difficult to identify the individual natural frequencies because the vibrations can be transmitted between the different elements (pump rotor, case, and support structure). The impact technique is useful for separating the structural natural frequencies because each element can be impacted individually. Individual resonances can generally be identified by analyzing the vibrational mode shapes and comparing the transfer function amplitudes.

The transfer function amplitude is an indication of the relative stiffness of a structure. For example, a rigid structure that is difficult to excite or resonate would have a low amplitude transfer function (response/force). A highly responsive structure would have a higher amplitude transfer function because less force is required to obtain the response. This technique was used to determine the bearing housing resonance of a pump that is described in the case histories that follow.

Various types of data presentation formats are in common use and are described in the case histories. The data format is dependent upon the type of instrumentation available; however, for convenience of comparison to criteria or specifications, some formats may be more convenient than others.

Troubleshooting

As examples of the types of diagnostic procedures that are used to identify the causes of some typical pump problems, several field case histories will be presented.

Bearing Housing Resonance of Feed Charge Pumps. High vibration amplitudes and seal failures occurred on one of a pair of centrifugal pumps that are the feed charge pumps for a cat feed hydro-treater plant. The pumps were driven by electric motors at a speed of approximately 3,587 rpm (59.8 Hz). The pumps have five impellers; the first two impellers have six vanes and the last three impellers have seven vanes.

Vibration data recorded on both pumps indicated that the vibration levels were significantly higher on the inboard bearing of Pump A. The maximum vibration occurred at seven times running speed (418 Hz) which is the vane passing frequency of the last three impellers.

The high vibration levels and seal failures on the inboard bearing of the pump could be caused by several problems, including:

- Pulsation at the vane passing frequency that can increase the forces on the impeller and shaft.
- Mechanical or structural natural frequencies that amplify the vibration levels.
- Misalignment between the motor and the pump or internal misalignment between the pump bearings.

A field test was performed to measure the vibrations and pulsations in the system. The vibration amplitudes were considerably higher on the inboard bearing of Pump A which experienced the seal failures. The vibration amplitude increased significantly between the case and the bearing housing. This differential vibration could be the cause of the seal failures. The bearing housing vibrations of Pump A were primarily at seven times running speed (418 Hz).

Vibration measurements were taken at four locations between the end of the case and the end of the bearing to illustrate the vibration mode shapes. The vibration amplitudes on the inboard bearing were considerably higher near the end of the bearing housing compared to near the case. The horizontal vibrational mode can be defined from the vibration spectra at the four points shown in Figure 18-30. As shown, there was significant differential motion between the case and the end of the flange and across the bolted flange. The vibration amplitude on the bearing housing was approximately 6 g peak-peak at 418 Hz (0.44 in./sec peak). Most allowable vibration criteria would consider these amplitudes to be excessive. However, the actual differential displacement was only 0.13 mils peak-peak. The amplitudes at the outboard bearing were lower compared to the amplitudes on the inboard bearing.

The discharge pulsation amplitudes were 12 psi at seven times running speed and 3 psi at fourteen times running speed. The discharge pulsations were higher on Pump B (without any seal failures), which indicated that the pulsations were not the cause of the increased vibrations.

Impact Tests

Impact tests on the bearing housings indicated that the inboard bearing housings were very responsive compared to the outboard bearing hous-

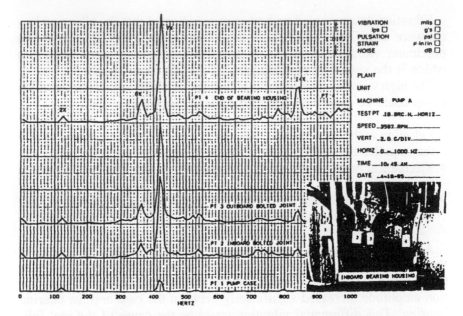

Figure 18-30. Vibrational response of inboard bearing housing.

ings. The transfer function (Figure 18-31) was plotted with a full scale amplitude equal to 0.4 (0.05 per major division on the graph paper).

The inboard bearing of Pump A had a major response at 420 Hz with a transfer function amplitude of 0.24. This frequency was almost coincident with seven times running speed (418 Hz) which would amplify the vibration levels at seven times running speed and appeared to be the primary cause of the high vibration at seven times running speed. The flange bolts were tightened and the frequency increased to 430 Hz which indicated that the response near 420 Hz was primarily associated with the bearing housing and its attachment stiffness to the case.

A similar major response near 432 Hz occurred in the vertical direction. The transfer function amplitude was lower and the frequency was higher than measured in the horizontal direction because the bearing was slightly stiffer in the vertical direction compared to the horizontal direction. When the flange bolts were tightened, the response near 432 Hz was increased to 440 Hz.

The outboard bearing was more massive and stiffer and not as responsive as the inboard bearing. There were no major responses in the horizontal direction near the excitation frequencies of interest. A vertical response was measured near 580 Hz; however, because it was above the frequencies of interest, it should not cause an increase in vibration on the bearing housings.

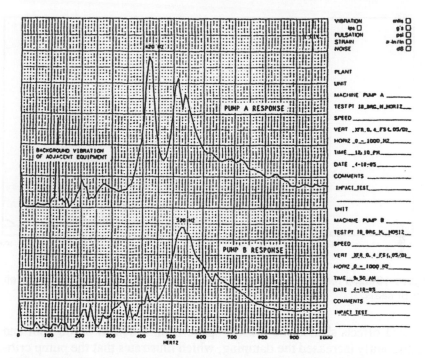

Figure 18-31. Structural natural frequency of inboard bearing housing.

Impact tests were performed on Pump B (no seal failures). The responses on the inboard bearing were 530 Hz in the horizontal direction (Figure 18-31) and 600 Hz in the vertical direction. Although visually the two pumps appeared to be identical, detailed measurements revealed slight differences in the cross-sectional thickness and lengths of the inboard support between the flange and the end of the cases. Apparently these slight dimensional differences caused the mechanical natural frequencies to vary considerably between the two pumps.

The obvious solution to the problem would be to replace the bearing housing casting or modify the inboard bearing housing support of Pump A. The mechanical natural frequency could be increased by stiffening the support with gussets. For improved reliability, the mechanical natural frequency should be increased approximately 10% above the primary excitation frequency at seven times running speed.

Pump Critical Speed Problem [46]. A critical speed analysis was performed on a centrifugal pump used as a jetting pump on a pipe laying barge. The predicted unbalanced vibration response of the centrifugal pump showing the effects of the seals is shown in Figure 18-32. Note that

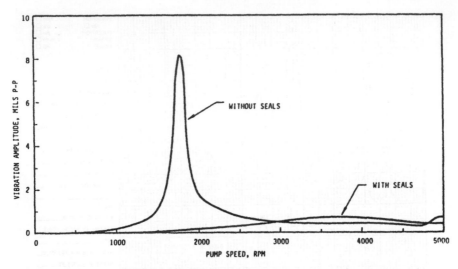

Figure 18-32. Predicted vibration of three-stage centrifugal pump showing effect of seals.

the seal effects shifted the critical speed from 1,800 to 3,700 rpm and significantly increased the damping, which illustrates that the pump critical speed should be sensitive to seal stiffness effects. When the seals were considered, the predicted amplitudes were reduced by a factor of more than 10 to 1.

The vibrations on the pump were measured (Figure 18-33) using the peak-store capabilities of a real-time analyzer. The vibrations were low until the pump reached 3,300 rpm and then sharply increased to 6 mils peak-peak at 3,600 rpm. The design speed was 3,600 rpm; however, the pump could not run continuously at that speed. The pump speed was kept below 3,400 rpm so that the vibrations were less than 2 mils peak-peak. Even with the reduced pump speed, the pipe-laying barge was able to set pipe-laying records.

After a year of operation, the pump vibrations began to increase until the vibrations at 3,400 rpm were unacceptable. The pressure breakdown bushing had worn which reduced its effective stiffness and the critical speed had dropped to 3,400 rpm. It was recommended that the pump be operated at a speed above the critical speed. This was tried and the pump operated at 3,600 rpm with vibration levels less than 2 mils peak-peak. The barge remained in service and reset the pipe-laying records during the next season.

The data analysis technique used to determine this critical speed was to use the peak-store capabilities of the real-time analyzer. An alternate method would be to use a tracking filter to determine the critical speed

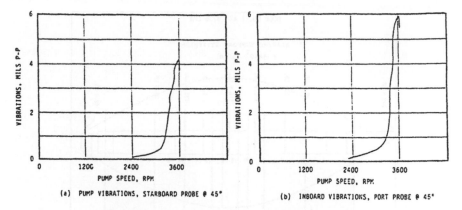

Figure 18-33. Jetting pump measured vibrations.

response because both the amplitude and phase data could be made available. Although it is generally better to have both the Bode and Nyquist plots, for this case they were not required to define the cause of the vibration plot.

High Vibrations of a Centrifugal Pump. Critical speed calculations were performed on a three-stage centrifugal pump to determine if a critical speed existed near running speed. The mode shape calculations for the first and second critical speeds are summarized in Figures 18-34 and 18-35. As discussed, the liquid seals can significantly affect the critical speeds of a pump; however, for this rotor, the seals only increased the critical speed by about 15%. This critical speed analysis considered a shaft with two bearings and the eight seals, or ten sets of stiffness and damping coefficients. Each of these coefficients varied as a function of speed and was included in the unbalanced response analysis (Figure 18-36). A comparison of predicted responses to measured test stand data is given in Figure 18-37. The agreement with the measured data was considered good, indicating that the rotordynamic model of the rotor and the bearing and seal clearances was acceptable.

A data acquisition system with a tracking filter was used to plot the amplitude and phase angle versus speed (Bode Plot). The lack of a significant phase shift through the critical speed could not be explained without a more detailed analysis. As in many field studies, other priorities prevailed, and additional tests were not possible. Although the critical speed was in the running speed range, the vibration amplitudes were low and any wear of the seals should move it further away from the rated speed of 3,600 rpm.

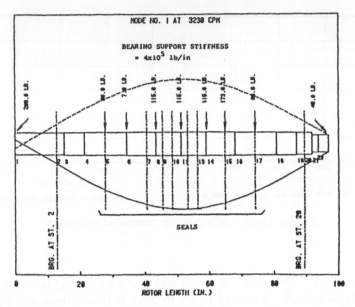

Figure 18-34. Three-stage pump—first mode response—with seals.

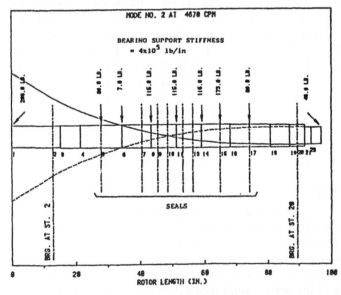

Figure 18-35. Three-stage pump—second mode response—with seals.

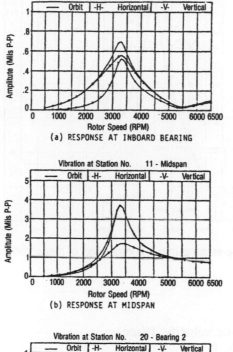

UNBALANCE RESPONSE - CENTRIFUGAL PUMP
NEW SEALS - DESIGN CLEARANCE
5SLOP BEARINGS - DESIGN CLEARANCE   CLAD = 6 MILS (NOMINAL)
UNBALANCE AT STATION NO. 11 - IMPELLER = 4.610 IN-OZ AT 0 DEGREES

Figure 18-36. Three-stage pump—unbalanced response—with seals.

Pulsation Induced Vibrations [14].

A four-stage centrifugal pump suffered repeated failures of the splitter between pump stages. A detailed field study revealed the cause of the problems to be an acoustic resonance of the long cross-over that connected the second-stage discharge with the third-stage suction (Figure 18-38). The resonant frequency was a half-wave acoustic resonance.

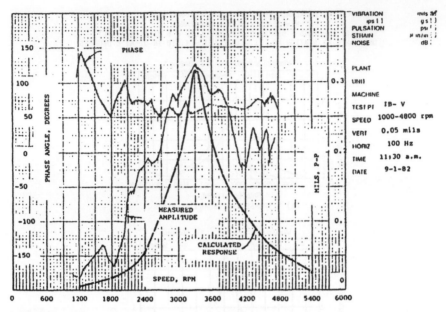

Figure 18-37. Three-stage pump—comparison of measured and calculated response.

$$f = \frac{c}{2 \times L}$$

where: c = acoustic velocity, ft/sec
        L = length, ft

The speed of sound in water is a function of the temperature, and at 310°F was calculated to be 4,770 ft/sec. The length of the cross-over was 5.75 ft. The acoustic natural frequency was

$$f = \frac{4770}{2 \times 5.75} = 415 \text{Hz}$$

The acoustic resonant frequency was excited by the vane passing frequency (seven times running speed). Coincidence occurred at (415) (60)/ 7 = 3,560 rpm.

Dynamic pressure measurements in the center of the cross-over showed pulsation amplitudes of 100 psi peak-to-peak. The pulsations at the suction and discharge flanges were less than 10 psi peak-to-peak, which agreed with the mode shape of the half-wave acoustic resonance.

There were two possible changes that could eliminate the coincidence of vane passing frequency with the resonant frequency and reduce the

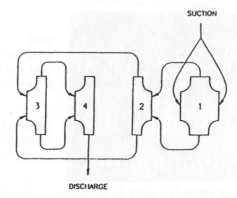

Figure 18-38. Flow schematic of a four-stage pump.

vibration levels that occurred at a speed of 3,560 rpm. One possible change was to increase the vane tip clearance to minimize the turbulence that produces the pulsations. This modification would require trimming the diameter of the impellers and operating the pump at a higher speed to attain the capacity. Another possibility was to change to a six- or eight-vane impeller to alter the vane passing frequency. The impeller diameter modification was the quickest and most economical and was carried out in the field, and the splitter failures were eliminated.

This example also illustrates the importance of selecting proper test points when measuring pulsations. For example, if an acoustical resonance is expected in the cross-over, then the pressure transducer should be installed near the center of the cross-over length rather than at the ends.

Sometimes the cross-over or cross-under passage can have an acoustical resonance that can be so severe that it can excite the shaft and result in high shaft vibrations [46]. This is illustrated for a different pump problem (Figure 18-39) which had shaft vibrations at operating speed (3,400 rpm) of 1 mil peak-peak and vibrations at five times speed of 0.5 mils peak-peak. Pulsation levels in the cross-under were over 250 psi and caused approximately 0.5 mils peak-peak of shaft vibration at the acoustic natural frequency.

Shaft Failures Caused by Hydraulic Forces. Repeated shaft bending fatigue failures were experienced in a single stage overhung high-pressure water pump. The failures exhibited the classical fatigue beach marks with the failures occurring straight across the shaft at the sharp corner at the change in diameter. The pump was instrumented with proximity probes, pressure transducers, and accelerometers. Measurements were made

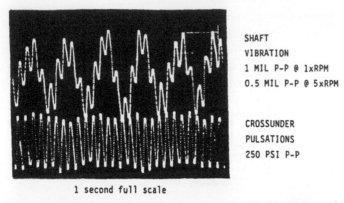

SHAFT
VIBRATION
1 MIL P-P @ 1xRPM
0.5 MIL P-P @ 5xRPM

CROSSUNDER
PULSATIONS
250 PSI P-P

1 second full scale

Figure 18-39. Shaft vibrations caused by acoustic resonance.

over a wide range of startup and flow conditions. The pumps had a double-volute casing that was supposed to balance the radial forces on the impeller; however, measurements made of the shaft centerline by measuring the DC voltage on the proximity probes showed that the impeller was being forced upward against the casing. This caused a large bending moment on the shaft as it rotated. The pump shaft center location near the bearings was displayed on an oscilloscope (Figure 18-40) during startup and during recycle. A differential movement of 6 mils occurred across a short distance along the shaft. The wear patterns on the impellers were consistent with the major axis of the orbit and the direction that the shaft moved. The large movement in the bearing journals only occurred under certain start-up conditions; therefore, it was possible to modify the start-up procedures to ensure that the large hydraulic forces would not cause shaft failures.

Hydraulic forces can cause shaft failures as illustrated by this example; therefore, it is good practice to determine if the shaft is properly aligned in its journals under all operating conditions.

Pump Instability Problem. A high speed pump experienced high vibrations in the process of startup at the plant site. Originally, the problem was thought to be due to a lateral critical speed causing increased synchronous vibration levels when full speed was reached.

Analysis of this problem was particularly difficult due to the extremely short startup time of the motor-driven pump and the even more rapid rate at which the vibration levels increased as the pump approached rated speed. To analyze the problem, an FM recording of a startup was analyzed while running the recorder playback at 1/8 of recorded speed, which in effect, caused the startup period during playback to be eight

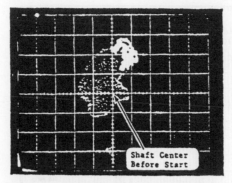

PUMP SHAFT MIDSPAN
STARTUP AND RECYCLE

DURING RECYCLE THE
SHAFT WAS OFFSET 6
MILS FROM THE INITIAL
POSITION AT REST

2 mils/div

Shaft Center
Before Start

(a) PUMP SHAFT MIDSPAN SPATIAL POSITION

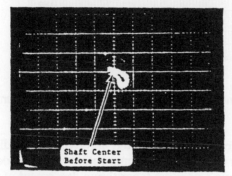

PUMP SHAFT INBOARD
STARTUP AND RECYCLE

THE VIBRATION AMPLITUDE
AND DC POSITION OF THE
ORBIT WAS MUCH LESS
THAN AT THE SHAFT
AT MIDSPAN

2 mils/div

Shaft Center
Before Start

(b) PUMP SHAFT INBOARD SPATIAL POSITION

Figure 18-40. Pump shaft vibration orbits.

times as long. A cascade plot of the vibration data (Figure 18-41) showed that just before trip of the unit at 21,960 rpm, an instability vibration component occurred near 15,000 cpm. From this and other data, it was determined that the high vibrations were caused by:

- A sudden increase in nonsynchronous vibration as the unit approached full speed resulting in shaft bow.
- A sudden increase in unbalance due to the shaft bow and as a result, a rapid increase in synchronous vibration levels as the nonsynchronous components disappeared.

After the problem source was identified using the above data analysis technique, computer simulation of the rotor led to a solution consisting of bearing modifications. The stability analysis of the pump rotor predicted that the pump had an unstable mode at 15,000 cpm with a negative loga-

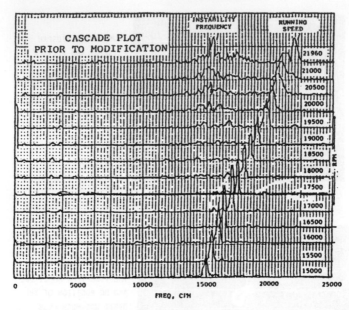

Figure 18-41. Nonsynchronous instability vibration of high speed pump.

rithmic decrement of 0.01 for a simulated fluid aerodynamic loading of 1,000 lb/in. at the impellers [36, 47]. The pump rotor with the modified bearings was predicted to have a positive logarithmic decrement of 0.10. The rotor vibrations after the bearing modifications were made are shown in Figure 18-42. The nonsynchronous vibration component was no longer present and the unit has since operated successfully.

Appendix
Acoustic Velocity of Liquids

The acoustic velocity of liquids can be written as a function of the isentropic bulk modulus, K_s and the specific gravity:

$$c = 8.615 \sqrt{\frac{K_s}{sp\ gr}} \tag{1}$$

where c = acoustic velocity, ft/sec
      sp gr = specific gravity
       K_s = isentropic (tangent) bulk modulus, psi

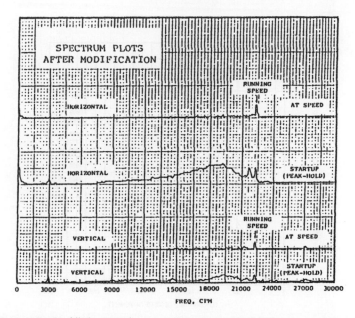

FREQ. CPM

Figure 18-42. High speed pump vibrations after bearing modification.

At low pressure, water can be considered to be incompressible and the acoustic velocity is primarily dependent upon the change of density with respect to temperature. The acoustic velocity of water at low pressures [48] is given for various temperatures from 32° to 212°F in Figure 18-43. For elevated pressures, the acoustic velocity must be adjusted for pressure effects and the function using the bulk modulus is convenient.

The bulk modulus of water [49] can be calculated with the following equation for pressures up to 45,000 psig and various temperatures.

$$K_s = 1000 \, K_o + 3.4P \tag{2}$$

where K_o = constant from Table 18-6
K_s = isentropic tangent bulk modulus, psi
P = pressure psia

The calculation accuracy is ±0.5% for the isentropic bulk modulus of water at 68°F and lower pressures. The error should not exceed ±3% at elevated pressures (greater than 44,000 psi) and temperatures (greater than 212°F).

The bulk modulus for petroleum oils [50] (hydraulic fluids) can be determined for various temperatures and pressures by using Figures 18-44 and 18-45 which were developed by the API. The isothermal secant bulk modulus for petroleum oils at 20,000 psig is related to density and tem-

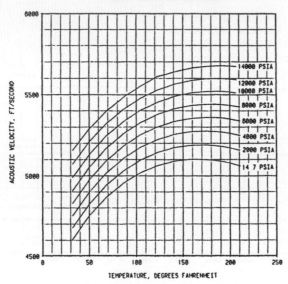

Figure 18-43. Acoustic velocity of distilled water at various temperatures and pressures.

Table 18-6 [49]
Constant K_o for Pressure Correction for
Bulk Modulus of Water

$$K_s = 1000\ K_o + 3.4P$$
P = pressure, psia
(valid up to 45000 psia)

Temperature (°F)	Constant K_o
32	289
50	308
68	323
86	333
104	340
122	345
140	348
158	348
176	341
194	342
212	336

perature in Figure 18-44. The isothermal tangent bulk modulus has been shown to be approximately equal to the secant bulk modulus at twice the pressure within $\pm 1\%$. Pressure compensation for the isothermal secant bulk modulus can be made by using Figure 18-45.

The isothermal bulk modulus, K_T, and isentropic bulk modulus, K_s, are related by the following equation:

$$K_s = K_T \frac{c_P}{c_V} \tag{3}$$

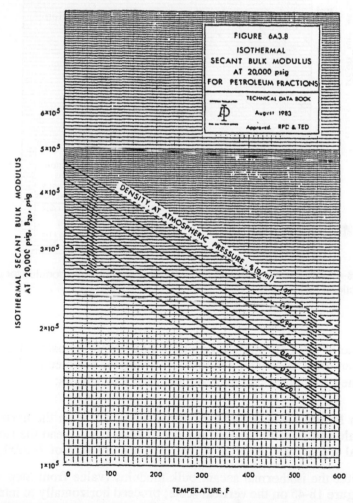

Figure 18-44. Isothermal secant bulk modulus at 20,000 psig for petroleum fractions [50].

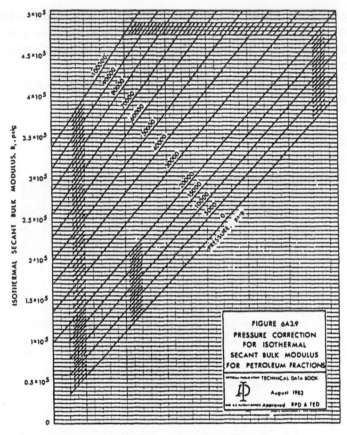

Figure 18-45. Pressure correction for isothermal secant bulk modulus for petroleum fractions [50].

The value of c_p/c_v for most hydraulic fluids is approximately 1.15.

The isentropic tangent bulk modulus for common petroleum oils can be determined from the Figures 18-44 and 18-45 as follows:

1. On Figure 18-44, enter the desired temperature on the horizontal scale and go to the proper specific gravity line to read the isothermal secant bulk modulus at the reference pressure of 20,000 psig.

2. Enter the isothermal secant bulk modulus (value from Step 1) on Figure 18-45 on the vertical scale; proceed horizontally to intersect the 20,000 psi reference pressure line; move vertically to the line corresponding to twice the desired pressure. Move horizontally to

the left and read the equivalent isothermal tangent bulk modulus on the scale.

3. Multiply the isothermal tangent bulk modulus by 1.15 (to compensate for ratio of specific heats) to determine the value of the isentropic tangent bulk modulus.

The value of the isentropic tangent bulk modulus obtained in Step 3 can be used in Equation 1 to calculate the acoustic velocity.

Piping systems with incompressible fluids (liquids) have an apparent effect on the acoustic velocity because of the pipe wall flexibility. The classical Korteweg correction can be used for thin wall pipe (wall thickness < 10% diameter) to adjust the acoustic velocity.

$$c_{adjusted} = c \sqrt{\frac{1}{1 + \dfrac{DK_s}{tE}}} \tag{4}$$

where D = pipe diameter, in.

  t = pipe wall thickness, in.

  E = elastic modulus of pipe material, psi

References

1. Florjancic, D., Schoffler, W. and Zogg, H., "Primary Noise Abatement on Centrifugal Pumps," Sulzer Tech., Rev. 1:24, 1980.
2. Szenasi, F. R. and Wachel, J. C., Section 8.4, "Pump Noise," *Pump Handbook*, 2nd ed., McGraw-Hill, 1986, pp. 8.101–8.118.
3. Jones, A. B., "The Mathematical Theory of Rolling-Element Bearings," *Mechanical Design and Systems Handbook*, McGraw-Hill, 1964, pp. 13-1 to 13-76.
4. *Centrifugal Pump Hydraulic Instability*, CS-1445 Research Project 1266-18, Electric Power Research Institute, 3412 Hillview Ave., Palo Alto, California, 1980.
5. Wachel, J. C. and Tison, J. D., *Field Instrumentation and Diagnostics of Pump Vibration Problems*, Presented at Rotating Machinery and Controls Short Course, University of Virginia, June 6–7, 1983.
6. Mancuso, J. R., *Couplings and Joints*, Marcel Dekker, Inc., New York, 1986, pp. 340–350.
7. API Standard 610 "Centrifugal Pumps for General Refinery Services," Section 2.8, 7th ed., American Petroleum Institute, Washington, D.C., 1989.

8. Bloch, H. P., "Improving Machinery Reliability," *Practical Machinery Management for Process Plants*, Vol. I, Gulf Publishing Company, Houston, Texas, 1982, pp. 72–162.

9. Atkins, K. E., Tison, J. D. and Wachel, J. C., "Critical Speed Analysis of an Eight Stage Centrifugal Pump," *Proceedings of the Second International Pump Symposium*, April 1985, pp. 59–65.

10. Wood, D. J. and Chao, S. P. "Effect of Pipeline Junctions on Waterhammer Surges," *Proceedings of the ASCE, Transportation Engineering Journal*, August 1978, pp. 441–457.

11. Fraser, W. H., "Flow Recirculation in Centrifugal Pumps," in *Proceedings of the Tenth Turbomachinery Symposium*, 1981.

12. Fraser, W. H., "Recirculation in Centrifugal Pumps: Materials of Construction of Fluid Machinery and Their Relationship to Design and Performance," paper presented at winter annual meeting of ASME, Washington, D.C., November 16, 1982.

13. Wachel, J. C. and Szenasi, F.R., et al., "Vibrations in Reciprocating Machinery and Piping," *EDI Report 85–305*, Chapter 2, October, 1985.

14. Sparks, C. R. and Wachel, J. C., "Pulsations in Liquid Pump and Piping Systems," *Proceedings of the Fifth Turbomachinery Symposium*, October 1976, pp. 55–61.

15. Wylie, E. B. and Streeter, V. L., *Fluid Transients*, FEB Press, Ann Arbor, Michigan, 2nd Ed., pp. 213–224, 1983.

16. *Technical Data Handbook-Petroleum Refining*, Vols. I, II, III, 4th Ed., American Petroleum Institute, Washington, D.C., 1983.

17. Kinsler, L. E. and Frey, A. R., *Fundamentals of Acoustics*, John Wiley & Sons, New York, 3rd Ed., 1982, Chapters 9–10.

18. Wachel, J. C., "Design Audits," *Proceedings of the Fifteenth Turbomachinery Symposium*, November 1986, pp. 153–168.

19. Massey, I. C., "Subsynchronous Vibration Problems in High-Speed Multistage Centrifugal Pumps," *Proceedings of the Fourteenth Turbomachinery Symposium*, October 1985, pp. 11–16.

20. Childs, D. W. "Finite Length Solution for Rotordynamic Coefficients of Turbulent Annular Seals," ASLE Transactions, *Journal of Lubrication Technology*, 105, July 1983, pp. 437–445.

21. Karassik, Igor, "Centrifugal Pump Construction," Section 2.2, *Pump Handbook*, 2nd Ed., McGraw-Hill, 1986, pp. 2–69.

22. Childs, D. W. and Scharrer, J. K., "Experimental Rotordynamic Coefficient Results for Teeth-on-Rotor and Teeth-on-Stator Labyrinth Gas Seals," ASME Paper No. 86-GT-12, 1986.

23. Iwatsubo, T., Yang, B. and Ibaraki, R., "An Investigation of the Static and Dynamic Characteristics of Parallel Grooved Seals," The Fourth Workshop of Rotordynamic Instability Problems in High-Per-

formance Machinery, Turbomachinery Laboratories, Texas A&M University, College Station, Texas, June 1986.

24. Black, H. F. and Jenssen, D. N., "Dynamic Hybrid Properties of Annular Pressure Seals," ASME Paper 71-WA/FF-38, 1971.
25. Wachel, J. C., and Szenasi, F. R., et al, "Rotordynamics of Machinery," *EDI Report 86-334*, April 1986, Chapters 3–5.
26. Allaire, P. E. and Flach, R. D., "Design of Journal Bearings for Rotating Machinery," *Proceedings of 10th Turbomachinery Symposium,* December, 1981.
27. Childs, D. W., "Dynamic Analysis of Turbulent Annular Seals based on Hirs' Lubrication Equation," ASME Transaction, *Journal of Lubrication Technology*, 105, July 1983, pp. 429–436.
28. Childs, D. W. and Kim, C. H., "Analysis and Testing for Rotordynamic Coefficients of Turbulent Annular Seals with Different, Directionally-Homogeneous Surface-Roughness Treatment for Rotor and Stator Elements," ASME paper, ASME-ASLE Joint Lubrication Conference, October 1984.
29. Black, H. F., "Effects of Hydraulic Forces in Annular Pressure Seals on the Vibrations of Centrifugal Pump Rotors," *Journal of Mechanical Engineering Science*, 11, (2), 1969.
30. Childs, D. W. and Moyer, D. S., "Vibration Characteristics of the HPOTP of the SSME," ASME Paper 84-GT-31,1984.
31. Von Pragenau, G. L., "Damping Seals for Turbomachinery," NASA TP-1987, March 1982.
32. Bolleter, U., Frei, A. and Florjancic, D., "Predicting and Improving the Dynamic Behavior of Multistage High Performance Pumps," *Proceedings of the First International Pump Symposium*, May 1984, pp. 1–8.
33. Wachel, J. C., "Rotor Response and Sensitivity," *Proceedings Machinery Vibration Monitoring and Analyses*, Vibration Institute, Houston, Texas, April 1983, pp. 1–12.
34. ASA STD2-1975 "Balance Quality of Rotating Rigid Bodies," Acoustical Society of American Standard, ANSI S2, 1975.
35. API Standard 617 "Centrifugal Compressors for General Refinery Services," Section 2.8, 5th Ed., American Petroleum Institute, Washington, D.C., 1988.
36. Wachel, J. C., "Rotordynamics Instability Field Problems," Second Workshop on Rotordynamic Instability of High Performance Turbomachinery, NASA Publication 2250, Texas A&M University, May 1982.
37. Black, H. F., "Calculation of Forced Whirling and Stability of Centrifugal Pump Rotor Systems," ASME Paper No. 73-DET-131, 1973.

38. Barrett, L. E., Gunter, E. J. and Nicholas, J. C., "The Influence of Tilting Pad Bearing Characteristics on the Stability of High Speed Rotor-Bearing Systems," *Topics in Fluid Film Bearing and Rotor Bearing Systems, Design and Optimization,* ASME Publication Book No. 100118, 1978, pp. 55–78.
39. Szenasi, F. R. and Von Nimitz, W., "Transient Analysis of Synchronous Motor Trains," *Proceedings of the Seventh Turbomachinery Symposium,* December 1978, pp. 111–117.
40. Szenasi, F. R. and Blodgett, L. E., "Isolation of Torsional Vibrations in Rotating Machinery," *National Conference on Power Transmissions,* 1975, pp. 262–286.
41. Military Standard 167, Bureau of Ships, 1964.
42. Peterson, R. E. *Stress Concentration Factors,* John Wiley and Sons, 1974, pp. 245– 253.
43. Frej, A., Grgic, A., Heil, W. and Luzi, A., "Design of Pump Shaft Trains Having Variable-Speed Electric Motors," *Proceedings of the Third International Pump Symposium,* May 1986, pp. 33–44.
44. Frank, K., "Frequency Converters for Starting and Speed Regulations of AC Motors for Turbomachine Applications," *Proceedings of the Tenth Turbomachinery Symposium,* December 1981, pp. 87–94.
45. Szenasi, F. R., "Diagnosing Machinery-Induced Vibrations of Structures and Foundations," ASCE Spring Meeting, April 29, 1986.
46. Wachel, J. C., "Prevention of Pump Problems," Worthington 2nd Technical Pump Conference, October 1981.
47. Allaire, P. E., et al., "Aerodynamic Stiffness of an Unbounded Whirling Centrifugal Impeller with an Infinite Number of Blades," *Workshop on Rotordynamic Instability,* Texas A&M University, May 1982.
48. Wilson, W. D., "Speed of Sound in Distilled Water as a Function of Temperature and Pressure," *Journal of the Acoustical Society of America,* Vol. 31, No. 8, August 1959, pp. 1067–1072.
49. Hayward, A. T. J., "How to Estimate the Bulk Modulus of Hydraulic Fluids," *Hydraulic Pneumatic Power,* January 1970, p. 28.
50. *Technical Data Handbook-Petroleum Refining,* Vol. I, 4th Ed., 8th Revision, American Petroleum Institute, Washington, D.C., 1983, pp. 6-65 to 6-82.

Part 4

Extending Pump Life

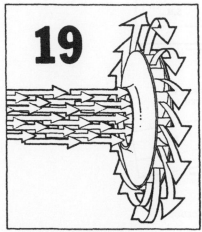

ALIGNMENT

by **Malcolm G. Murray, Jr.**
Murray and Garig Tool Works

Definitions

Internal Alignment. This type of alignment refers to machined fits and precise assembly that ensure that the pump incurs no excessive deviation from intended tolerances *prior to* installation on its pedestal and connection of piping and shaft coupling. Although of obvious importance, this type of alignment will not be emphasized here. It is covered in numerous general and specific manufacturers' manuals and shop practice standards. Further reference to it will be brief, and instructions for its accomplishment will not be given in this section.

External Alignment. This type of alignment refers to a procedure or condition that ensures the accuracy with which the centerline of one machine shaft coincides with the extension of the centerline of an adjacent machine shaft to which it is to be coupled. For horizontal machine elements, alignment figures are customarily given in vertical and horizontal planes along the shaft axis, both necessary for the complete picture. For vertical machine elements, similar information is given in two convenient planes perpendicular to each other. *Alignment* in this chapter will usually refer to external alignment. Other terms often used are *shaft alignment, coupling alignment, angular,* and *parallel alignment.* See Figure 19-1.

Why Bother With Precise Alignment?

This is a logical and legitimate question. Certainly, if pumps would run as well with poor alignment as with good alignment, this section of the

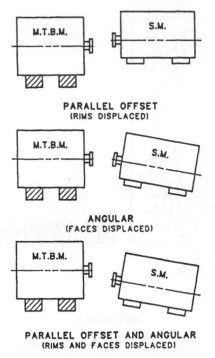

Figure 19-1. Basic conditions of misalignment.

book could be omitted. Experience has shown though, rather painfully at times, that excessive misalignment can have undesirable consequences. These include internal rubs, excessive vibration, increased static and cyclical stress. These factors can lead to premature failure of bearings, seals, couplings, and other parts. Service interruptions, fire, and personnel injury can result. Figure 19-2 shows a catastrophic pump failure caused by misalignment. Costs of such a failure will seldom be less than several thousand dollars and can easily run into the millions if the failure consequences go beyond the original pump.

Causes of Misalignment

Internal misalignment is generally caused by defective parts, poor machining, and/or improper fits and assembly. One other occasional cause is an improper field practice—varying the bolt torque or the gap at the pump head joint on a vertically split pump to make external alignment adjustments.

External (shaft/coupling) misalignment can be caused by a host of factors. These factors are described in the following pages.

Figure 19-2. Pump damage caused by bearing seizure. Bearing seizure was result of gear coupling damage from excessive misalignment caused by piping forces.

Management and Human Factors

Standards and Tolerances. Lack of suitable standards and tolerances, or lack of their enforcement, can make a mockery of any attempt to accomplish meaningful alignment. More specific recommendations will be given later.

Training. Inadequately trained alignment personnel are likely to do poor alignment, quickly, or good alignment, slowly. Well trained and equipped personnel, by contrast, can do good alignment systematically and efficiently without wasting time in a trial-and-error approach.

Time Skimping. In training people to do alignment systematically and efficiently, the most frequent comment from students goes something like this:

> I like the new tools and methods you have shown us. I would like to use them to do a better job. But, when I am given an alignment job to do, my supervisor wants a *fast* job, and quality is secondary. So,

I use a straightedge and take 30 minutes. If I use dial indicators, they are just for show, and nobody checks their readings.

Our answer to this is to suggest that the student politely tell his supervisor that a proper pump alignment job will typically require a half day rather than a half hour. Then ask, "Do you want me to do a sloppy job?" Few supervisors will answer this question in the affirmative. It all gets back, though, to standards and tolerances, and whether management is serious about their enforcement. If they are not, the training is a waste of time, and the only alignment tool needed is a crooked straightedge.

Tooling and Supplies. Even a well-trained field machinist or millwright, if inadequately equipped, will be severely handicapped in doing a good alignment job efficiently. For small machines, inexpensive tooling is available that can give adequate results if used properly. More typically, though, a process plant having a number of pumps of various sizes, can benefit greatly by acquiring several thousand dollars worth of alignment tools and supplies. Skimping here is false economy. If funds are tight for tool purchase, good alignment tools and tool kits can sometimes be rented. (See Murray [6] and [10] for more specific details on tools and examples showing return on investment for alignment tools.)

Physical Factors

Foundation. Occasionally, the foundation mass will prove insufficient for stability, resulting in an insufficient damping ability. To be assured of having adequate foundation mass, a rule of thumb calls for three times the machine weight for rotating machines and five times the machine weight for reciprocating machines, assuming a density of 150 lb/ft^3 for reinforced concrete. If soft soil conditions are present, spread footings should be considered. A flexible isolation joint between the foundation and the pavement slab can be helpful in avoiding undesired vibration transmission between adjacent foundations.

A common deficiency in pump foundation design is the insufficient extension of the foundation's perimeter beyond the outline of the pump baseplate. This can cause difficulties in leveling and grouting. Foundation extension three inches beyond the baseplate perimeter is recommended.

Foundation strength is related to the mix composition and rebar placement. The mix should be workable, but not soupy, giving at least 2,500 psi compressive strength when tested following a 28-day cure. It should not have shrinkage cracks. In addition to proper engineering design of the rebar size and layout, care should be taken to ensure at least two inches concrete cover to resist spalling.

The best designs fail if the executions are faulty. Taking pour samples for later testing (and making all concerned aware of this) helps dull the temptation to add excessive water to the mix. The actual rebar layout should be inspected for design compliance *before* the pour is made, for obvious reasons. A minimum three-day wet cure is desirable, although additives and/or plastic sheeting are sometimes used instead for moisture retention. Regardless of method, responsible followup is essential to be sure that what is called for really does get done.

Baseplate—Common Problems and Suggested Solutions. The baseplate may be excessively flexible. Another point to watch is shipping and in-plant transportation of baseplates. Electric motors are usually much heavier than the pumps they drive. For this reason, it is good to have motors over 100 horsepower shipped unmounted to avoid baseplate distortion caused by the motor weight acting on the ungrouted baseplate.

For welded steel baseplates, request oven stress relieving after welding and before final machining of the pad surfaces and before protective coating is applied. This will help avoid distortion due to gradual in-service stress relief.

High strength machine-to-pedestal hold-down bolts, such as SAE Grade 8, can be used as replacements, but should not bc used as original equipment in which their high strength is taken advantage of in determining bolt size and/or quantity. Sooner or later, somebody in the field will substitute one or more Grade 2 mild steel bolt(s), which will be vulnerable to overstressing.

Specify $3/16$ in. shim space beneath driver feet, rather than the $1/16$ in. or $1/8$ in. called for by API610. It is often necessary to remove more than $1/8$ in. of shim thickness, but seldom necessary to remove more than $3/16$ in.

To avoid restricting necessary horizontal alignment movement capability, specify a minimum of $1/8$ in. annular clearance between full-size (not undercut) SAE Grade 2 bolts and their boltholes after the driver has been aligned to the pump at the factory. Once satisfactory running alignment has been achieved in the field, install two Number 8 taper dowel pins at the pump for easy return of the pump to its pedestal to a *nearly aligned* position. Use an anti-seize compound when installing these pins. An exception would apply for a row of identical pumps with a *floating spare*. Here, doweling is useless. The driver should not be doweled because this will normally require some movement for aligning it to the pump each time the latter is reinstalled following shop repair. Three pairs of foot/pad punch marks can be used instead.

Specify rounded baseplate flange corners with approximately 2-inch radius. This will reduce the tendency for grout cracking at corners and will also improve safety in handling.

Baseplate surface protection should be specified for the dual purpose of corrosion protection and for a good grout bonding surface. One good system is an inorganic zinc silicate coating applied immediately after sandblasting to white metal per SSPC SP-5. This leaves a rough, dull surface with the coating having a molecular bond to the metal, and works well with both inorganic and epoxy grouts.

Preferably, one 4-inch grout filling hole should be provided at the center of each bulkhead section, with one 1/2-inch vent hole near each corner of the section. This allows controlled grout placement and verification that each section is filled with grout.

Vertical leveling screws should be specified. Shims and wedges should be avoided. If left in place after grouting, these may cause "hard spots" rather than the uniform support desired from the grout. Also, they may allow moisture penetration, followed by corrosion and grout spalling.

Mounting pads should be machined flat, parallel, and coplanar. This should be done after all welding and stress relieving are complete.

Individual requirements for the installation of alignment jackscrews should be examined closely. At times, as with an axially long boiler feed pump, it may be worthwhile to have jackscrews at the pump rather than or in addition to those at the driver. In the majority of cases, jackscrews can be omitted at electric motors if adequate access is available to allow the use of a soft-face deadblow hammer. This is a surprisingly accurate method for making horizontal alignment moves without damaging the motor. It will work even on large motors of several thousand horsepower and weighing several tons. For motors 500 horsepower and larger, it is worthwhile to ask for API 610's optional 2-inch minimum clearance beneath the motor center at each end of the motor. This will permit the use of hydraulic jacks for lifting or a pair of portable alignment positioners for easy adjustment in any direction, vertical and horizontal [9]. For turbine drivers, alignment jackscrews are nearly always worth having. Jackscrews and their mounting lugs should be resistant to corrosion and sized to avoid bending or breaking in use. For most installations 5/8-in. screws installed in 2-in. × 3/4-in. lugs are strong enough. Zinc plating protects against corrosion at low cost, or upgrading to 316L stainless steel may be considered. If the lugs are installed by welding, they should be placed in such a way to allow access for machine and shim installation and removal. If this is not feasible, they should be attached by drilling, tapping, and bolting. Tables 19-1 and 19-2 give further information on jackscrews and other alignment movement methods.

Grouting. This subject is too extensive to cover adequately here. We will instead make a few general remarks and mention some references having more information.

Table 19-1
Vertical Alignment Movement Methods

Method	Advantages	Disadvantages
Integral Jackscrews, Threaded into Machine Feet	No problem locating jacking surfaces. Inexpensive if included when the machine is purchased.	Shims require an extra hole or slot. Threads sometimes rust and cause difficulty in turning the jackscrew. If the screw breaks, removal of the remaining fragment may be difficult. Difficult and expensive to retrofit after machine is installed.
Prybar	Inexpensive. Readily available. Can get into low-clearance locations.	Access and clearance are sometimes a problem, limiting usefulness and load capacity. Safety can be compromised if the bar slips while someone is reaching beneath the machine feet to install or remove shims.
Portable Jacks and Wedges	Available in a variety of types and load ratings, most fairly inexpensive, with collapsed heights down to $3/16$ in.	Sometimes the one needed is somewhere else, and the one available is not suitable. This is not really a disadvantage for the method, however, but merely an indication of poor planning.
Crane or Hoist	The obvious solution if bottom clearance, access, or load bearing surfaces are unsuitable for jacking.	Expensive and bulky. Sometimes have access problems. Poor small movement control.
Portable Alignment Positioners	In a single device, they give vertical, precise horizontal transverse, and precise horizontal axial movement capability. Capacities up to 20 tons per pair are available. Used in groups of four, the maximum capacity increases to 40 tons. (For more details, see References 6, 9, and 14.)	Two inch minimum vertical clearance required. More expensive than regular jacks. Not usually worth bothering with for machines smaller than 500 hp.

(table continued on next page)

Table 19-1 Continued
Vertical Alignment Movement Methods

Method	Advantages	Disadvantages
Portable Alignment Positioners	Easy to use for movement of large electric motors. Can also be used on some turbines and gearboxes.	

Table 19-2
Horizontal Alignment Movement Methods

Method	Advantages	Disadvantages
Steel Hammers and Sledges	Readily available and inexpensive. No interference with shim installation and removal.	Poor movement control. Likely to damage machines being moved.
Soft-face, Hollow Head, Shot-loaded Deadblow (No-bounce) Hammers and Sledges	Fairly inexpensive. Good movement control (no bounce). Works similar to lead hammer, but far more durable. Does not cause machine damage. Available in a range of sizes and weights. Short handle 6 lb. and long handle 11 lb. will cover most alignment requirements.	Access sometimes insufficient for use.
Jackscrews	Good movement control if screws and mounting lugs are sufficiently large and rigid. Durable, if materials are chosen intelligently. Jackscrews do not cause machine damage. Easy to specify or install on new machines.	If mounting lugs are too flimsy, stick-slip action causes poor movement control. If not of suitable corrosion resisting material, screws may corrode and cannot be turned. Often interfere with direct shim installation and removal, requiring shims to be inserted and removed from foot ends or inside edges. This can become awkward at times. Bolting, rather than welding the jackscrew lugs in place, can solve this problem.

Table 19-2 Continued
Horizontal Alignment Movement Methods

Method	Advantages	Disadvantages
		Difficult to retrofit to existing machines. Not usually portable. Individual lugs and screws are normally needed at each machine. Portable clamp-on jackscrews are sometimes used, but usually it is difficult to find a good place and way to clamp them on.
Portable Jacks	Easy to use, where applicable. Fairly good movement control.	A nearby fixed point to jack against is required, and often this is not present.
Comealong or Horizontal Hoist	Can attach to fixed anchor point some distance away.	Poor control. Stick-slip action is likely.
Portable Alignment Positioners	(See previous description in Table 19-1.) These devices handle vertical and horizontal movement in the same overall step.	
Doweling	Returns a machine to its previous position on its support.	Dowels sometimes stick and break off. Access for drilling and reaming is sometimes insufficient.
Matching Punchmarks	Easy to apply, using automatic centerpunch, in three foot locations. Can be used on both machine elements, whereas doweling is normally used only on one.	Less precise than doweling. Does not move the machine, but gives an indication of where to put it to restore its previous position.

The purpose of grouting is to fill the baseplate cavity without leaving voids and to stiffen the baseplate, providing an even support to transmit load into the foundation.

Grout material is normally either organic (usually epoxy plus aggregate) or nonmetallic inorganic (Portland cement, sand, water, and nonmetallic anti-shrink additives). The choice is based on economics and environmental factors. More specifically, epoxy is more expensive and less resistant to high temperatures, but has higher early and ultimate strength and greater chemical and impact resistance than inorganics. It is also easier to place properly and to cure.

With either category of grout material, success depends on careful planning and preparation and attention to numerous details such as site preparation, equipment selection, and a stepwise procedure for proportioning and mixing, placement, curing, and protective coating. (For more thorough coverage of this subject, see Bloch [1] and Murray [12].)

Piping Fit and Support. Here, the object is to avoid excessive pump nozzle loading. Piping fit is a function of installation workmanship and effective construction inspection/quality control. Piping flexibility and support are functions of design, with due consideration given to weights, stiffnesses, and thermal effects. If not carefully designed to avoid such problems, heavy, hot piping can become "the tail that wags the dog," with the dog being the pump.

Flexible Coupling. Choice of flexible coupling can be important. Tradition and price are the two most often used criteria, and either can result in a poor choice. For most pumps, it is desirable to use a spacer coupling. This gives more parallel misalignment capability than a close coupling of the same type and also provides better access for maintenance.

If radial space is limited, a gear coupling may be required. This will transmit more horsepower at a given speed than any other type occupying the same space. It is, however, more sensitive to misalignment than other types and also requires lubrication. The latter factor is no longer as much of a disadvantage as it once was though, because non-separating coupling greases are now available. A gear coupling can handle limited end float applications for sleeve bearing motors, but doing this correctly can get complicated.

Gridmember couplings are another relatively inexpensive lubricated design used mostly for the smaller pumps.

Elastomeric couplings are popular for pumps in certain size and speed ranges. These accommodate fairly large misalignments, but are sensitive to high temperatures (including heat caused by rubbing against a coupling guard) and certain chemicals. This type of coupling may be unable to handle sleeve bearing motor limited end float applications well.

Dry metallic flexing element couplings are available in several designs—disk, diaphragm, and cantilever link. Chloride stress corrosion cracking is sometimes a problem, but can be minimized by judicious choice of element material, for example, type 316 stainless steel in preference to type 304. These couplings are a good choice for use with sleeve bearing motors because they have an elastic centering characteristic that is easy to apply to end float limitation and geometric centering of the motor rotating element.

Alignment Access. As in many fields, forethought is helpful when it comes to alignment access. A common problem, easily avoided in the specifying stage, is insufficient exposed shaft between front of bearing housing and back of coupling for concealed hub couplings. At least one inch of exposed shaft is desirable to allow alignment bracket attachment. It is helpful to visualize the radial space requirements of alignment brackets for full shaft rotation and to avoid placing piping or other interferences in this circular path. Access for moving and monitoring moves on the machine element to be adjusted for alignment (usually the driver) should also be considered. As mentioned earlier, a 2-inch vertical jacking access beneath each end of a motor will allow insertion of hydraulic jacks or alignment positioners [9]. Access to install and remove shims and place horizontal move monitoring dial indicators is also important. Access for applying and turning wrenches on hold-down bolts should not be overlooked. Aligning a machine that has one side immediately adjacent to a wall, or blocked by a maze of pipes, can require acrobatic maneuvers with a degree of difficulty directly proportional to the age, size, weight, and arthritic tendencies of the aligner.

Pre-Alignment Steps

It may seem to some readers that we are taking a long time to get into the ostensible subject—shaft/coupling alignment. This may be true, but there is good reason for it. One thing builds upon another. Without certain prealignment steps, the alignment itself is likely to be dangerous, difficult, or ineffective. Nobody knows exactly how the pyramids were built, but it is a pretty sure bet that they were not done from the top down. Pre-alignment steps will now be described.

Work Permit and Power Lockout. This is basic and is usually the first step. It is essential that all concerned agree on *which* pump/driver combination is to be aligned, and that it will be effectively prevented from starting under power during the alignment.

Personnel Coordination. Sometimes the work area is crowded, and several categories of work are to be done. Alignment is the type of work that suffers if non-alignment personnel, such as electricians and pipefitters, are doing their own jobs in the same vicinity and inadvertently lean against a jacked-up machine, kick a foot-move-monitoring dial indicator, etc. It is best in such cases to arrange for the various types of work to be done separately, either sequentially or on separate shifts.

Heaters and Circulation. Some electric motors have base or winding heaters that come on when the main power is deactivated. These can cause undesired motor growth and can also magnetize dial indicators and cause them to stick. It is usually desirable to get such heaters turned off before starting the alignment. The same is true for any hot fluid or steam circulation through pump or turbine.

Shading or Night Schedule. These are precautions more likely to be needed with large elevated compressor trains than with typical pumps. Nevertheless, it is worth mentioning that uneven sun heating can sometimes affect alignment and cause inconsistencies during the job. The usual ways of avoiding such problems are to erect a tarp sunshade or do the work at night.

Installation of Pump on Pedestal. The pump foot and pedestal surfaces should be clean and free of burrs.

Connection of Main Piping. Prior to connection of piping, stationary-mount vertical and horizontal dial indicators should be applied to the pump at its coupling end, set to mid range, and zeroed. The pump suction piping should then be connected, followed by the discharge piping. During flange bolt tightening, the indicators should be watched, and any indicated distortion should be kept less than 0.003 in. for each flange by selective tightening of flange bolts. If this cannot be achieved, it may be necessary to refit the pipe or use a dutchman. The latter is a tapered filler ring inserted between the flanges, with a gasket on each side to occupy the uneven gap. (See Murray [7] for more information on dutchmen.)

Coupling Surface Runout Check. This is sometimes necessary, but not always. It should be done at face and rim measurement surfaces of any coupling that will remain stationary while the dial indicators rotate, contacting these surfaces. If both coupling halves will be rotated in coordination or while coupled, the surface runout check can be omitted. To do the runout check, apply indicators mounted from a stationary point and rotate the coupling while watching for deviations from zero start. If these appear, mark their locations and set them away from quandrantal measurement locations (top-bottom-side-side).

Soft Foot Check and Shim Recommendations. Soft foot should be minimized to prevent machine frame distortion and inconsistent alignment results. It should be checked at electric motor drivers, which usually have four feet. Checking can often be omitted at turbines and pumps, which are frequently supported at two or three feet. To do the check, apply a vertical indicator mounted from a stationary base to each foot in turn, and measure the upspring when the hold-down bolt is loosened. If this exceeds 0.002 in., try shimming the indicated amount beneath the foot with the largest upspring. This should help, if the cause was improper shimming. Other causes may not be helped by this remedy though. These include heel-and-toe effect, tilting feet or support pads, and distorted or dirty shims. Remedies for these problems include remachining supports, using tapered or liquid epoxy shims, and replacement of bad shims with clean, flat shims. The best shim pack is a "sandwich" having thick shims top and bottom protecting a few thin shims in between. Measure the thickness of each shim. Markings are not always accurate, and thin shims often stick together, doubling their marked thickness. Stainless steel is generally the best material, and pre-cut shims are desirable if available in the outline size required. Pre-cut shims 0.050 inch and thicker should be checked with a straightedge for excessive edge distortion.

Axial Gap and End Float Check. For ball bearing motors, end float is zero and axial gap depends on coupling geometry. For elastic centering couplings (for example, metal disk), the coupling should be at zero axial deflection with the motor in the center of its float, or manufacturer's cold-spring recommendations should be followed. These are the easy ones.

For those who enjoy messy mathematics, Murray [6] has some examples showing how to calculate dimensions for achieving desired axial gaps and end floats with sleeve bearing motors having gear couplings. For most situations, however, it is easier to get these by intelligent trial and error. By knowing our objectives, the task becomes fairly easy. We wish to accomplish three things:

1. The motor shaft, when coupled, must be capable of passing through its float center.
2. It must be restrained by the connected coupling so that it cannot rub the outboard bearing stop.
3. It must be restrained by shaft contact, coupling thrust plates, or thrust buttons from rubbing the inboard bearing stop.

In effect, the last two transfer thrust forces to the pump thrust bearing through the coupling. Unlike the relatively weak motor thrust surfaces, it is capable of handling these forces.

By marking float extremities and float center on the shaft of the uncoupled motor, then coupling it and applying thrust in both directions while rotating on the oil film, it becomes easy to see whether our objectives have been met. If they have not, the solution becomes fairly obvious—move the coupling in or out on the shaft, move the motor frame in or out, add thrust buttons of necessary thickness, or a combination of these.

Bracket Sag Check. In most cases, particularly with spacer couplings, a sag check should be made on the indicator bracket to be used for the alignment. In doing this, rim sag should always be checked. Face sag, if face readings are to be taken, is usually negligible and can be ignored. An exception could occur with a long dogleg face mount. Procedures for handling this are shown in Murray [6], but will not be repeated here because this situation does not arise often with pumps. For vertical pumps, bracket sag may be ignored. It may exist, but it is constant everywhere and does not require correction. Vertical pumps seldom require field alignment because they normally achieve their alignment by machined fits.

To do a rim sag check, mount the bracket on a stiff pipe, zero the indicator at mid range in the top position, and roll the pipe from top to bottom on sawhorses, and note the bottom reading. This is "total sag," which is twice the bracket sag. Do this several times and "jog" or depress the indicator contact point and let it reseat at top and bottom reading.

To keep less than 0.001 in. error due to calibration pipe sag, using Schedule 40 calibration pipe, keep the following span limits:

Pipe	Span Between Supports
2 in.	2 ft. 6 in.
3 in.	3 ft.
4 in.	3 ft. 6 in.
6 in.	4 ft. 4 in.

For a face mounted bracket, do the same type of check between lathe centers or mount inside a pipe capped at one end. The latter method will usually require a sign reversal, but the numbers will be correct.

Measurement of Linear and Face Dimensions. This should be done to the nearest 1/8 in. for small machines and the nearest 1/4 in. for large machines, and the dimensions entered into a prepared alignment data form. Examples of such forms can be found in Murray [6]. Measurements should be based on indicator contact point centers and foot support centers.

Determination of Tolerances. This is necessary in order to know when the alignment is adequate and work can be stopped. Murray [6] and Bloch [1] contain a number of criteria for alignment tolerances based on coupling and machine type, distance between coupling flex planes, and rotating speed. Results vary with the different approaches, from very sloppy to very precise. There is general agreement, however, that coupling manufacturers' advertised tolerances are too lax for the machines being coupled, even though they may be all right for the couplings themselves. A simple rule of thumb that works well for most horizontal pumps is to allow 0.003 in. maximum centerline offset at measurement planes and the same amount for maximum face gap difference. For vertical pumps, in the infrequent cases requiring their manual alignment, 0.001 in. maximum face gap difference is recommended.

Methods of Primary Alignment Measurement

By "primary" measurement, we mean that thermal growth is not determined thereby, although it may be incorporated into the alignment data if known from some other source. Table 19-3 compares various primary alignment measurement methods. Figures 19-3A and B and 19-4 illustrate the two most common methods—face and rim and reverse indicator (rim and rim).

Table 19-3
Primary Alignment Measurement Methods

Method	Advantages	Disadvantages
Straightedge, with or without feeler gauge	Simple, inexpensive, easy to use if good coupling rim surfaces are available. Commonly used for preliminary alignment. Also used for final alignment of low speed and/or intermittently operated machines.	Less precise than most other methods.
Taper gauge; small hole gauge and outside micrometer; inside dial caliper	Simple and inexpensive. No bracket required. Fairly accurate with care. Used for measuring face gaps of close-coupled machines.	Limited application. Does only part of the alignment job.

(table continued on next page)

Table 19-3 Continued
Primary Alignment Measurement Methods

Method	Advantages	Disadvantages
Inside micrometer	Simple and inexpensive. No bracket required. Used for measuring face gaps. Not limited to close-coupled machines. Fairly accurate with care.	Limited application. Does only part of the alignment job.
Dial indicators, in any of several arrangements, as follows:	Precise measurement capability with relatively inexpensive equipment.	Costs more than straightedge and requires more careful handling.
Face and Rim (see Figure 19-3A)	Traditional arrangement familiar to most alignment personnel. Usable with large, heavy machines having one shaft difficult to rotate for alignment measurement. Geometrically more accurate than other methods for machines with large measurement diameters and short axial spans. Usually the easiest arrangement to apply to small, close-coupled machines.	If used with a sleeve bearing machine, shaft float can cause errors in face readings. There are at least three ways to eliminate these errors, but they make the procedure more complex. Reference 6 has further details. Geometrically less accurate than reverse-indicator arrangement for situations in which axial span exceeds face measurement diameter. This will usually be true for pumps having spacer couplings. Brackets are often face mounted, requiring coupling spacer removal, and are more difficult to calibrate for sag. Brackets are often custom built for a particular machine rather than being "universal" variable geometry type.
Reverse-Indicator (Rim and Rim). (See Figures 19-3B and 19-4)	Sleeve bearing shaft float does not cause error or require special procedure to avoid error. Geometrically more accurate	Inapplicable to machines having one or both shafts impractical to rotate for alignment measurement. Often impractical to use on

Table 19-3 Continued
Primary Alignment Measurement Methods

Method	Advantages	Disadvantages
	than face-and-rim arrangement, for spacer coupling applications. Face gap difference can be derived easily from rim measurements without the need for direct measurement. Reverse-indicator brackets are usually variable geometry type, clamped to hub rim or shaft, applicable to a range of machine sizes, and easy to calibrate for sag on a horizontal pipe. Both shafts turn together, so eccentricity and surface irregularities do not cause errors.	small, close-coupled machines.
Face-Face-Distance (See Reference 6)	Good for long spans. Applicable mainly to cooling tower drives, but sometimes used for pumps with long coupling spacers. Can also be used when spacers pass through a wall or bulkhead, as on some shipboard applications. Bracket sag is usually negligible.	Less accurate than other dial indicator methods. May require precautions to avoid axial float errors.
Partial Projection Reverse-Indicator (See Reference 14)	Good for long spans, mainly cooling tower drives with metal disk couplings. Avoids axial float error. More accurate than face-face-distance arrangement if used with metal disk couplings provided bracket sag is accounted for.	Must calibrate and account for bracket sag. Subject to diametral tooth clearance error if used with gear-type couplings.
Laser-Optic (OPTALIGN®)	Fast. Accurate.	Basic equipment. Will not handle two-element

(table continued on next page)

Table 19-3 Continued
Primary Alignment Measurement Methods

Method	Advantages	Disadvantages
	Laser beam has zero sag even over long spans. Measurement and calculating capability are combined in one system. Easy to use for basic alignment problems. Can also handle certain more complex problems, but procedures are more complex, or extra equipment is required. Requires about 7/8 in. exposed shaft for mounting, but this can be reduced to 9/16 in. by using non-original-equipment brackets. Metric and inch systems both present in same calculator. Can be used to measure soft foot originating from common causes. Can be used for progressive error reduction horizontal moves similar to the Barton method used with dial indicators. Can be used to accurately measure sag and its effect for long spacers and pipes.	or multi-element optimum move problems without supplemental calculations. Supplemental equipment available at extra cost can handle such problems. Both shafts must be turned together. This requirement can sometimes be met by using a roller bracket with inclinometer on a shaft that remains stationary. Subject to error if coupling backlash cannot be eliminated during rotation. Requires careful handling to avoid damage. Electrical. Requires periodic battery replacement, and gas test for permit to use in hazardous areas. Shielding may be required in the presence of bright sunlight, steam, or heat waves. Expensive. Recalibration recommended every two years. Severe misalignment can sometimes exceed range of system, requiring preliminary alignment by other means. For basic equipment, computer gives numbers and their orientation on a display representing the machine elements but no graphical shaft centerline relationship display. The latter is available in supplemental equipment at extra cost.

The OPTALIGN® laser-optic alignment system is manufactured by Prueftechnik Dieter Busch + Partner GmbH & Co., Ismaning, Germany. Represented in the U.S. by Ludeca, Inc., Miami, Florida.

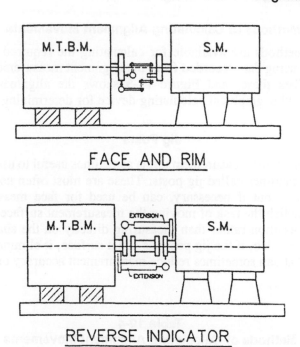

FACE AND RIM

REVERSE INDICATOR

Figure 19-3 A & B. Two most common methods of indicator alignment.

Figure 19-4. Reverse indicator alignment measurement set-up. Note the locking chain wrenches for ease of turning, and the inclinometers for accurate quadrantal rotation.

Methods of Calculating Alignment Movements

Several methods are available for calculating the required alignment movements using data recorded from the alignment measurements. Table 19-4 describes these, and Figure 19-5 shows the alignment plotting board, a popular graphical calculating device for determining alignment movements.

Jig Posts

For dial indicator measurements, it is sometimes useful to use auxiliary surfaces, sometimes called jig posts. These are most often used for rim measurements, but if necessary, can be used for face measurements. They accomplish the task of moving the measurement surface to a more convenient location rather than measuring directly on the shaft or coupling surface. Also, if the jig post has a flat surface, it eliminates curvature error that can sometimes reduce measurement accuracy on small diameters.

Table 19-4
Methods of Calculating Alignment Movements

Method	Advantages	Disadvantages
Nelson's Method	Simple pencil and paper math, using face-and-rim data. Easy to visualize and figure. Accurate.	Applicable only to face-and-rim measurement. A four-move method, less efficient than two-move methods. The second and fourth moves may reverse some of the corrections applied in the first and third moves.
Two-Step Mathematical Methods	Pencil and paper mathematics, using any type of measurement data. More efficient than Nelson's method. Accurate.	Harder to visualize than Nelson's method. More likelihood of calculation error due to incorrect choice of sign.
Graph Paper Plot	Pencil, straightedge, and graph paper, using any type of measurement data. Easier to visualize than straight mathematical methods. Can use on multi-element trains for optimum move calculations.	Slightly less accurate than straight mathematical methods, but not usually enough so to matter. Awkward to use in field.

Table 19-4 Continued
Methods of Calculating Alignment Movements

Method	Advantages	Disadvantages
Plotting Board with Overlay	Graphical method for any type of measurement data using plastic laminated graph and marked overlay (See Figure 19-35 and References 6 and 8.) More durable than graph paper and easier to use under field conditions. Easy visualization of shaft centerline relationships. Can use on two-element optimum-move problems.	More expensive than paper. Not practical to use for simultaneous optimum moves of three or more elements. Not as easy to learn as electronic calculator methods.
Barton Method	Progressive error reduction method. (See Reference 6.) Calculations can be done in one's head. No formal data recording required. Fast and simple, if machine can be adjusted easily and precisely with jackscrews.	Practical only for horizontal corrections. Requires a series of moves so good jackscrews must be present. Also requires reverse-indicator bracket arrangement with shafts that turn easily.
Electronic Calculators	Fast, easy to use, and accurate, for basic and some advanced calculations.	Most will not handle two or multiple element optimum move problems without supplemental calculations. Expensive. Electrical, usually battery operated. May require gas test for use in hazardous areas. In some models, programming can be lost while changing batteries, or if batteries go dead. Can be damaged if not handled carefully.
Combination Measurement/ Calculation Methods— OPTALIGN®	Fast. Easy to use. Accurate. (See Table 19-3 for more details.)	Expensive. Careful handling required. Electrical. (See Table 19-3 for more details.)

Figure 19-5. Machinery alignment plotting board used to determine corrective alignment movements.

Jig posts, although useful, can also cause problems if not handled properly. Their flat measurement surfaces can introduce an error in the case of a gear or gridmember coupling turned in such a way that backlash is present. The same problem would be present in a jig post with a curved measurement surface, if the surface is not concentric with the shaft center.

Another common source of jig post error is transverse inclined plane effect. To avoid this, level the rim measurement surfaces of both posts in coordination and rotate precise 90° quadrants using an inclinometer. With sleeve bearing machines, axial inclined plane effect may also be present. To eliminate this, the post surfaces must be axially parallel to the machine shafts. For face measurements on jig posts, the face surfaces must be parallel to the coupling faces, that is, perpendicular to the shafts in two 90° planes.

There are two ways to eliminate these last two inclined-plane errors. One is by using precisely machined jig posts that give the desired result automatically when mounted on shafts. The other way is to use tri-axially adjustable jig posts, with a leveling procedure that achieves the desired parallelism/perpendicularity. The $T + B$ vs. $S + S$ test, which is described in Murray [6], will detect inclined-plane error if this is present.

Numerical Examples

Numerical examples will not be given here due to lack of space to give proper coverage. Those wishing to examine a variety of such examples are urged to see Murray [6].

Thermal Growth

When machines are operating, their temperatures usually change from that at which they were aligned. In most cases they get hotter, although the opposite is occasionaly true. In the case of pumps, turbines, and compressors, the temperature of the connected piping also changes. These temperature changes cause expansion or contraction in the metal. In most cases, this expansion or contraction is not uniform and differs significantly for driver and driven machines. Their relative positions therefore change during operation, causing changes in the alignment relationship. In addition to temperature effects, torque, hydraulic effects, and oil film thickness may contribute to alignment changes. Because of these factors, a coupled machine train with good alignment in its "cold" or non-operating state may have less precise alignment under operating conditions.

For a pump that operates at conditions subject to little temperature variation and aligned at an ambient temperature that also varies little, the relationship between cold and hot alignment will be constant and repeatable. If the amounts and directions of growth can be determined, they can be used in the form of thermal offsets to include a deliberate misalignment when the machines are aligned. Then, under operating conditions, the growth will act equal and opposite to this misalignment, causing the machines to become aligned during operation.

So much for the theory. We live in an imperfect world, and things do not usually behave exactly as we might wish. The best laid plans of mice and men oft go astray, and thermal growth offsets are a good example of this.

There are two common approaches. One, favored by certain authorities, is to use formulas or rules of thumb involving fluid or metal temperatures, coefficients of thermal expansion, and machine geometric measurements to calculate predicted growths. These are then converted to offsets and incorporated into the initial machine alignment. If the machine runs well, with low vibration and without premature failure, the offsets are assumed to be correct, and are used thereafter. If the machine does not run satisfactorily, and the problem is attributed to misalignment caused by thermal growth other than that used to derive the offsets, the growth is then measured by some means, and the offsets are revised accordingly.

The other approach, favored by this author for most pumps (but not necessarily for large turbomachinery), is simpler. This is to say that formulas and rules of thumb are unreliable because they fail to account for piping growth effects that may influence machine movement more than the growth of the machines themselves. The formulas also fail to account for uneven temperature rises in different parts of the machine frames, which is quite common. In effect, formulas are unlikely to give good figures for amounts of relative growth and are equally unlikely to get the *directions* of such relative growths correct. This being the case, our most conservative course of action, as well as the easiest, is to align everything to zero offset, cold. Then, at least, the machines are starting from a good alignment condition, and relative growth in one direction is no more harmful than in another. If alignment-related vibration and reliability prove satisfactory, as they do about 90% of the time, we have simplified our alignment task by eliminating offsets. In the other 10% of the cases, the growths should be *measured*, not calculated, and offsets applied. Several methods of growth measurement are described in Table 19-5.

Table 19-5
Methods for Determining Thermal Growth Offsets for Machinery Alignment

Method	Advantages	Disadvantages
Manufacturers' Recommendations	When available, these require little effort to obtain. Sometimes they give good results.	Often not available, and often unreliable when they *are* available.
Calculations based on temperatures, thermal expansion coefficients, and machine geometry, using formulas and rules of thumb.	Easy to obtain, and very scientific-looking.	Unreliable. Likely to do more harm than good by getting the relative growth direction wrong and thus increasing the misalignment. Formulas fail to account for nonuniform temperature rise in machines and effects of temperature rise in connected piping.
Shut down and "quickly" take alignment measurements.	Makes the measurer feel he is doing something worthwhile.	Utterly useless and self-deceiving. The majority of the thermal movement occurs during the first

Table 19-5 continued
Methods for Determining Thermal Growth Offsets for Machinery Alignment

Method	Advantages	Disadvantages
		minute after shutdown—too quickly to obtain good measurements.
Use extension gauges to take triangulation measurements on tooling balls at several planes—Essinger/ Acculign (See Reference 3.)	Consistent results obtained if lower tooling balls are mounted on foundation or sole plates rather than on baseplate. Easy to apply, with straightforward calculations based on measurements taken at hot running and cold shutdown conditions. Electronic or non-electronic as desired, and non-optical; fairly rugged tooling, not overly expensive.	Piping congestion can sometimes make access difficult. Some loss of accuracy due to inability to measure to center of shaft—usually must stop short of this by about 15%.
Eddy current proximity probes mounted in water stands. (See Reference 5.)	Good for ground mounted machines. Accurate results if applied properly. Can also be applied to elevated foundation machines. The mount bases should be on foundations or soleplates, not on baseplates or structural steel.	Fairly bulky. Electronic.
Similar probes mounted in Dodd/Dynalign® bars. (See Reference 2.)	Good for both ground mounted and elevated machines, since measurements are relative, not ground based. Accurate results if applied properly.	Electronic. May run into span limitations with long coupling spacers.
Similar probes used in Indikon system. (See Reference 4.)	Effectively measures to center of rotating shafts, giving highly accurate results. No coupling span limitations. Good for both ground-mounted and	Electronic. Expensive. Built into coupling. Not easy to retrofit to existing machines.

(table continued on next page)

Table 19-5 Continued
Methods for Determining Thermal Growth Offsets for Machinery

Method	Advantages	Disadvantages
	elevated machines. Similar system can measure machine torque using strain gauges mounted in coupling.	
Optical Measurement	Non-electronic. Accurate if done properly. Can be used for both ground-mounted and elevated machines. No temperature limits.	Delicate instruments, fairly expensive. Considerable skill required.
Laser Measurement (See Reference 13.)	Accurate.	Expensive. Special mounts improve economics for multiple installations and accuracy.
Vernier-Strobe (See Reference 13.)	Special Vernier scales attached to coupling are read or photographed with aid of strobe light. Highly accurate and fairly inexpensive.	Limited to use with certain types of couplings, mainly metal disk and grease lubricated gear types.
Alignment adjustment while running in response to vibration.	Growth measurement bypassed, going instead to effects on running vibration.	Risk of machine damage if moves are not well controlled.

® *Dymac, a unit of SKF USA, Inc., San Diego, California.*

Considerably more could be said about machinery alignment than we have room for here. Those wishing to learn more are urged to investigate the references listed.

References

1. Bloch, H. and Geitner, F., *Practical Machinery Management for Process Plants, Vol. 3, Machinery Component Maintenance and Repair,* Houston: Gulf Publishing Company, 1985, pp. 159–224 and 106–114.
2. Dodd, V. R., *Total Alignment,* Tulsa: PennWell Publishing Company, 1975, pp. III-1–39.

3. Essinger, J. N., *Benchmark Gauges for Effective Hot Alignment of Turbomachinery*, Houston: Acculign, Inc., undated.
4. *Indikon Measuring Systems—Misalignment*, Cambridge, Massachusetts: Indikon Company, Inc., 1977.
5. Jackson, C., "Shaft Alignment Using Proximity Transducers—ASME Paper 68-PET-25," New York: The American Society of Mechanical Engineers, 1968.
6. Murray, M. G., *Alignment Manual for Horizontal, Flexibly-Coupled Rotating Machines*, 3rd Edition, Baytown, Texas: Murray & Garig Tool Works, 1983.
7. _____, "Flange Fitup Problems? Try a Dutchman," *Hydrocarbon Processing*, Vol. 59, No. 1, January 1980, pp. 105–106.
8. _____, *Machinery Alignment Plotting Board*, Baytown, Texas: Murray & Garig Tool Works, 1980
9. _____, *Machinery Alignment Postioners*, Baytown, Texas: Murray & Garig Tool Works, 1978.
10. _____, "Machinery Shaft/Coupling Alignment Strategy Cuts Manhours and Downtime," *Oil & Gas Journal*, Vol. 81, No. 28, July 11, 1983, pp. 55–61.
11. _____, "Out of Room? Use Minimum Movement Machinery Alignment," *Hydrocarbon Processing*, Vol. 58, No. 1, January 1979, pp. 112–114.
12. _____, "Practical Considerations for Pump Baseplates, Grouting and Foundations," *3rd International Pump Symposium Proceedings*, 1986, pp. 131–136.
13. _____, "Two New Systems for Measuring Alignment Thermal Growth," *Vibrations*, Vol. 6, No. 4, Willowbrook, Illinois: The Vibration Institute, December 1990, pp. 9–15.
14. Piotrowski, J., *Shaft Alignment Handbook*, New York: Marcel Dekker, Inc., 1986.

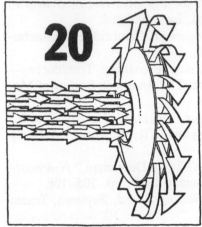

20

Rolling Element Bearings and Lubrication*

by **Heinz P. Bloch, P.E.**
Consultanting Engineer

Next to mechanical seal failures, bearing failures are most often responsible for pump outages and repair events. Both rolling element (antifriction) bearings and sliding element (plain) bearings are used in centrifugal pumps. Each of these two bearing categories is found in small as well as large and in single as well as multistage pumps. However, very large or very high speed pumps tend to favor sliding element bearings over rolling element bearings.

Because the preponderance of centrifugal pumps is small, say less than 200 hp (~150 kW), most pumps are equipped with rolling element bearings. Avoiding or reducing rolling element bearing failures should thus be a prime topic whenever pump life extensions are being discussed.

A rolling element bearing is a precision device and a marvel of engineering. It is unlikely that any other mass-produced item is machined to such close tolerances. While boundary dimensions are usually held to tenths of a thousandth of an inch, rolling contact surfaces and geometries are maintained to millionths of an inch. It is for this obvious reason that very little surface degeneration can be tolerated [12].

The life of a rolling element bearing running under good operating conditions is usually limited by fatigue failure rather than by wear. Under optimum operating conditions, the fatigue life of a bearing is determined by the number of stress reversals and by the cube of the load causing these stresses. As example, if the load on the bearing is doubled, the the-

---

* Material on lubrication of rolling element bearings adapted by permission from TRW Bearings Division, Jamestown, New York.

oretical fatigue life is reduced to one-eighth. Also, if speed is doubled, the theoretical fatigue life is reduced to one-half.

Friction Torque

The friction torque in a rolling element bearing consists essentially of two components. One of these is a function of the bearing design and the load imposed on the bearing. The other is a function of the lubricant type, lubricant quantity, and the speed of the bearing.

It has been found that the friction torque in a bearing is lowest with a quantity of the correct viscosity oil just sufficient to form a film between the contacting surfaces. This is just one of the reasons why oil mist (to be explained later) is a superior lubrication method. The friction will increase with greater quantity and/or higher viscosity of the oil. With more oil than just enough to separate the rolling elements, the friction torque will also increase with the speed.

Function of the Lubricant

A bearing lubricant serves to accomplish the following:

- To lubricate sliding contact between the cage and other parts of the bearing.
- To lubricate any contact between races and rolling elements in areas of slippage.
- To lubricate the sliding contact between the rollers and guiding elements in roller bearings.
- In some cases, to carry away the heat developed in the bearing.
- To protect the highly finished surfaces from corrosion.
- To provide a sealing barrier against foreign matter.

Oil Versus Grease

The ideal lubricant for rolling element bearings is oil. Grease, formed by combining oil with soap or non-soap thickeners, is simply a means of effecting greater utilization of the oil. In a grease, the thickener acts fundamentally as a carrier and not as a lubricant.

Although realtively few centrifugal pump bearings are grease lubricated, greases are in fact used for lubricating by far the largest overall number of rolling bearings. The extensive use of grease has been influenced by the possibilities of simpler housing designs, less maintenance, less difficulty with leakage, and better sealing against dirt. On the other hand, there are limitations that do not permit the use of grease. Where a lubricant must dissipate heat rapidly, grease should not be used. In many cases, associated machine elements that are oil lubricated dictate the use of oil for rolling element bearings. Listed below are some of the advantages and disadvantages of grease lubrication.

Advantages

- Simpler housing designs are possible; piping is greatly reduced or eliminated.
- Maintenance is greatly reduced since oil levels do not have to be maintained.
- Being a solid when not under shear, grease forms an effective collar at bearing edges to help seal against dirt and water.
- With grease lubrication, leakage is minimized where contamination of products must be avoided.
- During start-up periods, the bearing is instantly lubricated whereas with pressure or splash oil systems, there can be a time interval during which the bearing may operate before oil flow reaches the bearing.

Disadvantages

- Extreme loads at low speed or moderate loads at high speed may create sufficient heat in the bearing to make grease lubrication unsatisfactory.
- Oil may flush debris out of the bearing. Grease will not.
- The correct amount of lubricant is not as easily controlled as with oil.

Oil Characteristics

The ability of any oil to meet the requirements of specific operating conditions depends upon certain physical and chemical properties. In cold ambients, the lube oil must have a low enough pour point to ensure that the oil remains in the liquid condition, whereas in hot ambients, the lube oil must have a high enough viscosity to ensure that the rolling elements are coated with an oil film of sufficient thickness to prevent metal-to-metal contact.

Viscosity

The single most important property of oil is viscosity. It is the relative resistance to flow. A high viscosity oil will flow less readily than a thinner, low viscosity oil.

Viscosity can be measured by any of a number of different instruments that are collectively called viscosimeters. In the United States, the viscosity of oils is usually determined with a Saybolt Universal Viscosimeter. It simply measures the time in seconds required for 60 cc of oil to drain through a standard hole at some fixed temperature. The common temperatures for reporting viscosity are 100°F and 210°F (38°C and 99°C). Viscosities are quoted in terms of Saybolt Universal Seconds (SUS).

Generally, for ball bearings and cylindrical roller bearings, it is a good rule to select an oil that will have a viscosity of at least 13.1 cSt or 70 SUS at operating temperature. It is especially important to observe this guideline in hot climates where it would be prudent to opt for ISO Grade 100 lube oils whenever possible.

Viscosity Index

All oils are more viscous when cold and become thinner when heated. However, some oils resist this change of viscosity more than others. Such oils are said to have a high viscosity index (V.I.). Viscosity index is most important in an oil that must be used in a wide range of temperatures. Such an oil should resist excessive changes in viscosity. A high V.I. is usually associated with good oxidation stability and can be used as a rough indication of such quality.

Pour Point

Any oil, when cooled, eventually reaches a temperature below which it will no longer flow. This temperature is said to be the pour point of the oil. At temperatures below its pour point, an oil will not feed into the bearing, and lubricant starvation may result. In selecting an oil for rolling element bearings, the pour point must be considered in relation to the operating temperature. In other words, equipment operating at 10°F should be lubricated with oils that have a pour point of zero or perhaps even −10°F. It should be noted that many diester or polyalpha-olefin synthetic lubricants have pour points in the vicinity of −60°F and are thus well suited for low temperature operation.

Flash and Fire Point

As an oil is heated, the lighter fractions tend to become volatile and will flash off. With any oil, there is some temperature at which enough vapor is liberated to flash into momentary flame when ignition is applied. This temperature is called the flash point of the oil. At a somewhat higher temperature, enough vapors are liberated to support continuous combustion. This is called the fire point of the oil. The flash and fire points are significant indications of the tendency of an oil to volatilize at high operating temperatures. High V.I. oils generally have higher flash and fire points than lower V.I. oils of the same viscosity.

Oxidation Resistance

All petroleum oils are subject to oxidation by chemical reaction with the oxygen contained in air. This reaction results in the formation of acids, gum, sludge, and varnish residues that can reduce bearing clearances, plug oil lines, and cause corrosion.

Some lubricating fluids are more resistant to this action than others. Oxidation resistance depends upon the fluid type, the methods and degree of refining used, and whether oxidation inhibitors are used.

There are many factors that contribute to the oxidation of oils and practically all of these are present in lubricating systems. These include temperature, agitation, and the effects of metals and various contaminants that increase the rate of oxidation.

Temperature is a primary accelerator of oxidation. It is well known that rates of chemical reaction double for every 18°F (10°C) increase in temperature.

Below 140°F (60°C), the rate of oxidation of oil is rather slow. Above this temperature, however, the rate of oxidation increases to such an extent that it becomes an important factor in the life of the oil. Consequently, if oil systems operate at temperatures above 140°F (60°C), more frequent oil changes would be appropriate. Figure 20-1 shows recommended oil replacement frequencies as a function of oil sump or reservoir capacity.

The oxidation rate of oil is accelerated by metals such as copper and copper-containing alloys and to a much lesser extent by steel. Contaminants such as water and dust also act as catalysts to promote oxidation of the oil.

Emulsification

Generally, water and straight oils do not mix. However, when an oil becomes dirty, the contaminating particles act as agents to promote emulsification. In rolling element bearing lubricating systems, emulsification is undesirable. Therefore, the oil should separate readily from any water present, which is to say, it should have good demulsibility characteristics.

Rust Prevention

Although straight petroleum oils have some rust protective properties, they cannot be depended upon to do an unfailing job of protecting rust-

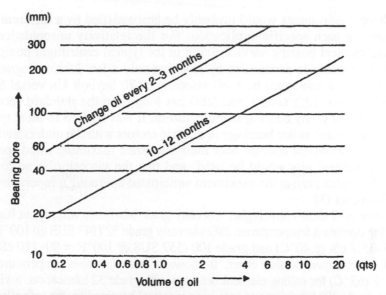

Figure 20-1. Recommended lube oil change frequency as a function of bearing bore diameter and lube oil sump capacity (courtesy of FAG Bearing Corporation, Stamford, Connecticut).

susceptible metallic surfaces. In many instances, water can displace the oil from the surfaces and cause rusting. Rust is particularly undesirable in a rolling element bearing because it can seriously abrade the bearing elements and areas pitted by rust will cause rough operation or failure of the bearing.

Additives

High grade lubricating fluids are formulated to contain small amounts of special chemical materials called additives. Additives are used to increase the viscosity index, fortify oxidation resistance, improve rust protection, provide detergent properties, increase film strength, provide extreme pressure properties, and sometimes to lower the pour point.

General Lubricant Considerations

The proper selection of a grease or oil is extremely important in high-speed bearing operation. Proper lubrication requires that an elastohydrodynamic oil film be established and maintained between the bearing rotating members. To ensure the build-up of an oil film of adequate load-carrying capacity, the lubricating oil must possess a viscosity adequate to withstand the given speed, load, and temperature conditions.

These requirements would probably be best satisfied by a different lubricant for each specific application. For the relatively unsophisticated rolling element bearing service found in the typical centrifugal pump, it is often possible to economize by stocking only a few lubricant grades. Maintaining a minimum base oil viscosity of 70 Saybolt Universal Seconds (SUS) or 13.1 centistokes (cSt) has long been the standard recommendation of many bearing manufacturers. It was applied to most types of ball and some roller bearings in electric motors with the understanding that bearings would operate near their published maximum rated speed, that naphthenic oils would be used, and that the viscosity be no lower than this value even at the maximum anticipated operating temperature of the bearings [3].

Figure 20-2 shows how higher viscosity grade lubricants will permit higher bearing operating temperatures. ISO viscosity grade 32 (147 SUS @ 100°F or 28.8–35.2 cSt @ 40°C) and grade 100 (557 SUS @ 100°F or 90–110 cSt @ 40°C) are shown on this chart. It shows a safe allowable temperature of 146°F (63°C) for rolling element bearings with grade 32 lubrication. Switching to grade 100 lubricant and requiring identical bearing life, the safe allowable temperature would be extended to 199°F (93°C). If a change from grade 32 to grade 100 lube oil should cause the bearing operating temperature to reach some intermediate level, a higher oil viscosity would result and the bearing life would actually be extended. Most ball and roller bearings can be operated satisfactorily at temperatures as high as 250°F (121°C) from the metallurgy point of view. The only concern would be the decreased oxidation resistance of common lubricants, which might require more frequent oil changes. However, the "once-through" application of oil mist solves this problem. Therefore, oil mist lubricated anti-friction bearings are often served by ISO grade 100 naphthenic oils. Where these oils are unavailable or in extremely low ambients, dibasic ester-based synthetic lubes have been very successfully applied. These synthetics eliminate the risk of wax plugging that has sometimes been experienced with mineral oils at low temperatures.

When applying greases to lubricate pump bearings, certain precautions are in order. Soft, long-fibered type greases, or excessively heavy oils, will result in increased churning friction at higher speeds, causing bearing overheating due to the high shear rate of these lubricants. Excessive amounts of lubricant will also create high temperatures.

Using oils of adequate film strength, but light viscosity, or using channeling or semi-channeling greases has the benefit of substantially reducing the heat-generating effects of lubricants. The advantages of these greases rest in their ability to "channel" or be pushed aside by the rotating ball or roller elements of a bearing and to lie essentially dormant in the side cavities of the bearing or housing reservoir. Channeling greases normally are "short-fibered" and have a buttery consistency that imparts

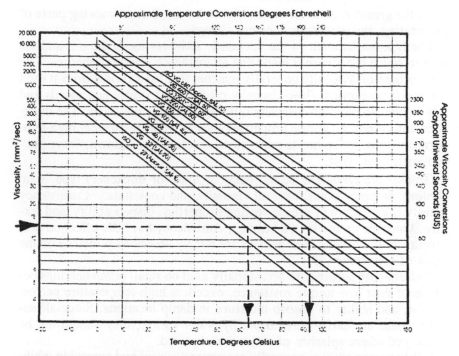

Figure 20-2. Temperature vs. viscosity relationships for various lube oil grades. Note that centrifugal pumps are typically lubricated with ISO grades 32 through 100.

a low shear rate to the lubricant. This low temperature aids an operating bearing to establish a temperature equilibrium, even if a lubricant is applied having a slightly higher viscosity than the application demands. Higher fluid friction increases the temperature of the lubricant until the viscosity is reduced to the proper level. It should be noted, however, that short-fibered greases may lead to "false brinelling" damage in applications subject to vibration without equipment rotation.

Greases generally are applied where oils cannot be used, for example, in places where sealing does not exist or is inadequate, dirty environments, in inaccessible locations, in places where oil dripping or splashing cannot be tolerated, and often, where "sealed-for-life" lubrication is desired.

Greases are fine dispersions of oil-insoluble thickening agents—usually soap—in a fluid lubricant such as a mineral oil. When a bearing lubricated with grease starts to move, the grease structure (created by the thickening agent) is affected by the shearing action, and lubrication is provided by the oil component of the grease. As the bearing slows to a stop, the grease regains its semi-solid structure. In non-moving parts of the bearing, this structure does not change.

The type and amount of the thickener, additives used, the oil, and the way in which the grease is made can all influence grease properties. The chosen base-oil viscosity generally matches that for a fluid lubricant used for the same service—low-viscosity oil for light loads, fast speeds and low temperatures; high-viscosity oils for differing conditions. The thickener will determine grease properties such as water resistance, high-temperature limit, resistance to permanent structural breakdown, "stay-put" properties, and cost.

Greases are classified on the basis of soap (or thickener) type, soap content, dropping point, base oil viscosity, and consistency. Consistency mainly is a measure of grease-sealing properties, dispensability, and texture. Once the grease is in a bearing, consistency has little effect on performance. But despite this, greases are widely described primarily on the basis of consistency [5].

Sodium-soap greases are occasionally used on small pump bearings because of their low torque resistance, excellent high-temperature performance, and ability to absorb moisture in damp locations. Since all sodium soaps are easily washed out by water sprays, they should not be employed where splashes of water are expected.

Lithium-soap greases generally are water resistant and corrosion inhibiting, and have good mechanical and oxidation stability. Many automobile manufacturers specify such grease—often with additives to give wide protection against problems caused by shipment, motorist neglect, and the extended lubrication intervals now popular. Widely used in centralized lubrication systems, these versatile greases are also favored in both sliding and rolling element bearings.

Simple calcium-soap greases resist water-washout, are noncorrosive to most metals, work well in both grease cups and centralized lubrication systems, and are low-cost lubricants. They are, dependent on manufacture and ingredients, limited to services < 160°F to 200°F (71°C to 93°C).

Complex calcium-soap greases, wisely applied, can provide multi-purpose lubrication at a fraction of the cost of a lithium-soap grease; however, misapplication of these greases will more likely cause more difficulty than the same error committed with lithium greases.

Special-purpose greases are available for food processing (both the thickener and oil are nontoxic), fine textile manufacture (light colors for nonstaining, or adhesive grades to avoid slingoff), rust prevention, and other special services.

The user is well advised to observe the grease compatibility constraints indicated in Table 20-1. Experience shows that many pumps and electric motor drivers are initially factory-lubricated with polyurea greases to

Table 20-1
Results of Grease
Incompatibility Study

	Aluminum Complex	Barium	Calcium	Calcium 12-hydroxy	Calcium Complex	Clay	Lithium	Lithium 12-hydroxy	Lithium Complex	Polyurea
Aluminum Complex	X	I	I	C	I	I	I	I	C	I
Barium	I	X	I	C	I	I	I	I	I	I
Calcium	I	I	X	C	I	C	C	B	C	I
Calcium 12-hydroxy	C	C	C	X	B	C	C	C	C	I.
Calcium Complex	I	I	I	B	X	I	I	I	C	C
Clay	I	I	C	C	I	X	I	I	I	I
Lithium	I	I	C	C	I	I	X	C	C	I
Lithium 12-hydroxy	I	I	B	C	I	I	C	X	C	I
Lithium Complex	C	I	C	C	C	I	C	C	X	I
Polyurea	I	I	I	I	C	I	I	I	I	X

B = Borderline Compatibility
C = Compatible
I = Incompatible

Derived from a paper given by E. H. Meyers entitled "Incompatibility of Greases" at the NLGI's 49th Annual Meeting.

prevent oxidation damage during storage. Obviously, this grease should not be mixed with many of the typical greases used by modern process plants.

Application Limits for Greases

Bearings and bearing lubricants are subject to four prime operating influences: speed, load, temperature, and environmental factors. The optimal operating speeds for ball and roller type bearings—as related to lu-

brication—are functions of what is termed the DN factor. To establish the
DN factor for a particular bearing, the bore of the bearing (in millime-
ters) is multiplied by the revolutions per minute, that is:

$$75\text{mm} \times 1000 \text{ rpm} = 75,000 \text{ DN value}$$

Speed limits for conventional greases have been established to range
from 100,000 to 150,000 DN for most spherical roller type bearings and
200,000 to 300,000 DN values for most conventional ball bearings.
Higher DN limits can sometimes be achieved for both ball and roller type
bearings, but require close consultation with the bearing manufacturer.
When operating at DN values higher than those indicated above, use either special greases incorporating good channeling characteristics or circulating oil. Typical relubrication intervals are given in Figure 20-3.

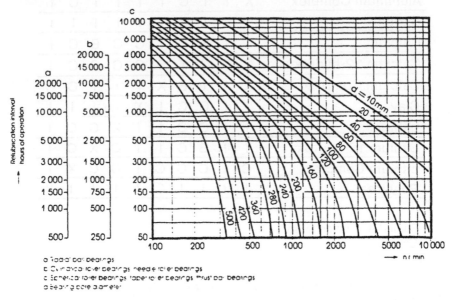

a radial ball bearings
b cylindrical roller bearings needle roller bearings
c spherical roller bearings tapered roller bearings thrust ball bearings
d bearing bore diameter

Figure 20-3. Relubrication intervals recommended by SKF Bearing Company.

Life-Time Lubricated, "Sealed" Bearings

As shown in Table 20-2, "lubrication for life" ranks last in order of
preference [8]. Lubed-for-life bearings incorporate close-fitting seals in
place of, or in addition to shields. These bearings are customarily found
on low horsepower motors or on appliances that operate intermittently.

Table 20-2
Influence of Lubrication On Service Life
(Source: FAG Bearing Corporation)

Oil	Oil	Grease	Dry lubricant
Rolling bearing alone	Rolling bearing with gearwheels and other wearing parts	Rolling bearing alone	Rolling bearing alone
Circulation with filter, automatic oiler Oil-air Oil-mist Circulation without filter*	Circulation with filter Oil-air Oil-mist	Automatic feed	
Sump, regular renewal	Circulation without filter*	Regular regreasing of cleaned bearing	
	Sump, regular renewal* Rolling bearing (a) in oil vapour (b) in sump (c) oil circulation	Regular grease replenishment	
Sump, occasional renewal			
	Sump, occasional renewal Rolling bearing (a) in oil vapour (b) in sump (c) oil circulation	Regular renewal Occasional renewal Occasional replenishment Lubrication for-life	Regular renewal
			Lubrication for-life

↕ Decreasing service life**

* By feed cones, bevel wheels, asymmetric rolling bearings.
** Condition: Lubricant service life ≪ Fatigue life.

However, at least one large petrochemical company in the United States has expressed satisfaction with sealed ball bearings in small centrifugal pumps as long as bearing operating temperatures remained below 150°C (302°F) and speed factors DN (mm bearing bore times revolutions per minute) did not exceed 300,000 [15]. Studies showed, on the other hand, that close-fitting seals can cause high frictional heat and that loose fitting seals cannot effectively exclude atmospheric air and moisture that will cause grease deterioration. It should be assumed that these facts preclude the use of lubed-for-life bearings in installations that expect "life" to last more than three years in the typical petrochemical plant environment. Moreover, we believe this to be the reason why some bearing manufacturers advise against the use of sealed bearings at DN values in excess of 108,000. At least one bearing manufacturer considers the DN range between 80,000 and 108,000 the "gray" area, where either sealed or open (shielded) bearing would be acceptable. The DN range below 80,000 is generally considered safe for sealed bearings. This would generally exclude sealed bearings from all but the smallest centrifugal pumps.

Oil Viscosity Selection

The Oil Viscosity Selection Chart (Figure 20-2) may be used as a guide for the selection of the proper viscosity oil. The chart may be used for applications where loads are light to moderate with moderately high conditions of speed and temperature.

Applications of Liquid Lubricants in Pumps

The amount of oil needed to maintain a satisfactory lubricant film in a rolling element bearing is extremely small. The minimum quantity required is a film averaging only a few micro-inches in thickness. Once this small amount has been supplied, make-up is required only to replace the losses due to vaporization, atomization, and seepage from the bearing surfaces [12].

How small a quantity of oil is required can be realized when we consider that 1/1000 of a drop of oil, having a viscosity of 300 SUS at 100°F (38°C) can lubricate a 50 mm bore bearing running at 3,600 RPM for one hour. Although this small amount of oil can adequately lubricate a bearing, much more oil is needed to dissipate heat generated in high speed, heavily loaded bearings.

Oil may be supplied to rolling element bearings in a number of ways. These include bath oiling, oil mist from an external supply, wick feed, drip feed, circulating system, oil jet, and splash or spray from a slinger or nearby machine parts.

One of the simplest methods of oil lubrication is to provide a bath of oil through which the rolling element will pass during a portion of each revolution. It can be shown that only a few drops of oil per hour would satisfy the lubrication requirements of a typical rolling element bearing in a centrifugal pump. However, unless frictional heat generated by the bearing is readily dissipated, the lubricating oil must serve also as a coolant.

The MRC Bearing Division of TRW calculates the theoretical oil flow required for cooling from the expression.

$$Q = \frac{.000673 \times Fr \times P \times PD \times RPM}{H_s \times (T_o - T_i)} \text{ lbs/minute}$$

where   Fr = Coefficient of friction referred to PD

.00076 = cylindrical roller bearings
.00089 = pure thrust ball bearings
.00103 = radial ball bearings
.00152 = angular contact ball bearings
.002-.005 = tapered roller bearings

P = imposed equivalent load in pounds
PD = pitch diameter in inches
RPM = operating speed in revolutions per minutes
T_o = outlet oil temperature in °F
T_i = inlet oil temperature in °F ($T_o - T_i$, generally about 50°F)
H_s = specific heat of oil in BTU/LB/°F (usually .46 − .48)
= .195 + .000478 (460 + T_i)

Conversion: (pounds of oil per minute) × (0.135) = gal/min

Hence, where cooling is required in high speed and heavily loaded bearings, oil jets and circulating systems should be considered. If necessary, the oil can be passed through a heat exchanger before returning to the bearing.

Oil Bath Lubrication

A simple oil bath method is satisfactory for low and moderate speeds. The static oil level should not exceed the center line of the lowermost ball or roller. A greater amount of oil can cause churning that results in abnormally high operating temperatures. Systems of this type generally employ sight gauges to facilitate inspection of the oil level.

Figure 20-4 shows a constant level arrangement for maintaining the correct oil level. Figure 20-5 shows a similar arrangement that includes a pressure-balanced constant level oiler.

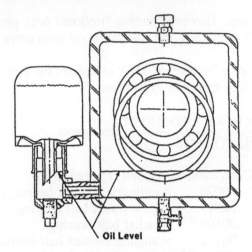

Figure 20-4. Constant level oilers are typically used for conventional lubrication of centrifugal pumps.

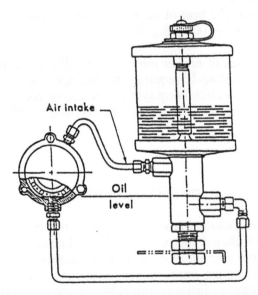

Figure 20-5. Pressure-balanced constant level oilers are required for bearing housings with excessive back pressure or vacuum. They are ideally suited for purge oil mist applications (courtesy of Oil-Rite Corporation, Manitowoc, Wisconsin).

Drip Feed Lubrication

Although not usually found on centrifugal pumps, this system is widely used for small and medium ball and roller bearings operating at moderate to high speeds where extensive cooling is not required. The oil, introduced through a filter-type, sight feed oiler, has a controllable flow rate that is determined by the operating temperature of the bearings.

Figure 20-6 illustrates a typical design and shows the preferred location of the oiler with respect to the bearings.

Forced Feed Circulation

This type of system uses a circulating pump and is particularly suitable for low to moderate speed, heavily loaded applications where the oil has a major function as a coolant in addition to lubrication. If necessary, the oil can be passed through a heat exchanger before returning to the bearing. Entry and exit of the oil should be on opposite sides of the bearing. An adequate drainage system must be provided to prevent an excess accumulation of oil. Oil filters and magnetic drain plugs should be used to minimize contamination.

In applications of large, heavily loaded, high speed bearings operating at high temperatures, it may be necessary to use high velocity oil jets. In such cases the use of several jets on both sides of the bearing provides more uniform cooling and minimizes the danger of complete lubrication loss from plugging. The jet stream should be directed at the opening between the cage bore and inner ring O.D., see Figure 20-7. Adequate scavenging drains must be provided to prevent churning of excess oil after the oil has passed through the bearing. In special cases, scavenging may be required on both sides of the bearing.

At extremely high speeds, the bearing tends to reject the entry of sufficient oil to provide adequate cooling and lubrication with conventional oil jet and flood systems. Figure 20-8 shows an under-race lubrication system with a bearing having a split inner ring with oil slots. This method ensures positive entrance of oil into the bearing to provide lubrication and cooling of the inner ring and may well become a standard on the high-performance pump of the future.

Oil Mist Lubrication

Oil mist lubrication, as shown schematically in Figure 20-9, is a centralized system that utilizes the energy of compressed air to supply a continuous feed of atomized lubricating oil to multiple points through a low pressure distribution system operating at $\approx$ 20 in ($\approx$ 500 mm) H_2O. Oil

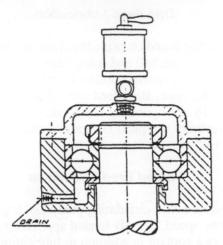

Figure 20-6. Drip feed lubrication occasionally used on small vertical pumps.

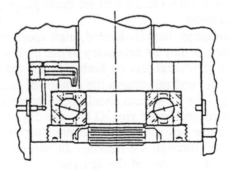

Figure 20-7. Oil jet lubrication for heavily loaded high speed bearings.

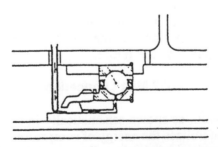

Figure 20-8. Under-race oil jet lubrication applied through split inner ring with oil slots ensures positive entrance of oil into bearing.

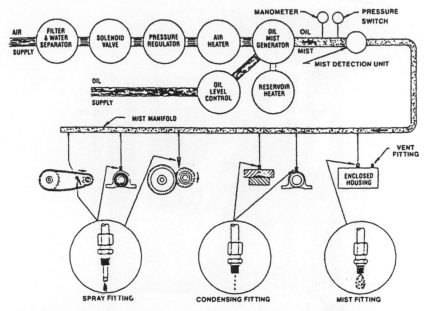

Figure 20-9. Oil mist lubrication schematic (courtesy of Alemite Division of Stewart-Warner Corporation, Chicago, Illinois).

mist then passes through a reclassifier nozzle before entering the point to be lubricated. This reclassifier nozzle establishes the oil mist stream as either a mist, spray, or condensate, depending on the application of the system.

Since the mid-1950s, the oil mist lubrication concept has been accepted as a proven and cost-effective means of providing lubrication for centrifugal pumps. Typical petrochemical plant pump applications are illustrated in Figures 20-10 through 20-13. Centralized oil mist systems have also been highly successful on electric motors, gears, chains, and horizontal shaft bearings such as on steam turbines and steel plant rolling mill equipment.

The actual method of applying oil mist to a given piece of equipment is governed to a large extent by the type of bearing used. For sliding bearings, oil mist alone is not considered an effective means of lubrication because relatively large quantities of oil are required. In this case, oil mist is used effectively as a purge of the oil reservoir and, to a limited extent, as fresh oil make-up to the reservoir. Purge mist lubricated pumps generally use a constant level oiler with a balance line between the bearing housing and the oiler lower bowl. This type of installation, which was shown earlier in Figure 20-5, is required for bearing housings having an excessive back pressure or vacuum and is ideally suited for purge oil

Figure 20-10. Oil mist console suitable for lubricating 30–60 pumps in a refinery (courtesy of Lubrication Systems Company, Houston, Texas).

mist applications. A constant level can be maintained in spite of pressure or vacuum in the housing as the equalizing tube provides static balance of pressure between the bearing housing and oiler bowl.

Rolling element bearings, on the other hand, are ideally suited for dry-sump lubrication. With dry sump oil mist, the need for a lubricating oil sump is eliminated. If the equipment shaft is arranged horizontally, the lower portion of the bearing outer race serves as a mini-oil sump. The bearing is lubricated directly by a continuous supply of fresh oil condensation. Turbulence generated by bearing rotation causes oil particles suspended in the air stream to condense on the rolling elements as the mist passes through the bearings and exits to atmosphere. This technique offers four principal advantages:

- Bearing wear particles are not recycled back through the bearing, but are washed off instead.

Figure 20-11. Multi-stage pump and electric motor connected to oil mist tubing (courtesy of Lubrication Systems Company, Houston, Texas).

- The need for periodic oil changes is eliminated.
- Long-term oil breakdown, oil sludge formation, and oil contamination are no longer factors affecting bearing life.
- The ingress of atmospheric moisture into the motor bearing is no longer possible and even the bearings of standby equipment are properly preserved. Without oil mist application, daily solar heating and nightly cooling cause air in the bearing housing to expand and contract. This allows humid, often dusty air to be drawn into the housing with each thermal cycle. The effect of moisture condensation on rolling element bearings is extremely detrimental and is chiefly responsible for few bearings ever seeing their design life expectancy in a conventionally lubricated environment.

It has been established that loss of mist to a pump or motor is not likely to cause an immediate and catastrophic bearing failure. Tests by various oil mist users have proven that bearings operating within their load and temperature limits can continue to operate without problems for periods in excess of eight hours. Furthermore, experience with properly maintained oil mist systems has demonstrated outstanding service factors. Because there are no moving parts in the basic oil mist components, and

Figure 20-12. Single-stage overhung pump using oil mist lubrication (courtesy of Lubrication Systems Company, Houston, Texas).

Figure 20-13. Between-bearing pump bracket receiving continuous oil mist lubrication (courtesy of Lubrication Systems Company, Houston, Texas).

because the system pressure is very low, oil mist is a reliable lubrication method. Proper lubrication system operation can be interlocked with pump operation or an alarm system, assuring adequate lubrication.

The savings due to lower preventive maintenance labor requirements, equipment repair cost avoidance, and reductions in unscheduled production outage events have been very significant and cannot possibly be overlooked by a responsible manager or cost-conscious manufacturing facility. Oil mist systems have become extremely reliable and can be used not only to lubricate operating equipment, but to preserve stand-by, or totally deactivated ("mothballed") equipment as well.

The majority of centrifugal pump installations using oil mist lubrication have, as of 1991, applied open, once-through mist flow. However, closed systems with virtually no losses to the environment do exist and should be the preferred configuration in the future [4, 2].

Selecting Rolling Element Bearings for Reduced Failure Risk

Less than 10% of all ball bearings run long enough to succumb to normal fatigue failure, according to the Barden Corporation [13]. Most bearings fail at an early age because of static overload, wear, corrosion, lubricant failure, contamination, or overheating. Skidding (the inability of a rolling element to stay in rolling contact at all times), another frequent cause of bearing problems, can be eliminated by ensuring the bearing will always be loaded. With pairs of angular contact bearings, preloading may be necessary.

Actual operations have shown that better bearing specification practices will avert the majority of static overload problems. Other problems caused by wear, corrosion, and lubricant failure, contamination and overheating can be prevented by the proper selection, application, and preservation of lubricants. Oil viscosity and moisture contamination are primary concerns, and higher-viscosity lubricants are generally preferred [3]. The detrimental effects of moisture contamination are described in Table 20-3.

Unlike API pump bearings, which petrochemical companies often specify for a B-10 life of 40,000 hours, ANSI pump bearings are selected on the basis of an expected 24,000-hour life. Nominally, this means that 90% of the ANSI pump bearings should still be serviceable after approximately three years of continuous operation. However, the failure statistics quoted by MacKenzie [13] indicate that conventionally lubricated ANSI pump bearings do not even approach this longevity. Lack of lubrication, wrong lubricants, water and dirt in the oil, and oil-ring debris in the oil sump all cause bearing life expectancies to be substantially less. It must be assumed that similar findings by other major users of ANSI

Table 20-3
Fatigue Life Reduction of Rolling Element Bearings Due to Water Contamination of Lubricant (Mineral Oil)

Base Oil Description	Water Content of Wet Oil, %	Fatigue-Life Reduction, %	Test Equipment and Hertzian Stress
Mineral oil,	0.002	48	Rolling 4-Ball
dried over	0.014	54	test, 8.60
sodium	3.0	78	GPa (1.25 ×
	6.0	83	10^4 psi)

pumps prompted the search for "life-time lubricated" rolling element bearings that we had alluded to earlier, but which nevertheless have their own particular vulnerabilities.

Problem incidents caused by dirt and water have been substantially reduced by oil mist lubrication. However, serious failure risk can also be introduced by certain specification practices, including some contained in API 610. Without going into the many possible factors that could influence bearing life, several items must be considered in the mechanical design of reliable centrifugal pumps. First among these would be that deep-groove Conrad-type radial bearings with loose internal clearance* are more tolerant of off-design pump operation than bearings with precision tolerances. Also, it should be recognized that centrifugal pumps which undergo signficant bearing temperature excursions are prone to cage failures. Phenolic cages are best suited for operation of radial bearings below 225°F (107°C). New cage materials, such as Polyamide 66, provide a much higher temperature limit and excellent lubricity at these temperatures. Metallic cages are, of course, equally suitable and are sometimes preferred in thrust bearings.

The pump designer and pump user must also come to grips with questions relating to preload values and contact angles. The API requirement to utilize duplex 40° contact angle thrust bearings was prompted by the desire to obtain maximum load capacity from a given size bearing [11]. Similarly, the requirement of installing these bearings back-to-back with a light axial preload was aimed at reducing axial shuttling of rotors to prevent brinelling of contact surfaces (raceways) and to prevent ball skidding. Occasionally, an effective way to prevent this destructive skidding is to install matched, preloaded sets of angular contact bearings with *dif-*

---

* The AFBMA (Anti-Friction Bearing Manufacturers Association) calls this recommended clearance a "C3" clearance as opposed to looser or tighter clearances (C1 or C5, for instance).

ferent contact angles. The MRC Division of SKF markets these bearings under the trade designation "Pumpac." Experience shows these bearings are capable of extending the mean time between pump repairs, especially in single-stage, overhung pumps with unidirectional thrust load, in sizes above 20 hp and at speeds in excess of 1,800 rpm [7].

As regards quantification of preload, Figure 20-14 will prove very enlightening. It shows that for a given bearing (FAG 7314 B.UO, 70 mm bore diameter) a shaft interference fit of 0.0003 in. will produce an al-

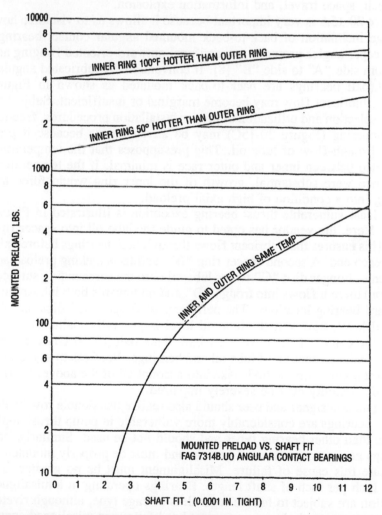

Figure 20-14. Mounted preload vs. shaft fit for a specific 70mm bore diameter bearing (courtesy of FAG Bearing Corporation).

most insignificant preload of approximately 22 lbs, whereas an interference fit of 0.0007 in. would result in a mounted preload of 200 lbs. A much more significant preload would result from temperature differences between inner and outer bearing rings. Such differences could exist in pumps if heat migrated from high temperature pumpage along the shaft or if the pump design incorporated cooling provisions that might artificially cool the outer ring and would thus prevent it from expanding. By far the worst scenario would be for a pump operator to apply a cooling water stream from a firehose. It is sad to see this done today, in the age of high tech, space travel, and information explosion.

Figure 20-15A is very important because it allows us to visualize how the cage inclination of back-to-back mounted angular contact bearings with steeper angles promotes a centrifugal outward-oriented flinging action from side "A" to side "B" [8]. If conventionally lubricated angular contact ball bearings are back-to-back mounted as shown in Figure 20-15B, lubricant flow may become marginal or insufficient. Subject to proper selection and utilization of proper installation procedures, face-to-face mounting (Figure 20-15C) may be advantageous because it promotes through-flow of lube oil. This presupposes that the temperature difference between inner and outer race is minimal. If the temperature difference were substantial, growth of the inner ring would force the bearing into a condition of high axial preload.

The least vulnerable thrust bearing execution is illustrated in Figure 20-16. Here, the vendor has opted to guide the lube oil into spacer ring "A." This ensures that lubricant flows through both bearings before exiting at each end. A second spacer ring "B" facilitates making preload adjustments. Flinger disc "C" tosses lube oil onto the surrounding surfaces and from there it flows into trough "D" and on towards both inboard and outboard bearing locations. The periphery of flinger "C" dips into the lube oil level; however, the lube oil level is generally maintained well below the center of the lowermost ball. This reduces oil churning and friction-induced heat-up of lube oil and bearings. Needless to say, unless lubricant application methods take into account all of the above, bearing life and reliability may be severely impaired.

The pump designer and user should also realize that double row "filler notch" bearings are considerably more vulnerable in pump thrust applications than other bearing types and should not be used. Similarly, ball bearings are sensitive to misalignment and must be properly mounted to eliminate this cause of failure. Misalignment must be no greater than 0.001 inch per inch of shaft length. Bearings operating in a misaligned condition are subject to failure regardless of cage type, although riveted cages seem particularly prone to rivet head fatigue in misaligned condition.

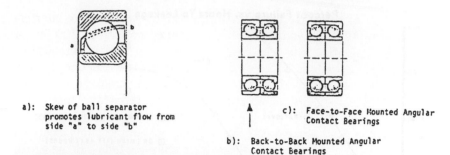

a): Skew of ball separator
   promotes lubricant flow from
   side "a" to side "b"

c): Face-to-Face Mounted Angular
   Contact Bearings

b): Back-to-Back Mounted Angular
   Contact Bearings

Figure 20-15. Angular contact bearings commonly used for thrust take-up in centrifugal pumps.
a) Skew of ball separator promotes lubricant flow from side "a" to side "b."
b) Back-to-back mounted angular contact bearings.
c) Face-to-face mounted angular contact bearings.

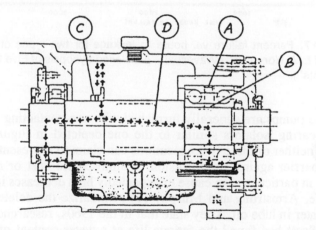

Figure 20-16. Advantageous lubrication arrangement ensures optimized delivery of lube oil to all bearings (courtesy of KSB Pump Company, Kaiserslautern, Germany).

Magnetic Shaft Seals in the Lubrication Environment

Most pump shaft seals are generally inadequate. ANSI pumps are usually furnished with elastomeric lip seals. When these seals are in good condition, they contact the shaft and contribute to friction drag and temperature rise in the bearing area. After 2,000 to 3,000 operating hours, they are generally worn to the point at which they no longer present an effective barrier against contaminant intrusion. Percent failure vs. hours to leakage for two types of lip seals is shown in Figure 20-17 [10].

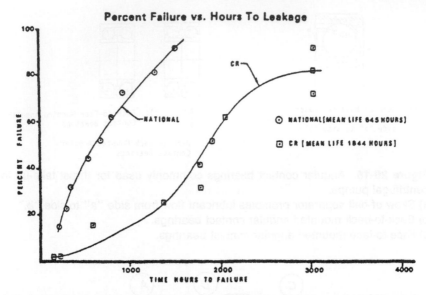

Figure 20-17. Percent failure vs. hours to leakage for two types of lip seals (courtesy of L. A. Horve, "CR Waveseal—A Detailed Synopsis of Five Comparative Life Tests").

API-type pumps are generally furnished with non-contacting labyrinth seals or bearing isolators similar to the one depicted in Figure 20-18. However, neither of these two executions can claim to represent a totally effective barrier against the intrusion of atmospheric dust or moisture. Moisture, in particular, can cause unexpectedly high decreases in bearing fatigue life. Armstrong and Murphy [1] summarize the deleterious effects of water in lube oil. They state that in the 1960s, researchers Grunberg and Scott had found the fatigue life at a water content of 0.002% reduced 48% and, at 6.0% water, found it to be reduced 83%. Others found a fatigue life reduction of 32% to 43% for squalene containing 0.01% water, and a third group of researchers detected about an 80% drop with a moist air environment contacting dried mineral oil (see Table 20-3). While the detailed mechanism for the reduction of fatigue life by water in a lubricant is not completely understood, it is thought to relate to aqueous corrosion. There is much evidence that the water breaks down and liberates atomic hydrogen. This causes hydrogen embrittlement and increases the rate of cracking of the bearing material by a significant margin.

A solution to the problem of sealing pump-bearing housings can be found in aircraft and aerospace hydraulic pumps, which make extensive use of magnetic face seals. This simple seal consists of two basic compo-

Figure 20-18. "Bearing isolators" are non-contacting bearing housing seals which are sometimes used in centrifugal pumps (courtesy of INPRO Seal Company, Rock Island, Illinois).

nents as shown in Figure 20-19. These components are: (1) a magnetized ring, having an optically flat sealing surface that is fixed in a stationary manner to the housing and sealed to the housing by means of a secondary O-ring, and (2) a rotating ring having a sealing surface that is coupled to the shaft for rotation and sealed to the shaft with an O-ring.

The rotating ring, which is fabricated from a ferromagnetic stainless steel, can be moved along the shaft. When no fluid pressure exists, the sealing surfaces are held together by the magnetic force, which is reliable and uniform, creating a positive seal with minimum friction between the sealing faces and ensuring the proper alignment of the surfaces through the equal distribution of pressure.

Magnetic seals are very reliable. Many have operated continuously for 40,000 hours without repair or adjustment under conditions considerably more severe than those to which chemical process pumps are typically exposed.

Finally, if sealed-bearing housings must accommodate changes in internal pressure and vapor volume, an expansion chamber similar to that shown in Figure 20-20 can be used. This small device, which incorpo-

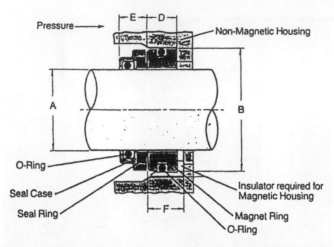

Figure 20-19. Magnetic face seals find extensive application in aerospace pumping services (courtesy of Magnetic Seal Corporation, West Barrington, R.I.).

rates an elastomeric diaphragm and constitutes a completely enclosed system, is screwed into the housing vent opening. It accommodates the expansion and contraction of vapors in the bearing housing without permitting moisture and other contaminants to enter. Carefully selected from a variety of plain and fabric-supported elastomers, the diaphragm will not fail prematurely in harsh chemical environments [14].

The upgraded medium-duty bearing housing shown in Figure 20-21 incorporates the various bearing life improvement features that have been discussed: (1) a deep-groove Conrad-type bearing with loose internal clearance (C3); (2) a duplex angular contact (40°), lightly preloaded back-to-back-mounted thrust bearing; (3) a vent port that remains plugged for dry-sump oil-mist lubricated bearings and that can be fitted with an expansion chamber if conventionally lubricated; (4) a magnetic seal; and (5) a bearing housing end cover made to serve as a directed oil-mist fitting [6].

If oil mist is not available at a given location, the conventional lubrication method shown earlier in Figure 20-16 would be the second choice for engineers experienced with pump operation and maintenance. Third choice would be as indicated in Figure 20-22, with the oil level at "A" or reaching to the center of the lowermost ball. The oil ring, Item 1, serves only to keep the oil in motion. If the oil level is allowed to drop to point "B," the oil ring is expected to either feed oil into the bearings or generate enough spray to somehow get adequate amounts of lubricant to the various bearings. Unfortunately, there is no real assurance of this happening. An oil ring with a given geometry will be fully effective only if it

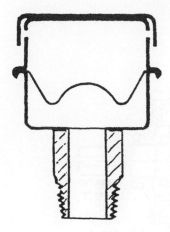

Figure 20-20. A small expansion chamber accommodates vapor expansion and contraction in hermetically sealed bearing housings (courtesy of Gits Manufacturing Company, Bedford Park, Illinois).

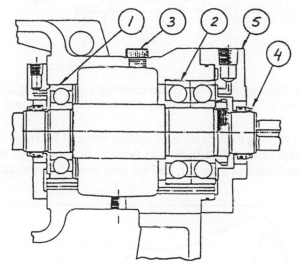

Figure 20-21. Bearing housing for so-called upgraded medium duty (UMD) pump marketed in the U.S. since 1985 (courtesy of Carver Pump Company, Muscatine, Iowa).

is immersed to the proper working depth, turns at the proper speed, and is surrounded by lube oil of the proper viscosity. This is graphically illustrated in Figure 20-23, which is derived from Heshmat and Pinkus [9]. In this context, *fully effective* would mean that the oil ring should deliver a sufficient amount of oil even as the shaft speed increases. This is achieved by the ring provided with grooves on the inside diameter, whereas the ring with the conventionally executed inside diameter appears to deliver sufficient amount only at low shaft speeds.

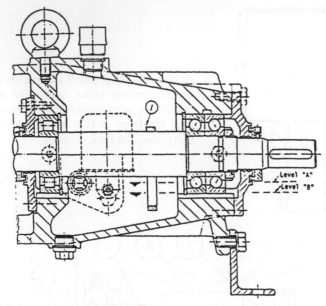

Figure 20-22. Conventionally lubricated centrifugal pumps using oil rings must pay close attention to oil levels (courtesy of Sulzer-Weise, Bruchsal, Germany).

In summary, proper lubrication of rolling element bearings in centrifugal pumps depends on such factors as bearing preload, cage inclination, point of introduction of lube oil, and oil ring design, immersion, oil vis-

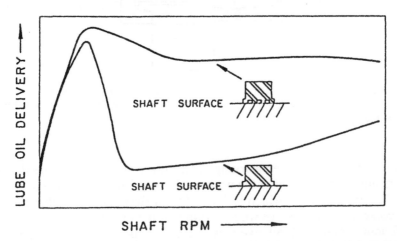

Figure 20-23. Amount of oil delivered by grooved and non-grooved rings (courtesy of H. Heshmat and O. Pinkus, "Experimental Study of Stable High-Speed Oil Rings").

cosity, and oil cleanliness. Unless these factors are understood and unless both user and designer take them into full account, bearing life may be erratic or perhaps even consistently too low.

References

1. Armstrong, E. L., Murphy, W. R., and Wooding, P. S., "Evaluation of Water-Accelerated Bearing Fatigue in Oil-Lubricated Ball Bearings," *Lubrication Engineering.* Vol. 34, No. 1, pp. 15–21.
2. Bloch, H. P., "Executing Oil Mist Lubrication in the 90s," *Hydrocarbon Processing,* October 1990, p. 25.
3. _____, *Improving Machinery Reliability,* Second Edition, Gulf Publishing Company, Houston 1988.
4. _____, *Oil Mist Lubrication Handbook,* Gulf Publishing Company, Houston, 1987.
5. Bloch, H. P. and Geitner, F. K., *Machinery Component Maintenance and Repair,* Second Edition, Gulf Publishing Company, Houston, 1990.
6. Bloch, H. P. and Johnson, D. A., "Downtime Prompts Upgrading of Centrifugal Pumps," *Chemical Engineering,* November 25, 1985.
7. Elliott, H. G. and Rice, D. W.; "Development and Experience with a New Thrust Bearing System for Centrifugal Pumps," Proceedings of 4th International Pump Users Symposium, Texas A&M University, 1987.
8. Eschmann, Hasbargen, Weigand, *Ball and Roller Bearings,* John Wiley & Sons, New York, 1983.
9. Heshmat, H. and Pinkus, O., "Experimental Study of Stable High-Speed Oil Rings," Transactions Of The ASME, Journal of Tribology, January 1985.
10. Horve, L. A., "CR Waveseal—A Detailed Synopsis of Five Comparative Life Tests," CR Industries, Elgin, Illinois, 1977.
11. James, R., Jr., "Pump Maintenance," *Chemical Engineering Progress,* February 1976, pp. 35–40.
12. "Lubrication of Anti-Friction Bearings," Form 446-1, TRW Bearings Division, Jamestown, New York 14701, January 1983.
13. MacKenzie, K. D., "Why Ball Bearings Fail," Product Data Bulletin No. 5, The Barden Corporation, Danbury, Connecticut.
14. Nagler, B., "Breathing: Dangerous to Gear Case Health," *Power Transmission,* January 1981.
15. Shelton, W. A., "Lubed-For-Life Bearings on Centrifugal Pumps," *Lubricating Engineering,* June 1977.

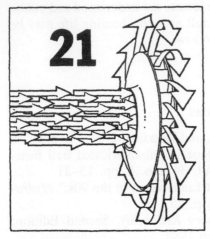

21

Mechanical Seal Reliability

by **Gordon S. Buck**
John Crane Inc.

In spite of recent advancements in mechanical sealing technology, excessive leakage is the most common cause of "pump repairs." In fact, seal-related repairs represent approximately 60% to 70% of all centrifugal pump maintenance work. Therefore, seal-related repairs are an excellent area to concentrate pump reliability improvement programs. Unfortunately, because of the many facets of seal reliability, this chapter can only serve as an overview and introduction to mechanical seal reliability.

Although a mechanical seal may be small enough to hold in your hand and simple in concept (see Chapter 17, Figure 17-1), it is actually a complex device. The successful operation of mechanical seals in centrifugal pumps calls for careful attention to detail in several areas. These areas are:

- Seal hardware (including sleeve, gland, and gaskets).
- Seal installation.
- Pump hardware (including piping).
- Pump repair and installation.
- Pump operation (including flush to seal).

Of course, these areas are not independent, but often seal reliability can be improved by concentrating on individual areas to solve particular problems. Naturally, the pertinent area must be selected or the effort may be entirely wasted. Therefore, the first step in reducing seal failures is to establish a seal failure analysis program.

Failure Analysis

The purpose of a mechanical seal is to prevent, or at least limit, leakage. Normally, the leakage from a mechanical seal is so slight that it is difficult to measure. For example, a 2-inch seal containing water at 100 psig should leak less than 2 ml/hr. Current federal regulations require that the concentration of certain volatile organics be less than 10,000 ppm near the seal leakage point. In some areas, this limit is 1000 ppm. It is likely that none of these examples would show any visible leakage. Because most seals are wearing devices, they will fail, that is, leak *excessively*, at some point in time. The machinery engineer must therefore be prepared to learn from each seal failure and use that information to improve future designs and applications so that seal life is improved. In order to do this, detailed records of each failure should be maintained.

Failure analysis is simply allowing the seal to tell why it did not perform as expected. This means acting the part of a detective. The following simple guidelines will make failure analysis easier and more consistent:

- Know what a new seal looks like.
- Know what a successfully applied seal looks like.
- Examine the failed seal carefully.
- Write down the differences between the new, successful, and failed seals.
- Formulate a consistent explanation about the differences.

Although manufacturers and mechanics are familiar with new seals, sometimes a machinery engineer may not see the unused item. He only sees the failures. Similarly, he may not see many successfully applied seals because they are not called to his attention. He is too busy looking at failures. On the other hand, some failure modes are so common that a mechanic may come to accept them as "normal wear." For failure analysis to be useful, all differences in the new and failed seal should be noted.

These differences must be written down in a consistent format. The form shown as Table 21-1 is a good starting point and can be modified to suit particular needs. There is a great temptation to take mental notes and write them later. This temptation must be overcome, and written notes must be made while examining the seal. If these notes become soiled, they can be rewritten later.

For many refineries and chemical plants, the statistics in Table 21-2 have been developed from analysis of many pump seal failures.

Statistics show that most seals fail prematurely, that is, prior to wearing out. They also show that these early failures can be prevented by sim-

Table 21-1
Mechanical Seal Failure Analysis Checklist

Machine Identification _____

Reason for Service Request _____

Last Known Failure _____

Summary of Analysis _____

_____

Seal Identification

Manufacturer _____   Type _____

Face Materials _____

Gasket Materials _____

Mounting _____   Balance Ratio _____

Operating Conditions

Liquid _____   Specific Gravity _____

Temperature, F _____   Sealed Pressure, psig ____

Seal RPM _____

Inspection of Seal

Stationary Face	(OK)	Rotating Face	(OK)
Work out	( )	Worn out	( )
Grooved	( )	Grooved	( )
Heat checked	( )	Heat checked	( )
Warped	( )	Warped	( )
Corroded	( )	Corroded	( )
Gasket, Stationary Face	(OK)	Gasket, Rotating Face	(OK)
Hardened	( )	Hardened	( )
Chemical attack	( )	Chemical attack	( )
Broken/cut	( )	Broken/cut	( )
Gasket, Gland	(OK)	Gasket, Sleeve	(OK)
Hardened	( )	Hardened	( )
Chemical attack	( )	Chemical attack	( )
Broken/cut	( )	Broken/cut	( )
Gland	(OK)	Drive Collar	(OK)
Corroded	( )	Corroded	( )
Bushing rubbed	( )	Turned on shaft	( )
Number of bolts	____	Setscrews	____
Antirotation Pins	(OK)	Sleeve/Shaft	(OK)
Fretted	( )	Fretted	( )
Corroded	( )	Corroded	( )
Flush Plan _____		Springs/Bellows	(OK)
Location _____		Broken	( )
Hole Size _____		Corroded	( )

Table 21-2
Causes of Seal Failures

Operating problems	30% to 40%
Improper installation	20% to 30%
Pump design/repair	10% to 20%
Misapplied seal	10% to 20%
Worn out	1% to 5%
Miscellaneous	15% to 20%
Total	100%

ply selecting the proper seal, installing it correctly, and operating the pump carefully.

Seal Hardware Failures

Another way to look at seal reliability is to insist that the seal must survive in spite of any problems with the pump or its operation. In this case, failures are examined from a hardware or design point of view. In other words, if a pump requires repair because the seal is leaking, the corrective measures are directed towards the seal hardware. This may not always be correct; however, the statistics presented in Table 21-3 seem to apply to hardware failures in refinery and chemical plant pump seals. Naturally, seals or pumps that were not properly installed or repaired are omitted from Table 21-3.

Table 21-3
Mechanical Seal Hardware Failures

Rotating or stationary face	40% to 70%
Loss of axial movement (hangup)	10% to 20%
Fretting	10% to 20%
Corrosion	10% to 20%
Miscellaneous	5% to 20%
Total	100%

Seal Faces. Table 21-3 shows that, in pump repairs caused by hardware failures, the seal faces are the apparent problem 40% to 70% of the time. These faces were severely worn or damaged. In some instances, an immediate improvement was made by simply changing face materials. For example, a stellite or ceramic face might be replaced with a solid tungsten carbide face. In other instances, the face materials could not be changed or improved and efforts were focused on other areas.

Because mechanical seal failures can be expensive, there is no excuse for using second rate materials. In many refinery and chemical plant services, premium grade carbon and tungsten carbide or silicon carbide faces have proven reliable. Springs are frequently Alloy 20. Glands, sleeve, shells, adapters, etc. are made from 316SS. In order to reduce inventories and avoid costly mix-ups, these same materials are also used in easy services—even cold water.

Hangup. Loss of axial movement, or seal hangup, is caused when leakage accumulates and hardens beneath the flexible element (see Figure 21-1). In most seals, this is the rotating element. Hangup is a particularly troublesome problem in high temperature seals because any leakage tends to form "coke." There are four approaches to solving hangup problems:

- Change the liquid (external flush that will not solidify).
- Reduce leakage (higher balance ratio and/or narrow face seal).
- Remove leakage (steam quench).
- Remove opportunity to hangup (change to a stationary seal).

Some stationary seals are designed so that there is a relative motion between the "floating" element and the rotating shaft. This relative motion makes it difficult for solids to accumulate and cause hangup. Because stationary metal bellows seals include this feature as well as narrow faces and are rated for high temperatures, they are often used in services that tend to cause hangup of conventional seals.

Fretting. Fretting is the damage caused by the constant rubbing of the flexible element against its drive lugs or the dynamic gasket against the sleeve. The movement may be less than a thousandth of an inch, but eventually the sleeve or drive lug is damaged. The damage may cause leakage directly or may restrict axial movement so that leakage occurs between the faces.

The solution to fretting problems usually involves one of the following approaches:

- Improve seal alignment during installation.
- Reduce shaft axial float (adjust bearings).
- Use O-rings for gaskets.
- Change to a bellows seal.

Corrosion. Most mechanical seals are made of relatively corrosion-resistant materials. For example, a typical off-the-shelf seal might be made of 316 stainless steel, carbon graphite and tungsten carbide with fluoroelastomeric gaskets. Metal parts are also available in Alloy 20, Hastelloy, and other corrosion resistant materials. Carbon graphites are available

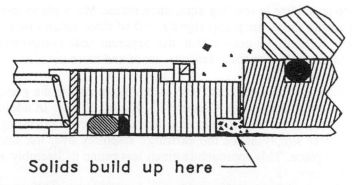

Solids build up here ———

Figure 21-1. Seal hangup from accumulated solids (courtesy John Crane Inc.).

with many different types of resin and metallic binders. Hardfaces include both nickel and cobalt bound tungsten carbide, various silicon carbides and other ceramics. Gaskets may be made from many materials including graphite foil, Teflon, nitrile, Kalrez, etc.

It is most likely that a suitable set of materials may be found for even the most corrosive seal environment. Still, while searching for those materials, corrosion problems do occur. Even after the proper materials have been found, the wrong material may be inadvertently substituted at the next repair.

Other Problems. There are other hardware problems that occur in addition to face damage, hangup, fretting, and corrosion. The most perplexing problem occurs when there is no obvious damage. This often happens when sealing light hydrocarbons. For these products, the liquid may flash to a vapor between the seal faces. Since most seals are designed for liquids, the resulting force imbalance from flashing causes the faces to "pop open." The solution may involve changing the balance ratio or seal type or installing an external flush that will not vaporize so readily.

Seal Failures from Installation Problems

Proper installation is so important that all seal manufacturers conduct extensive training courses on how to install their seals correctly. Most of the larger seal users have their own internal training courses as well. Consultants are available to teach proper seal installation. Still, the statistics show that 20% to 30% of all seal failures are caused by a poor or incorrect installation. Many installation problems seem to stem from improper assembly of seal components and internal misalignment.

Improper Assembly. A large refinery may use several different seal types from four or five manufacturers. Some plants do not have technicians who specialize in rotating equipment repair. Many plants use independent repair shops for pump repairs. All of these factors make it increasingly difficult to assure that the separate seal components are assembled properly. For these reasons, more and more users are turning to cartridge seals.

A cartridge seal is a complete seal assembly, including the gland, that is mounted on a sleeve (see Chapter 17, Figure 17-26). In addition to being assembled, the seal is also adjusted to the correct operating position. The user simply slides the cartridge assembly over the shaft and fastens it in place. This approach is much less prone to assembly and installation errors.

In the larger plants, used cartridge seals may be reconditioned by a specialist and stocked for future installation. Smaller plants may rely on the seal manufacturer to both recondition and stock the cartridge units.

Internal Misalignment. Most mechanical seal designs incorporate one flexible element and one fixed element. The most common arrangement has the flexible element rotating and the fixed element stationary. Seal life is greatly improved by assuring that the fixed element is mounted perpendicular to the rotating shaft. Because the fixed element is contained in the gland and the gland is bolted to the stuffing box, several surfaces must be checked for perpendicularity. Furthermore, the shaft is mounted in the bearing housing that is bolted to the pump case. These mounting surfaces must also be checked.

For overhung process pumps, a precision dial indicator can be used to assure that internal misalignment is minimized. The bearing bracket should be assembled as a complete unit so that the following checks can be made:

- Shaft run-out < 2 mil TIR.
- Shaft lift < 3 mil.
- Shaft float < 5 mil (this author recommends < 1 mil).
- Bracket face square to shaft within 3 mil.
- Sleeve (mounted on shaft) run-out < 2 mil.

In addition, the stuffing box face must be checked to ensure that it is perpendicular to the shaft within 3 mils TIR. The most reassuring method of checking this is to measure it directly. On some pumps, this may be done by mounting the assembled bearing bracket to the pump without including the gland. A dial indicator that reads the stuffing box face is placed on the shaft. If there is not enough room to do this, an indirect method must be used. On these pumps, the cover should be checked to see that the face of the stuffing box and the mounting face for the bearing bracket are parallel within 3 mils TIR.

Seal Failures Related to Pump Hardware

Sometimes seal failures are so strongly related to the pump and its performance that reliability can be improved only by modifying the pump. The pump may have a true design problem or may simply have been misapplied. Some of the more common pump hardware related seal failures are:

• Excessive shaft deflection.
• Narrow annulus of stuffing box.
• Shaft axial movement.
• Excessive piping loads.

Excessive Shaft Deflection. If the shaft deflects, the seal must move axially each revolution to compensate. API Standard 610 for centrifugal pumps specifies a maximum of 0.002-inch shaft deflection at the location of the seal faces. Some older pumps and non-API pumps may not meet this specification. In particular, older pumps designed for packing may have excessive shaft deflection. Shaft deflection is reduced by increasing the shaft diameter and/or reducing the bearing span or shaft overhang.

Narrow Annulus. In spite of the fact that the vast majority of centrifugal pumps are equipped with mechanical seals, most older pumps are actually designed for packing. This is particularly true for heavy duty API type pumps manufactured prior to the seventh edition of API Standard 610.

A pump designed for packing has a cylindrical "stuffing box" through which the shaft passes. Packing is compressed within the annulus between the OD of the shaft and the ID of the stuffing box. This annulus may be from 3/8 inch to 1 inch in cross section. If the pump is actually equipped with a seal, the seal must fit inside this annulus. The result is that very little liquid surrounds the seal.

Seal manufacturers know that their products are more reliable when operated within a "seal chamber" that contains more liquid than a stuffing box. However, many pump users insist on the ability to convert from seals to packing. Also, pump manufacturers prefer to use existing designs and patterns that are interchangeable with older models. The result is that most pumps are designed for packing but use seals. It should be no wonder that the seal is the most common cause of a pump repair.

In actual laboratory tests, seal face temperatures have been observed to decrease when the stuffing box bore is increased. For example, a standard arrangement had about 1/16-inch radial clearance between the seal and the stuffing box bore. When this clearance was increased to 1/2 inch, the seal face temperature decreased by 38°F. In response to this sort of information and experience, many pump manufacturers are now supplying "oversized" stuffing boxes for their pumps.

Shaft Axial Movement. Whenever the shaft moves axially, one of the sealing elements must move in the opposite direction to maintain the existing gap between the faces. If this movement occurs repeatedly, fretting may occur between the seal and shaft/sleeve. Table 21-3 shows that fretting is a significant cause of seal failures.

For reasons that are not completely clear, many manufacturers assemble their pumps so that the shaft can "float" axially for 3 mils to 8 mils. This is a built-in opportunity for the seal to fret. To reduce this potential, many pump users go against the manufacturer's recommendation and modify the pumps to restrict "float" to less than 1 mil.

Excessive Piping Loads. A centrifugal pump is not a good pipe anchor. API 610 specifies maximum loads that must be used by both pump manufacturer and piping designer. Normally this is not a problem except for end suction pumps used in high temperature services.

Beginning with the sixth edition, API 610 specifies piping loads that are considerably higher than previous editions. While this makes the job of the piping designer easier, it has imposed a severe constraint on the pump manufacturer. Most pumps that were designed to meet the early editions of API 610 will not meet the more stringent requirements of the sixth and seventh editions. Because they are not as rigid, older pumps may require even more attention to installation and hot alignment than the newer models.

Seal Failures Caused by Pump Repair and Installation

The mechanical seal is sensitive to the environment in which it operates. In particular, the mechanical condition of the pump has a significant effect on seal reliability. Some of the major considerations include:

- Rotating imbalance.
- Alignment to driver.
- Flushing the seal.

Rotating Imbalance. Vibration reduces seal life and any rotating imbalance will create vibration. The frequency of this vibration will be the running speed of the pump. The best way to prevent this is to dynamically balance the rotating assembly. Although many technicians will work with an assembly until no further improvement can be detected, a good commercial standard is 0.0045 ounce-inch per pound of assembly.

Alignment to Driver. Misalignment between the pump and driver is another cause of vibration. This vibration occurs at two times the running speed of the pump. Both radial and axial components may be present. In order to reduce this vibration, many pump users take great pains to align pumps to drivers.

Flushing the Seal. Because a mechanical seal is a rubbing device that generates heat, it must be lubricated and cooled. Cooling and lubrication are normally accomplished by circulating a suitable liquid around the seal. This is called flushing the seal. The most common arrangement is the discharge bypass flush, sometimes called API Plan 11 (see Chapter 17, Figure 17-9), but there are many others (see Chapter 17, Figure 17-26).

Perhaps the most important feature of a good flushing system is that the liquid circulates around the rubbing interface of the seal. This is accomplished by injecting or withdrawing the liquid in close proximity to the faces.

In general, seal flush systems should be as simple as possible to avoid installation and operating errors. This means designing the system to minimize valves, orifices, filters, etc. A common problem with flushing systems is that liquid does not actually flow because a valve is closed, a filter is blocked, etc.

Seal Failures Caused by Pump Operation

Although mechanical seals should be designed to survive operational upsets, seal life can certainly be enhanced by observing a few simple precautions. Some of the more common operating problems include:

• Operation at low flow.
• Operation at no flow.
• Cavitation.

Low Flow. In spite of the performance curve, which shows operation from no flow to maximum, centrifugal pumps may not operate smoothly at less than about 50% of Best Efficiency Point (BEP) flow. Vibration and noise may increase markedly at less than this minimum stable flow. The result is a decrease in seal life. Resolution of this problem may require hydraulic modification by the pump manufacturer or a by-pass line to artifically increase flowrate.

At still lower flows (less than 20% of BEP), the liquid passing through the pump may be significantly raised in temperature. According to API 610, the minimum continuous thermal flow is "the lowest flow at which the pump can operate and still maintain the pumped liquid temperature below that at which net positive suction head available equals net positive suction head required."

No Flow. In spite of assurances to the contrary by the operating crew, sometimes pumps have obviously been running "dry." There are many ways this can happen—a broken level controller, a plugged suction strainer, a broken check valve or a discharge valve that was never opened. All of these and more can prevent liquid from entering or leaving the pump.

Some batch processes actually are designed so that the pump runs dry as it pumps out a reactor or supply vessel. Dissimilar pumps operating in parallel are a classic example of how one pump may force another to run at no flow.

In addition to the noise and vibration problems described for low flow, operation with no flow may generate enough heat to vaporize any liquid trapped in the pump. In particular, seals that are flushed from the pump discharge or suction may quickly become "gassed up." The obvious solution is to *avoid* the no flow situation. If the pump must run under these conditions, a separate flushing system may be required for the seal.

Cavitation. Cavitation has been rightly and wrongly blamed for many ills in both pumps and seals. Certainly cavitation increases pump vibration and vibration reduces seal life. Recent studies have shown that the simple 3% head loss rule that is used to define NPSHR may not adequately define the onset of cavitation problems. Also, operation at low flow sometimes produces symptoms similar to cavitation.

Cavitation problems are sometimes difficult to solve. Frequently, increasing the available NPSH is prohibitively expensive. Reducing the required NPSH is sometimes possible with special impellers or inducers. For these reasons, cavitation problems must be carefully addressed during the initial specification of the pumps.

Reliability

Once the potential problems of selection, installation, and operation are addressed, mechanical seals can provide reliable sealing for many years. Many large refineries and chemical plants have obtained a gross average mean time between pump failures that is greater than two years. (This statistic includes all causes of pump repairs and not just mechanical seals.)

Some services will always be more difficult than others and have a correspondingly lower reliability. Higher pressures, higher temperatures, higher solids content, and increased corrosion rates will generally result in a lower pump and seal reliability. Similarly, large seals and high shaft speeds may have lower reliability. Rather than accept lower reliability, many users elect to use special purpose seals and sealing systems that are engineered for these specific applications.

Considering the wide variety of designs, arrangements, and materials available, it is no wonder that the mechanical seal is the most popular device for sealing modern centrifugal pumps.

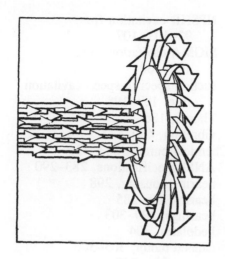

Index

Printed and bound by CPI Group (UK) Ltd, Croydon, CR0 4YY

03/10/2024

01040433-0017